ACKNOWLEDGMENTS

Much of the technical information that appears in this book originated from a textbook Bob and I wrote for professional mechanics several years ago. With the increase of mandatory emissions control testing, we felt it was time to give the average do-it-yourselfer some insights on current emissions-control technology. So here it is—a book written in plain English that clearly explains the mystery of emissions control.

I'd like to thank my wife Nina for putting up with the long hours I put into this manuscript. And, I'd like to thank Bob for shooting some great photos. I'd also like to thank the various manufacturers and other sources that provided us with the illustrations that appear throughout. –*Larry*

I want to thank all the professional automotive technicians and the engineers and trainers, both aftermarket and O.E., who have given generously of their time to help me in my research over the last two decades. I also want to thank Larry for being so easy to work with since we started our collaboration in the late seventies.–*Bob*

CONTENTS

	Introduction	V
1	From Air Injection to Zirconium	1
2	The Sky Is Falling	11
3	Rules, Regulations & Laws	20
4	Exhaust Emissions: HC, CO & NOx	31
5	Positive Crankcase Ventilation (PCV)	39
6	Evaporative Emissions Control (EEC)	44
7	Heated Air Intake & Early Fuel Evaporation (EFE)	52
8	Air Injection Systems	57
9	Catalytic Converters	64
10	Exhaust Gas Recirculation (EGR)	68
11	How Air/Fuel Ratio Affects Emissions	75
12	How Ignition Affects Emissions	86
13	Engine Wear & Other Factors	95
14	Computerized Engine Controls	103
15	Emissions Testing	112
16	Troubleshooting	124
17	Tuning Emissions-Controlled Engines	141
	Glossary	148
	Acronyms	162
	Index	165

UNDERSTANDING AUTOMOTIVE EMISSIONS CONTROL

**THEORY • TROUBLESHOOTING
MAINTENANCE • TESTING • TUNING • REPAIR**

LARRY CARLEY & BOB FREUDENBERGER

HPBooks

HPBooks

are published by
The Berkley Publishing Group
200 Madison Avenue
New York, New York 10016

First Edition: April 1995

© 1995 Larry Carley & Bob Freudenberger
10 9 8 7 6 5 4 3 2 1

Library of Congress Cataloging-in-Publication Data

Carley, Larry W.
    Understanding automotive emissions control : theory, troubleshooting, maintenance, testing, tuning, repair / Larry Carley & Bob Freudenberger.
       p. cm.
   Includes index.
   ISBN 1-55788-201-0 (trade pbk.)
   1. Automobiles—Pollution control devices. I. Freudenberger, Bob. II. Title.
TL214.P6C36974 1995   94-24720
629.25'28—dc20   CIP

Book Design & Production by Bird Studios
Interior photos by the authors unless otherwise noted

All rights reserved. No part of this publication may be reproduced, stored in a retrieval system, or transmitted in any form, by any means electronic, mechanical, photocopying, recording or otherwise, without the prior written permission of the publisher.

NOTICE: The information in this book is true and complete to the best of our knowledge. All recommendations on parts and procedures are made without any guarantees on the part of the authors or The Berkley Publishing Group. Tampering with, altering, modifying or removing any emissions-control device is a violation of federal law. Author and publisher disclaim all liability incurred in connection with the use of this information.

INTRODUCTION

This is a book about automotive emissions control and how it affects you. Whether you're a believer in strict emissions control or someone who would like to see a return to the good ol' days of unregulated cars and uncluttered engine compartments, the fact remains that emissions control is here to stay. So for better or worse, we have to accept the restrictions and limitations that have been imposed upon us by legislation. The purpose of this book is to help you better understand what those restrictions and limitations are as well as what emissions control is all about. To achieve that, we're going to give you a brief history of emissions control so you can understand how it has become the primary driving force of change in automotive technology today. We're also going to tell you about the basic pollutants that are produced by the internal combustion engine when it burns a fossil fuel, and how the various emissions-control systems and devices keep those emissions in check. Beyond that, we'll explain how emissions control fits into the overall scheme of late-model electronic engine management systems. And, finally, we're going to show you how you should maintain your own vehicle's emissions-control systems and what to do if you have an emissions problem.

WHY THINGS ARE THE WAY THEY ARE TODAY

The automobile that created our mobile society and gave us the freedom to travel where we want, when we want and with passengers of our choice, has now become the object of environmental scorn. The automobile today is often blamed for everything from air pollution, global warming and holes in the ozone layer to urban sprawl. Consequently, it has been singled out for regulatory crucifixion.

The media today cite all kinds of negative "facts" about the automobile's role in pollution. We hear statements like "nearly two-thirds of the total carbon monoxide and half the hydrocarbon and nitrogen oxides polluting our atmosphere come from motor vehicles." We also hear alarmist statements like "three out of five Americans face lung damage from ozone-polluted air."

We can't argue with the evidence that says automotive emissions are a major contributor to urban air pollution. You don't need a Ph.D. in chemistry to experience firsthand the nasty effects of exhaust pollution in a traffic jam. But we don't think many Americans are dropping dead in their tracks from breathing polluted air. It may be a contributing factor to lung disease and respiratory ailments in some people. But deadly? Not unless you're running an engine in a closed garage.

The fact is today's cars produce less pollution than ever before, and are getting even cleaner. Pre-1963 vehicles are the worst because they lack any emissions controls whatsoever. In 1963, positive crankcase ventilation (PCV) was added to recycle the blowby vapors in the crankcase back into the intake manifold so they could be reburned. This virtually eliminated crankcase emissions as a source of air pollution. Sealed fuel

Are the days of carbureted hot rod engines with little or no emissions-control devices coming to an end? Not quite . . . but regulations will force enthusiasts to rethink their definition of performance and adapt accordingly. Photo by Michael Lutfy.

systems and charcoal canisters showed up in 1971, which reduced evaporative emissions to zero. In 1973, exhaust gas recirculation (EGR) was added to reduce oxides of nitrogen (NOx) in the exhaust. An even greater change appeared in 1975 when the catalytic converter (which requires unleaded gasoline) was introduced. The converter greatly reduced unburned hydrocarbons (HC) and carbon monoxide (CO) levels in the exhaust by "reburning" the pollutants. The switch to unleaded gas was necessary because lead poisoned the catalytic agents. This change also eliminated lead as an exhaust pollutant (this metal has been linked to lead poisoning and various learning deficiencies in urban children). In 1981, onboard computers for closed-loop running and three-way oxidation/reduction converters were added to most cars, which cut CO and NOx emissions by another 50%.

The tailpipe emissions from late-model cars (1981 and later) with computerized engine controls and three-way catalytic converters are only a fraction of the older pre-emissions controlled cars. Today's cars produce 96% less HC and CO, and 76% less NOx than their pre-emission counterparts.

Additional reductions are being phased in on 1994 through 1998 model year cars and light trucks. The federal standards up through 1993 allowed no more than 0.41 grams per mile (gpm) of HC, 3.4 gpm of CO and 1.0 gpm of NOx for the first 50,000 miles. The new standards slash these limits by almost half: CO is cut to 0.25 gpm, while HC and NOx drop to 0.4 g/m.

So what do all these numbers really mean? It means a 1975 to 1979 model car puts out roughly as much HC and CO pollution as four new cars; a 1972 to 1975 model year car produces as much of these pollutants as seven new cars; a 1968 to 1971 model year car produces pollution equivalent to 10 new cars; and a '63 or earlier pre-emissions-controlled car pumps as much crud into the atmosphere as 25 or more new cars! These comparisons assume that a vehicle produces no more pollution than it did when new —which is usually NOT the case once an engine has accumulated high mileage. It's not unusual to find high mileage engines in older vehicles (1981 and earlier) that are belching out the pollution equivalent of 100 new cars!

The real problem is uninformed people making blanket statements that condemn all cars as being polluters. Older vehicles can be reasonably clean if their engines are maintained in good running condition. But there are those who see all motor vehicles as environmentally evil and would ban them from the face of the earth.

Another problem, of course, is the fact that we have too many automobiles crowded onto too few roads during peak rush hour in most urban areas. That, in turn, creates traffic gridlock and concentrated pollution. But that doesn't mean we have to abandon the automobile as a means of transportation. In fact, most of us have no real alternative to the automobile. Walking and bicycling are fine if you're not going very far or if the weather is mild. But neither is a practical option for the millions of people who commute considerable distances to their jobs, who have to travel to shop, haul their kids around, or simply need to get from point A to point B. The availability of public transportation is limited or virtually nonexistent outside most urban areas, and is expensive, inconvenient, and in many instances downright dangerous in many large cities. So the privately owned automobile (or light truck or sport utility vehicle) will continue to provide our primary means of transportation for the foreseeable future. The fuels we burn will evolve, or may someday be replaced by batteries or hydrogen, but the basic utility and versatility of the privately owned motor vehicle remains the best transportation option we have. ∎

This 5.0 Mustang illustrates the future of street-legal performance, and demonstrates that you can increase horsepower dramatically while controlling emissions. All items on this powerplant are legal and passed a California emissions test, the toughest in the nation. Photo by Michael Lutfy.

FROM AIR INJECTION TO ZIRCONIUM

Since the early 1970s, automotive emissions control has undergone an evolution that would make Darwin's head spin. Each new year has brought with it countless changes in the emissions-control systems on practically every new vehicle.

Prior to the beginnings of emissions control in the 1960's, basic automotive technology had remained relatively unchanged for decades. Engines and drivetrains had improved, but the fundamental components and systems that made the engine run had all remained essentially unchanged. If you could tune a 1939 Ford, you could probably tune a 1969 Ford. Carburetors had gained automatic chokes, but contact point ignition systems were still the same, engine compartments were relatively uncluttered and provided plenty of elbow room, and almost anyone with a little know-how and a few simple tools could call himself a mechanic. Then came emissions control. It was the end of an era and the end of ignorance.

The late 1960s and early 1970s were periods of great turmoil and social change in this country. People were taking a look at themselves, their society, and their traditional values. And many did not like what they saw: war, poverty, social injustice, and rampant pollution of our air and water. Like the other problems of that time, pollution was nothing new. The only thing new about it was that it seemed to be getting progressively worse as the population grew and urban areas became more congested.

The mood of the times was such that people wanted to change everything they

The millions of motor vehicles on our roads (nearly 200 million registered vehicles in the U.S. alone) amount to mobile pollution factories. Although the automobile has been blamed for much of the air pollution problem we have today, cars built within the last decade or so only produce a fraction of the pollution of their pre-control era ancestors.

The energy crisis of 1973 only served to hasten the demise of big-block, V8 American muscle cars. Gasoline consumption and pollution concerns far outweighed the interests of performance enthusiasts, and cars were immediately downsized and equipped with emissions controls. By 1975, the Z28 shown in the back of this photo would be gone. Chevy decided that reduced performance wasn't worthy of the once-proud Z28 name. It would be more than a decade before Chevy was able to devise emissions-legal performance that could wear the Z28 label. Photo by Michael Lutfy.

saw wrong with the world around them—immediately. They really believed that they could change things for the better, and so the environmentalist movement was born. People rallied around the message of Earth Day and supported legislation that would set in motion efforts to clean up the mess we had created. The result was the Clean Air Act of 1970 and the creation of the Environmental Protection Agency.

With the advent of emissions legislation, the automobile manufacturers were forced to rethink and redesign a lot of engine components they really didn't want to redo. But they had no choice. So each manufacturer came up with its own unique approach to meeting the government's new emissions standards as we started down the path to a cleaner future. Each new model year brought with it countless changes and "improvements" in ignition, carburetion and engine design all for the sake of reducing emissions. Consequently, we had to suffer through successive generations of emissions-controlled engines that provided progressively lower emissions—but also progressively poorer performance.

The first-generation cars of the early 1970s were a disaster. The add-on emissions controls of the era contributed to lousy fuel economy and virtually killed performance. A host of new emissions-control gadgets were tacked onto once uncluttered engines, including smog pumps, EGR valves, ported vacuum switches, vacuum amplifiers and vacuum delay valves. The result was a spaghetti-like tangle of tubes, hoses and wires that made routine maintenance and repairs a nightmare for professional mechanics and do-it-yourselfers alike. To make matters worse, "anti-tamper" plugs were installed over carburetor idle mixture screws to discourage any adjustments or fiddling that might have helped these engines run better.

Then came the second-generation cars in the mid-1970s, equipped with catalytic converters and designed to run on unleaded gasoline. The change to electronic ignition was a step in the right direction, but the addition of the converter and the low-octane unleaded gasoline requirement (which was more expensive) were seen as major hindrances to performance. At the same time, we had the so-called "energy crisis" thrust upon us as world oil supplies and prices were manipulated by the Organization of Petroleum Exporting Countries (OPEC). The 1973 oil embargo and the resulting panic it caused in this country forced Congress to mandate improved corporate average fuel economy (CAFE) standards—which meant an abrupt end to big V8 engines, muscle cars and performance. The President ordered a 55-mph national speed limit and America embarked on a crusade to save energy and reduce our dependence on foreign oil. Unnecessary driving was curtailed and there was serious talk of gasoline rationing, including possibly even restricting the days on which people could legally drive (odd-even day driving) to force conservation as well as car pooling. Those were dark days as far as most automotive enthusiasts were concerned.

CONVERTER CAPERS

The story of how the catalytic converter came to be sheds some light on the relationship that has evolved between the EPA and the automakers. When the EPA was formulating its emissions-control standards back in the early 1970s, it dutifully reviewed what the auto industry estimated it could do to reduce emissions by 1980—then made those estimates law for the 1975 model year, not 1980! This was done to spur the rapid development of new technology, or as one senator put it, "to keep the industry's

EMISSIONS CONTROL BENEFITS

Device	Emissions Function	Non-Emissions Benefit
EGR system	Reduces NOx	Prevents engine damaging detonation under load
PCV system	Recycles crankcase blowby gases	Prolongs oil and engine life by reducing sludge formation in crankcase
Evaporative emissions control	Prevents escape of fuel vapors from tank and carburetor	Conserves fuel
Catalytic converter	Reduces HC & CO in exhaust (also NOx)	Raises exhaust temperature to prolong life of muffler and pipes
Feedback electronic carburetion	Reduces CO & HC	Better fuel economy
Electronic Fuel Injection	Reduces CO & HC	Easier cold starting, better fuel economy and performance

feet to the fire"—a philosophy that continues to this day.

The automobile manufacturers cried foul, pleaded with the EPA to change its rules and used every legal maneuver their lawyers could think of to postpone or lower the standards. But their efforts were in vain. The standards dictated a 90% reduction in hydrocarbon and carbon monoxide emissions by 1975 and that was it. End of debate. If the car makers could not meet the standards, they would be out of the car business. The turn of events had put the industry in an uproar and everyone was making predictions of doom. But in June 1973, General Motors issued a statement saying that a catalytic converter had been developed that would meet the new EPA emissions requirements. The breakthrough no one thought possible had in fact been made.

As the 1975 model year drew nearer, there were dire predictions that catalytic converters might create more pollution than they would eliminate. The chemical reaction in the converter that changed hydrocarbon and carbon monoxide pollutants into harmless water vapor and carbon dioxide also transformed tiny amounts of sulfur in the exhaust into sulfate. The sulfates, in turn, reacted with water vapor to form corrosive sulfuric acid. Tests proved, however, that the amount of sulfuric acid produced by a converter was so small that it posed no significant danger or threat to the environment.

There were also fears that the high temperatures inside the catalytic converter might create a potential fire hazard. A test was conducted for the benefit of the press by the EPA where several cars equipped with catalytic converters and several without were parked in ankle-deep dry grass. The cars were allowed to idle while onlookers waited to see what happened. To the delight of EPA officials, only one grass fire resulted—and that was under a car that was not equipped with a catalytic converter.

A NEW TYPE OF "CARB"

While the EPA was fighting with the automakers in Detroit over nationwide clean air standards, the state of California decided to take matters into its own hands. Los Angeles has always had the worst air pollution of any major metropolitan area in the nation because of its geography and urban sprawl. Warm air over the Los Angeles basin forms an "inversion layer" which traps pollution underneath. The surrounding mountains prevent the wind from blowing the pollution away, so a huge dome of stagnant, polluted air hangs over Los Angeles like a cloud.

What many people don't know is that the problem existed long before Los

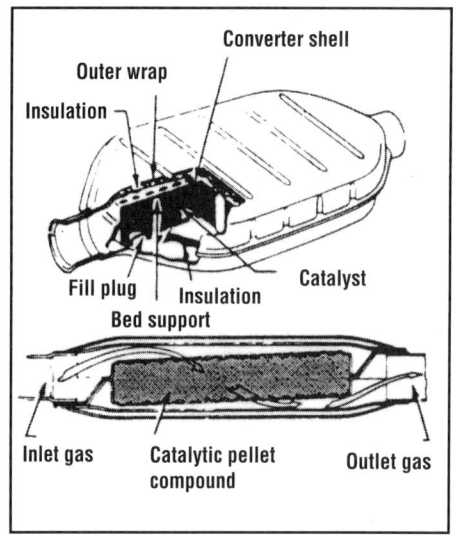

Skeptics predicted that the catalytic converter would actually create more pollution or start fires. However, the catalytic converter has proven to be one of the most effective components for reducing automotive exhaust pollution.

CALIFORNIA CLEAN AIR STANDARDS						
HC:	0.41	0.25	0.13	0.08	0.04	0.0 ppm
CO:	7.0	3.4	3.4	3.4	1.7	0.0 ppm
NOx:	0.4	0.4	0.4	0.2	0.2	0.0 ppm
Model Year:	1982	1993	1994	1996	1997	1998
Percent of Fleet:	100	40	10	25	2	2

California decided that it could not wait for the federal government to act, so it formed CARB (California Air Resources Board) to enact its own regulations governing automotive emissions. This chart illustrates how California hopes to reduce and eliminate automotive pollution.

Angeles became a city. The native Americans who lived in the area long before the first missionaries and settlers arrived were familiar with the perpetual haze created by natural sources of pollution.

In any event, Californians did not want to wait for federal emissions standards so they created the California Air Resources Board (CARB) to implement their own clean air standards. CARB became the nation's leader in establishing model legislation for future federal emissions standards. CARB required tougher emissions standards than those in the other 49 states. As a result, certain engines and drivetrains that were offered in the rest of the country (known as a "49-state legal" car) were not available in California and vice versa because of the different emissions standards. Consequently, California consumers soon found that they had to pay higher prices for new cars because of the added pollution equipment.

The situation in California created a serious problem for the automakers because they had to build cars capable of meeting one set of standards in California and another set of standards in the other 49 states. But they soon turned it to their advantage. Because California's standards were one to two years ahead of the rest of the country, it gave the car companies an opportunity to get the bugs out of their new emissions-control systems before the equipment was installed on the rest of their cars—which meant Californians sometimes had to deal with unproven and troublesome emissions-control systems.

Emissions Inspections

One of the approaches CARB undertook to ensure clean air for the Los Angeles basin was to implement a program of periodic motor vehicle inspection (PMVI). The idea behind PMVI (or inspection and maintenance (I/M) as it is also called), is to inspect vehicles once a year to make sure their emissions-control equipment is in good working condition (see Chapter 15 for more details). This is accomplished by subjecting the vehicle to a tailpipe emissions test and an underhood inspection. The tailpipe test certifies that the vehicle's emissions are within the limits set by law. The underhood inspection verifies that the pollution control equipment has not been disconnected or tampered with.

Though most people say they're for clean air, most do not like the inconvenience, hassle and expense of having to subject a vehicle to an annual emissions inspection. Even so, government regulators are committed to annual inspections as the best means of ensuring continued compliance with air

At first, enthusiasts responded to emissions controls by simply ignoring the EPA's mandates on emissions-control devices, by either unhooking, circumventing or eliminating them altogether. Photo by Michael Lutfy.

quality standards. The EPA wants every major metropolitan area of the country that doesn't meet federal air quality standards to eventually implement and strictly enforce automotive emissions testing programs. Two-thirds of the states have such programs either statewide or in major metropolitan areas, and it won't be long before it is in every state.

SIDE-STEPPING THE LAW

Restrictions in the 1970 Clean Air Act made it illegal for professional mechanics to remove, disconnect or tamper with factory emissions controls. But at the time there were no restrictions preventing vehicle owners from making a few of their own "improvements," such as removing or disconnecting various emissions-control devices and switching to less expensive leaded regular gasoline. So, many people replaced their catalytic converters with straight pipe "test tubes," unhooked a lot of the "pollution junk" and knocked out the restrictor in the fuel tank filler tube so they could switch to leaded gasoline. Although these modifications rarely improved performance unless done by a person who truly understands the subtleties of internal combustion and is willing to take a total systems approach, they became popular among enthusiasts. This raised the ire of the Environmental Protection Agency (EPA), which did not like the idea of people removing, disconnecting or tampering with the government mandated emissions controls. That's why CARB and the EPA found it necessary to push annual vehicle inspections.

In California, the annual inspection effectively discouraged tampering by making it very difficult for individuals to sneak vehicles with modified or missing emissions controls past the state inspectors. Inspection programs in other states, though not as tough as those in California, had a similar effect. The incidence of tampering and fuel switching decreased as motorists got the word that getting caught could be expensive. Missing emissions-control components had to be replaced, and those that were unhooked or made inoperative would have to be reconnected or fixed.

This period in emissions history could best be described as a game of cat-and-mouse between those who were dedicated to enforcing the emissions legislation and those who thumbed their noses at the law. "For off-road use only" labeling of aftermarket performance products was virtually a joke, except in California where the strict vehicle inspection program did a good job of discouraging violators.

Today, it's a different story. The revisions in the Clean Air Act of 1990 have now made it illegal for anyone to remove, disconnect or tamper with any emissions-control component on any vehicle that's driven on a public road. Off-road-only parts are still available for true race car applications, but are no longer street legal anywhere in the U.S. In reality, the only places where one is apt to run afoul of the law using off-road parts on the street are areas where

Today it's a different story. Tougher laws and annual inspections now make it impossible (except for certain exemptions for older cars) to get a vehicle registered in many states without compliance with emissions control laws. However, as shown on this emissions-legal but heavily modified 5.0-liter Mustang, it is still possible to have performance and remain within the law. Photo by Michael Lutfy.

vehicles have to be inspected annually. Even so, the law says that any aftermarket performance parts that are installed on a street-driven vehicle must be "emissions certified," which means they won't cause the engine to produce more emissions than it did in stock trim (see Chapter 3 for further details).

COMPUTER CARS

Back to our historical perspective. In 1980 in California, and 1981 in the remaining 49 states, the first generation of computerized cars arrived with "feedback" engine management systems and electronic carburetion. Fuel, ignition and emissions-control functions became more integrated than ever before, and for most people the new computer cars were "too complex" to work on, let alone modify, for fear of hopelessly screwing things up or damaging the expensive electronic components. So, most serious performance enthusiasts shunned the new cars and the new technology in favor of older, less high-tech models—at least initially.

By the mid-1980s, carburetors were

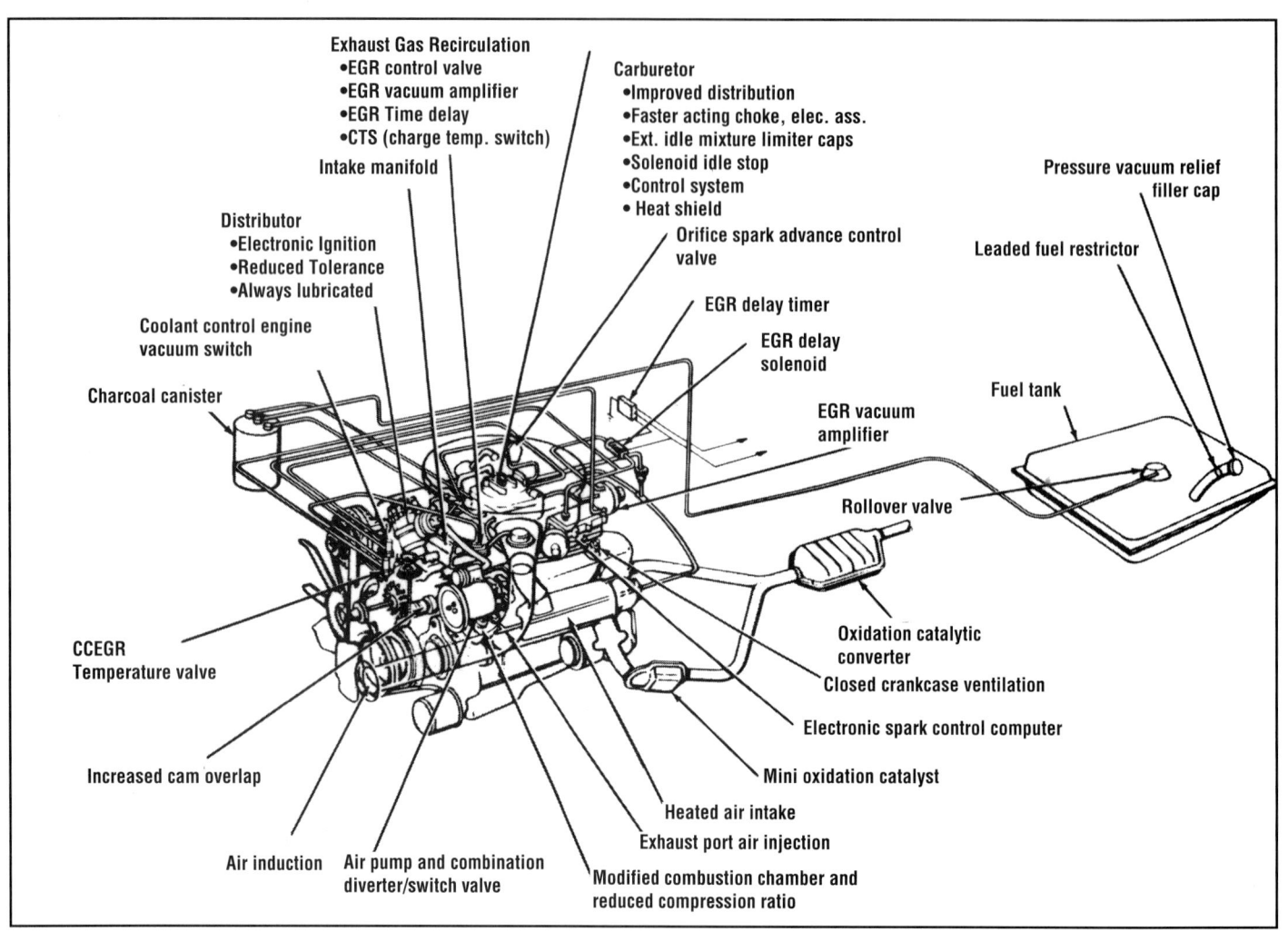

The emissions-control systems on today's cars are so complex and interconnected that any modifications you might make would almost surely hurt performance more than it would help it.

giving way to electronic fuel injection (EFI). Fuel injection was also initially met with skepticism but proved to be a welcome improvement over carburetion in terms of performance, driveability, fuel economy and emissions. First came throttle body injection (TBI) systems, which were followed by much better multi-point injection (MPI) systems where each cylinder is fed by its own individual fuel injector.

Once again we had to suffer through a period where technological changes outpaced the evolution of fuels. During this time, it was quite common for injectors to gum up with varnish because much of the gasoline that was being sold lacked sufficient detergent additives to keep the injectors clean. That problem is mostly history now, and the need for injector cleaning isn't what it once was.

Even so, cleaning a set of injectors can often make a noticeable improvement in performance as well as fuel economy and emissions.

Today we have computerized engine and drivetrain management, electronic multi-point fuel injection and sophisticated emissions-control strategies that actually result in the best of all worlds: low emissions, good fuel economy AND excellent performance and driveability. But the diversity of engine management systems, along with the complexity of the technology itself, has made it increasingly difficult for anyone who would like to tinker or modify anything under the hood. Add to this the new legal restrictions on vehicle modifications and you can appreciate why emissions diagnosis and repair has become a specialty unto itself. It takes a substantial amount of training and expertise, as well as up-to-date service literature and special tools, to perform any kind of serious diagnostic or repair work on a late-model powerplant with its computerized engine management system.

In an effort to make emissions diagnosis easier, the government is mandating a standardized approach called "onboard diagnosis" (OBD) for identifying and troubleshooting emissions performance problems. The OBD rules require a standard diagnostic connector on all 1995 model year vehicles so a common "scan tool" can be used to retrieve "trouble codes" from the vehicle's computer. The codes correspond to a specific fault that the computer has detected. The code or codes can then be used to begin a troubleshooting procedure that eventually allows you to pinpoint the exact cause of the problem—or so the theory goes. Trouble codes and onboard diagnostics have been around since the introduction of the first computerized cars in 1980, but they haven't necessarily made troubleshooting emissions and performance problems much easier. They are better than nothing, but they can also be misleading. A trouble code is not an end in itself, but a beginning. Unless an exact step-by-step diagnostic procedure is followed to rule out all the other various possibilities, you may or may not end up making the correct diagnosis and identifying what's really causing the problem. That's why computer cars with emissions and performance problems are often so hard to fix. Whether OBD will make any difference remains to be seen. For more about OBD, see p. 116.

THE FUTURE OF EMISSIONS CONTROL

In the future, we may get super-clean cars powered by hydrogen fuel, such as this prototype BMW. However, the problem is where to store the enormous tank. This BMW is experimenting with a cryogenic tank that keeps the hydrogen in a liquid state until it is used, therefore requiring less storage capacity.

Have we reached a plateau in automotive exhaust emissions? Hardly. California will require 2% of all the vehicles that are sold in that state to be "zero emissions vehicles" (ZEVs) by 1998, increasing to 5% by 2001 and 10% by 2003. ZEVs will have to be electrically powered since no other technology is currently capable of meeting a zero emissions requirement.

A certain percentage of California vehicles will also have to be "ultra-low emissions vehicles" (ULEVs). These will be a combination of super-clean gasoline-powered vehicles as well as those that run on alternative fuels such as propane, natural gas and alcohol.

Changes for the Near Future

One change that's recently been made is the introduction of "reformulated" gasoline in many major cities that don't meet federal clean air standards. Reformulated gasoline typically contains alcohol-based additives to provide extra oxygen for a slightly leaner, slightly cleaner burning mixture. In older cars, reformulated gasoline can make a measurable improvement in reducing carbon monoxide emissions.

I/M 240—Another change that's underway is the phase-in of the new "I/M

There's been a lot of press coverage of electric vehicles lately, but none is truly practical as all-around transportation yet (Chrysler).

240" inspection program. The new inspection program is being implemented primarily in cities with the dirtiest air. The new test is much more comprehensive than a simple idle emissions test in that it simulates the same urban driving test cycle that's used by the car makers to certify new vehicles

If the proposed clunker legislation becomes mandatory rather than voluntary, restoring cars such as this pristine Firebird convertible may not be possible. Such a law would virtually eliminate the specialty equipment high performance aftermarket, as well as the market for collector and classic cars. Get involved in stopping this legislation by contacting the Specialty Equipment Market Association (SEMA), 1575 South Valley Vista Drive, P.O. Box 4910, Diamond Bar, CA 91765-1647. Tel: 909/396-0289. Fax: 909/860-0184.

for emissions compliance. It's called I/M 240 because the test is 240 seconds in duration. It involves "loaded mode" testing with the vehicle being run at various speeds on a dynamometer to check emissions under various driving conditions. The new test also measures NOx emissions for the first time, and includes a check of the charcoal canister and fuel system to make sure fuel tank vapors are being properly contained. The new I/M 240 test is also much tougher to pass, which means more and more motorists will discover they have an emissions problem that requires fixing. At the same time, the "waiver" limit (the amount of money motorists have to spend on repairs before the government will allow their vehicles to pass inspection even if they still fail to meet the emissions standards) has been upped to $450! This is a big increase from the $15 to $70 waiver limits that were formerly commonplace. The bottom line is: emissions compliance is getting tougher and emissions repairs will get more expensive. For more details, see p. 116.

Big Brother—Although the EPA is committed to its I/M 240 program, some states and municipalities are considering the use of mobile roadside sniffers to catch "gross polluters." Such devices can single out vehicles that are producing extremely high levels of pollution so their owners can be cited for violating clean air standards. Sound like Big Brother?

Clunker Legislation

As for new legislation, one course of action that's being pursued at both state and federal levels is "clunker" legislation that would pay people to voluntarily scrap their older vehicles that emit higher levels of pollution. Such proposals are based on the notion that permanently removing older vehicles (1980 and earlier) from the road has a more beneficial impact on cleaning up overall air quality than passing stricter emissions standards and inspection programs. The idea is especially appealing to stationary polluters such as utilities and refineries whose clean-up costs are measured in hundreds of millions of dollars. In exchange for each older vehicle that's crushed, stationary polluters could purchase or be granted "pollution credits" to offset their own emissions.

Clunker legislation is viewed as a major threat by most aftermarket parts and service suppliers as well as trade associations because of the impact such a program would have on their industry. Junking significant numbers of older vehicles would reduce the number of such vehicles that currently need parts and service. A massive scrapping campaign might also destroy many potentially valuable collector cars and make it harder for the owners of older vehicles to find spare parts or older vehicles for restoration.

The biggest threat is if clunker legislation becomes mandatory rather than voluntary. If the government says you can't drive a car once it reaches a certain age, it would obviously limit your choice of vehicle. Such a rule would seriously limit the high performance automotive aftermarket and the market for collectible classic cars. ■

MAINTENANCE & REPAIRS: A QUESTION OF ENVIRONMENTAL RESPONSIBILITY

CHANGING OIL? Though the EPA has decided not to label used motor oil a hazardous waste (a move that would have imposed added paperwork and required expensive disposal procedures for service businesses that do a lot of oil changes), some municipal landfills will no longer accept used oil in normal household garbage. Some people still sneak the stuff into the garbage, pour it down the drain, dump it in a vacant lot or otherwise dispose of it in an environmentally unfriendly way. This sort of behavior can have serious consequences if you or your neighbors happen to rely on well water for drinking. You may also be risking a fine of up to several hundred dollars for illegal dumping if someone sees you in the act or can prove you did it.

One quart of used motor oil, if it seeps down through the soil and leaches into the ground water, can contaminate up to a million (yes, we said a million) gallons of drinking water, making it unfit for human consumption.

So what's the environmentally responsible way to get rid of used motor oil, transmission fluid or gear oil? Pour it into a sealed container and take it to a service station, quick lube shop or other facility that accepts used oil for recycling. Some auto parts stores as well as neighborhood recycling programs will also accept used motor oil. In most cases, the used oil is not recycled into oil again but is used to make other petroleum based products or is burned.

Do not mix anything else in with oil that is destined for recycling. Adding any type of chlorine-based solvent, paint thinner, degreaser, antifreeze, insecticide or other automotive or household chemical may transform your used motor oil into a hazardous waste. Recyclers won't want the stuff then because it will contaminate their storage tanks and require them to have the entire batch disposed of as hazardous waste (which isn't cheap!). For that reason, many service stations and quick lube shops won't accept used motor oil from someone they don't know.

CHANGING COOLANT? Disposing of used coolant or anti-freeze in a responsible manner is much harder than recycling used motor oil. Why? Because there are few facilities for recycling it. Some shops now have equipment for cleaning and recycling their own customers' anti-freeze, but few would want to accept used antifreeze from a do-it-yourselfer—unless you were willing to pay them to recycle it for you. One alternative here is to not change your own coolant. Take your vehicle to a shop that recycles anti-freeze and have professionals change and replenish your coolant.

Ethylene glycol anti-freeze is poisonous and will kill plants as well as animals. Most municipalities still allow you to dump the stuff down the drain where a combination of dilution and natural breakdown will

Changing your own oil? Then dispose of your used oil and filter responsibly. Dumping it into the ground not only pollutes our drinking water, but it also qualifies you for a hefty fine and in some cases, imprisonment.

eventually render the stuff harmless. But don't dump it on the ground or leave it sitting around in an open container where a pet or small child might find it and decide it tastes good. When you refill your cooling system with fresh antifreeze, you might consider using one of the new "low-toxic" anti-freezes that does not contain ethylene glycol but contains propylene glycol instead.

DOING A TUNE-UP? Whether you live in an area that requires annual emissions testing or not, you should follow the idle speed, timing and plug gap specifications on your underhood emissions decal when maintaining or tuning your engine. Timing, in particular, can have a major impact on emissions. Idle speed may not sound like a big deal, but setting it either too high or too low may cause enough of an increase in idle emissions to make your vehicle flunk an emissions test.

REPLACING A BATTERY? Batteries contain lead, a heavy metal (like mercury and cadmium) that is toxic. Many states now have laws prohibiting the disposal of used batteries in the trash or landfills. The battery must be returned to a battery retailer or dealer to be recycled. If you live in an area that does not have such legal restrictions on battery disposal, it's still a good idea to exchange your old battery when you buy a new one. The store will usually give you a few bucks for it, then sell it to a recycler.

REPLACING TIRES? Many municipal trash haulers and landfills will not accept used tires because they're bulky and don't bury well in a landfill. They tend to work their way back up to the surface, and provide unwanted habitats for various pests. In some places, used tires are being shredded and used as ground cover in parks. Ground-up tires are also used to add resiliency and durability to asphalt paving. In Taiwan, they make old tires into sandals. Tires can also be cooked in an airtight oven to recover the oil and other ingredients that are in them. They can even be burned in a municipal incinerator to make electricity. But for the individual who's trying to dispose of a set of used tires, short of selling them at a garage sale or trading them in when you buy new ones, there aren't a lot of options except for recycling or retreading. Most heavy-duty truck tires are retreaded several times before they end up in a scrap pile somewhere. But automotive retreads are not as popular because there's not as much difference in price between them and new ones.

RECHARGING YOUR AIR CONDITIONER? R-12 Freon automotive refrigerant contains CFCs, chemicals thought to be responsible for a thinning of the earth's protective ozone layer. If your A/C system is low on refrigerant, it means the refrigerant is leaking into the atmosphere. The environmentally responsible course of action, therefore, would be to find and fix the leak before recharging the system with refrigerant (which can no longer be purchased in auto parts stores).

To ensure your air conditioning system is not leaking, check it thoroughly and frequently for leaks using professional equipment such as this electronic leak detector.

THE SKY IS FALLING

Will the thinning of the Earth's protective ozone layer be "the end of life on Earth as we know it today" as some of the tabloid newspaper and television reports have claimed? Or is it an issue that's still (pardon the pun) up in the air?

For more than two decades, scientists have hotly debated the causes and possible effects of what appears to be a mysterious deterioration of the Earth's protective "ozone layer" that shields the planet from 95% to 99% of the sun's ultraviolet radiation. The ozone layer is actually a region of the upper stratosphere that extends from 12 to 21 miles above the Earth's surface. The air there is very thin and very cold (-70 degrees F on average). What makes it unique is that a special form of oxygen called "ozone" is formed by the interaction of sunlight and air.

WHAT IS OZONE?

Ozone is a pungent bluish gas that gives the sky its blue color on a sunny day. It is a special form of oxygen that contains three atoms linked together instead of the usual two. It forms in the upper atmosphere when ordinary oxygen molecules are split apart by ultraviolet radiation from the sun. Some of the molecules then recombine in triplicate instead of pairs, creating ozone. According to some estimates, 300 million tons of ozone are created and destroyed daily by natural processes in the stratosphere.

Ozone is also created in the lower atmosphere by lightning, electrical arcing (as when welding or running an electric motor), and by the interaction of sunlight with unburned hydrocarbons (HC) and oxides of nitrogen (NOx) from automobile exhaust.

Because of its unique makeup, ozone is a highly reactive chemical. As such, it has many beneficial uses, such as an antiseptic and bleaching agent. But when

The ozone layer in the upper atmosphere protects the earth from harmful ultraviolet radiation. But when ozone is created down low, by the reaction of automotive emissions and sunlight, and becomes part of the air we breathe, it becomes a harmful pollutant.

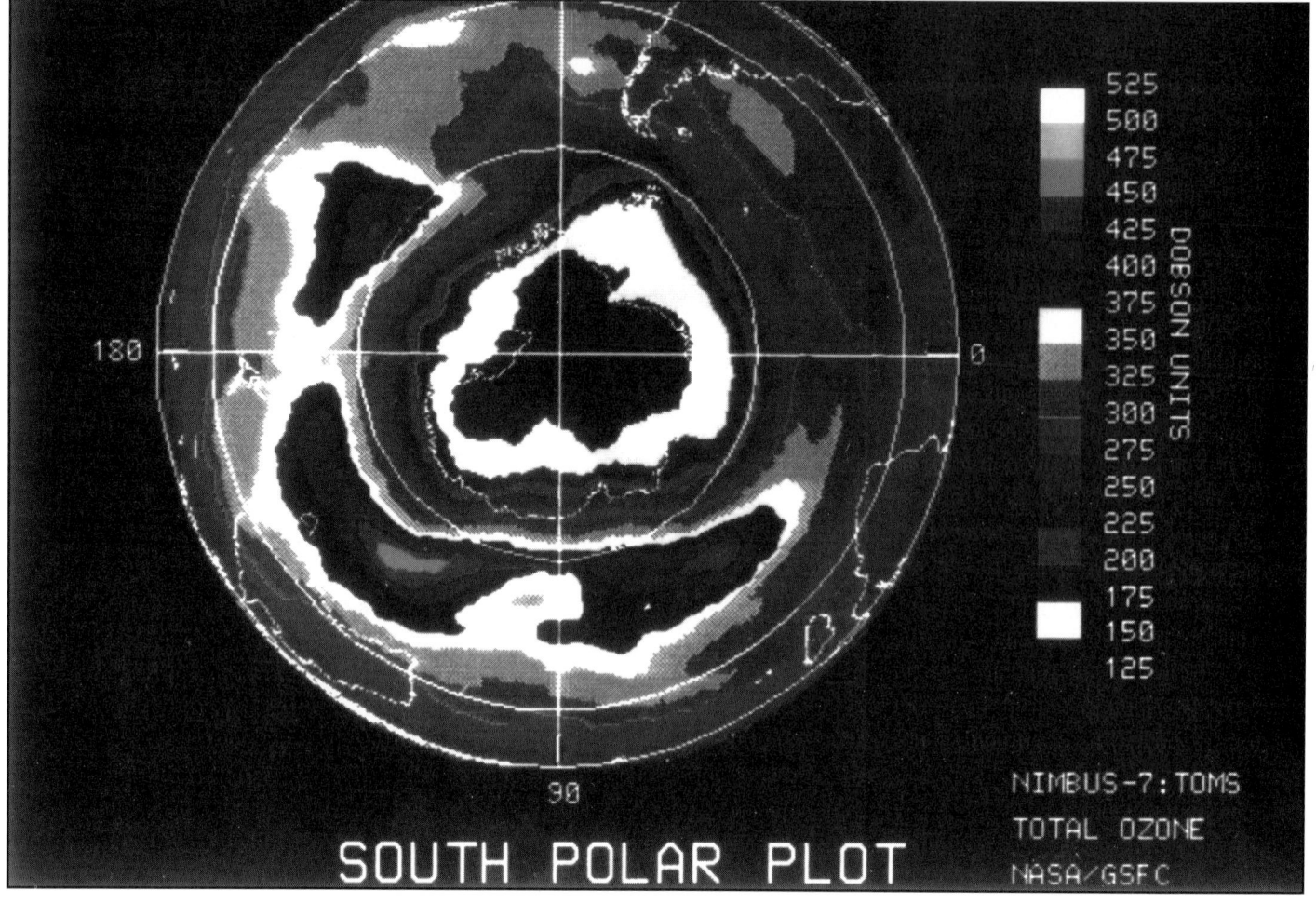

This satellite photo confirms that something is causing a huge hole in the ozone layer over the South Pole. The dark area indicates an absence of ozone in the upper atmosphere. Some scientists think it is a natural seasonal phenomenon rather than a manmade problem.

it's in the air we breathe, it's a pollutant. Ground level ozone is undesirable because it irritates the eyes and lungs, makes breathing difficult and aggravates heart and lung disease. It can also give you headaches, chest pains and make you cough. You can blame ozone for that tired feeling on hot summer days when thermal inversion layers form and trap ozone close to the ground. That's why the EPA issues "ozone alerts" when air pollution levels get too high.

Yet at higher altitudes, ozone is extremely beneficial because it blocks most of the sun's potentially harmful ultraviolet radiation. Without it, the northern latitudes would receive as much ultraviolet exposure as latitudes near the equator, while exposure near the equator would be even more intense.

Although the ozone layer itself is miles thick, the actual amount of ozone contained in the layer is quite small. There's only one part of ozone in nearly half a million parts of air! If all the ozone in the stratosphere could be brought down to ground level, it would form a layer only three feet thick. Yet this small amount of ozone is all it takes to shield life on Earth from harmful solar radiation.

Down to Earth

Why don't we just let the ozone pollution down low make up for the loss up high? Unfortunately it doesn't work that way. One reason is that ozone formed close to the ground never reaches the stratosphere. It stays down here and is "used up" before it ever gets very high. The other reason has to do with the complex chemistry of the atmosphere itself and the effects of heating and cooling on wind patterns and the weather.

Ozone has to be high up in the atmosphere to block the sun's radiation before it reaches the ground to provide any real protection. When ultraviolet light is allowed to penetrate deeper into the atmosphere, it forms ozone down close to the ground, where it pollutes the air and doesn't block the light.

Allowing solar radiation to penetrate deeper into the atmosphere also affects the altitudes at which air is heated, and consequently global wind patterns, which could cause hard-to-predict weather changes and increase global warming.

Disappearing Ozone

According to some scientific findings, chlorine molecules in the stratosphere, apparently from manmade CFCs (chlorofluorocarbons, including R-12 refrigerant from automotive air conditioning systems), appear to be eating up the ozone layer at an alarming rate. Chlorine interacts with ozone and breaks it down into ordinary oxygen. The chlorine is not consumed in the chemical reaction, so it remains free to repeat the process over and over again as long as there's sunlight and more ozone to destroy.

An enormous "hole" in the ozone layer, half the size of Antarctica, has opened up over the South Pole (see p. 12), and the fear is that the same thing may soon happen over the Northern latitudes. The ozone hole over Antarctica has been well documented by aircraft, balloon and satellite observations, but the debate continues as to whether this is a naturally occurring phenomenon or something that's due to manmade pollution.

At the present estimated rate of destruction, some scientists say there may be a 7% to 18% reduction of the Earth's protective ozone layer within the next 20 to 40 years. Nobody is sure of the exact timetable or the ultimate amount of damage that will eventually be done to the ozone layer, but based on present usage trends of CFCs, the outlook isn't good. Any reduction of the ozone layer allows more ultraviolet radiation from the sun to reach the Earth's surface. The increase in solar radiation that's predicted to occur based on the current rate of ozone destruction could have potentially devastating effects on the world's food supply, the ocean's food chain, global climate, prevailing weather patterns, human susceptibility to skin cancer and eye cataracts, and possibly even the planet's ability to maintain the ongoing balance between oxygen and carbon dioxide in the air we breathe. There are studies that claim an increase in the incidence of skin cancer in New Zealand due to a thinning of the ozone layer over that part of the world (though no one has yet shown concrete evidence to support such a link). Even so, some scientists predict the incidence of skin cancer will increase 4% for every 1% reduction in the ozone layer.

Of far greater concern is the effect increased ultraviolet radiation may have on life itself. Studies have shown that many important food crops give lower yields when subjected to excessive sunlight. There is also concern that increased ultraviolet solar radiation may disrupt the microscopic phytoplankton that lives near the surface of the world's oceans. Phytoplankton not only forms the basis for the entire marine food chain, but it plays a major role in regulating the composition of the atmosphere. Phytoplankton removes carbon dioxide (a gas that contributes to the "Greenhouse effect") from the air while replenishing the oxygen we need to survive.

So environmentalists are convinced that unless something is done to curtail the use and release of manmade CFCs such as R-12 automotive refrigerant into the atmosphere, ozone destruction will continue to accelerate with potentially catastrophic consequences for all of mankind as well as planet Earth.

On the Other Hand—In spite of all the research that's been done on the ozone layer, what's causing the problem and its possible consequences, there remains a fair amount of controversy over several unanswered questions. Among these are:

• How do manmade CFCs, which are several times heavier than air and settle to the ground when released into the atmosphere, end up in the stratosphere? The only explanation that's been given is that the CFCs apparently disperse in the atmosphere and don't break down until they reach the stratosphere.

• Are manmade CFCs the real culprit? Some scientists now think that chlorine from volcanoes may be causing the changes, not manmade CFCs.

• The hole in the ozone layer above

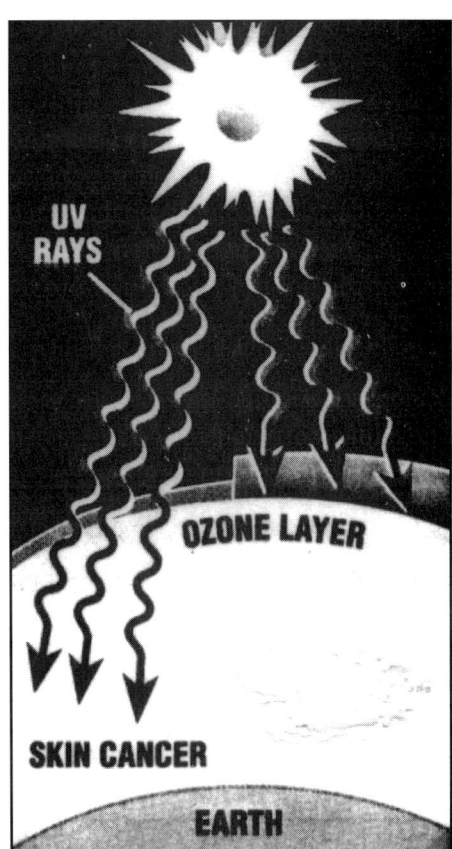

The ozone layer is a region of the stratosphere that shields us from the sun's harmful ultraviolet radiation. The scientific debate continues to rage as to whether or not the upper, protective ozone layer is really thinning as a result of pollution from manmade CFCs, such as the refrigerant used in automotive air conditioning systems.

Antarctica has occurred there because of the extreme cold and unique climatology over the pole. Could the hole be part of some naturally occurring long-term cycle? We've only been monitoring the ozone layer for about 30 years and nobody seems to know if the hole is part of a long-term cycle or not.

• Does a hole in the ozone layer over the South Pole necessarily mean holes will open up elsewhere? Nobody knows. A special space shuttle mission by NASA failed to turn up any evidence one way or the other.

• Would the 7% to 18% reduction in the ozone layer be as catastrophic as the alarmists predict? Until it actually occurs, there's no way to know for sure. Even the best atmospheric models that

TERMS YOU SHOULD KNOW

CFC—Abbreviation for chloro-fluorocarbons, a family of manmade chemical compounds that includes R-12 (Freon) automotive refrigerant.

FREON—A registered trademark of the DuPont Corporation for their family of CFC refrigerants, including R-12 and others.

GREENHOUSE EFFECT—A gradual warming of the Earth's climate as a result of increased heat retention due to atmospheric changes. Certain gases (primarily carbon dioxide from the burning of fossil fuel, but also CFCs) increase the retention of heat from the sun in the atmosphere. Global warming could cause major changes in established weather patterns and melt much of the polar ice caps.

OZONE DEPLETION—A thinning of the ozone layer that is taking place apparently because of manmade CFCs in the stratosphere, destroying ozone at a faster rate than it is being formed by natural processes. Without sufficient layers of ozone, increased skin cancer rate, changing weather patterns and global warming are the likely result.

OZONE LAYER—A region in the stratosphere 12 to 21 miles up where the air is very cold and thin, and ozone is found in high concentrations. The ozone layer is continually replenished by solar radiation and screens out about 95% to 99% of the sun's ultraviolet radiation. Ozone is a pungent bluish gas that gives the sky its blue color on a sunny day. It is a special form of oxygen that contains three atoms linked together instead of the usual two.

R-12—Dichlorodifluoromethane, commonly known as "Freon," the type of CFC-based refrigerant that's used in all 1992 and earlier automotive A/C systems. It will no longer be produced after 1995.

R-134a—The "ozone friendly" CFC-free replacement for R-12 refrigerant in automotive A/C systems. It is not a direct substitute, however, and requires various system modifications before it can be used.

REFRIGERANT RECOVERY AND RECYCLING—A mandatory requirement for any shop that performs A/C service work. Recycled R-12 and R-134a refrigerant must meet government standards for purity in terms of moisture content and other contaminants.

STRATOSPHERE—The region of the atmosphere that starts about 5 miles up and extends to about 50 miles up.

ULTRAVIOLET RADIATION—The "shorter" wavelengths of light from the sun that are invisible to the naked eye and cause sunburn. It is the job of the ozone layer to screen out the UV rays.

are run on supercomputers give conflicting results.

- Is it already too late to do anything to stop or reverse the problem? The life span of CFCs in the atmosphere can range from 22 to 110 years depending on the particular chemical (see CFC chart), so even a total ban on CFCs today would mean CFCs would still be destroying stratospheric ozone a century from now.

- Conspiracy theories abound. It's interesting to note that DuPont's patent on Freon (R-12) recently expired, which means any manufacturer with the desire to do so can now make R-12 without having to pay royalties to DuPont.

Such books as *Holes In The Ozone Scare* by Roger Maduro and Ralf Schauerhammer challenge current assumptions about the ozone issue and raise a number of important questions as to whether we should be banning R-12 at all. But debating the validity of the issue or the merits of various approaches to solving the so-called problem are beyond the scope of this book. Like it or not, R-12 is on its way out and "ozone-friendly" CFC-free refrigerants such as R-134a are on their way in.

CFCs

When a scientist named Sherry Rowland first proposed a possible threat to the ozone layer from manmade CFCs back in 1973, his theory was met with considerable skepticism—until subsequent measurements of ozone levels by other scientists seemed to confirm that something was indeed going on. The thinning of the ozone layer over the South Pole was initially blamed on the use of CFCs as propellants in aerosol products (hair spray, deodorant, etc., which accounted for nearly 50% of CFC usage

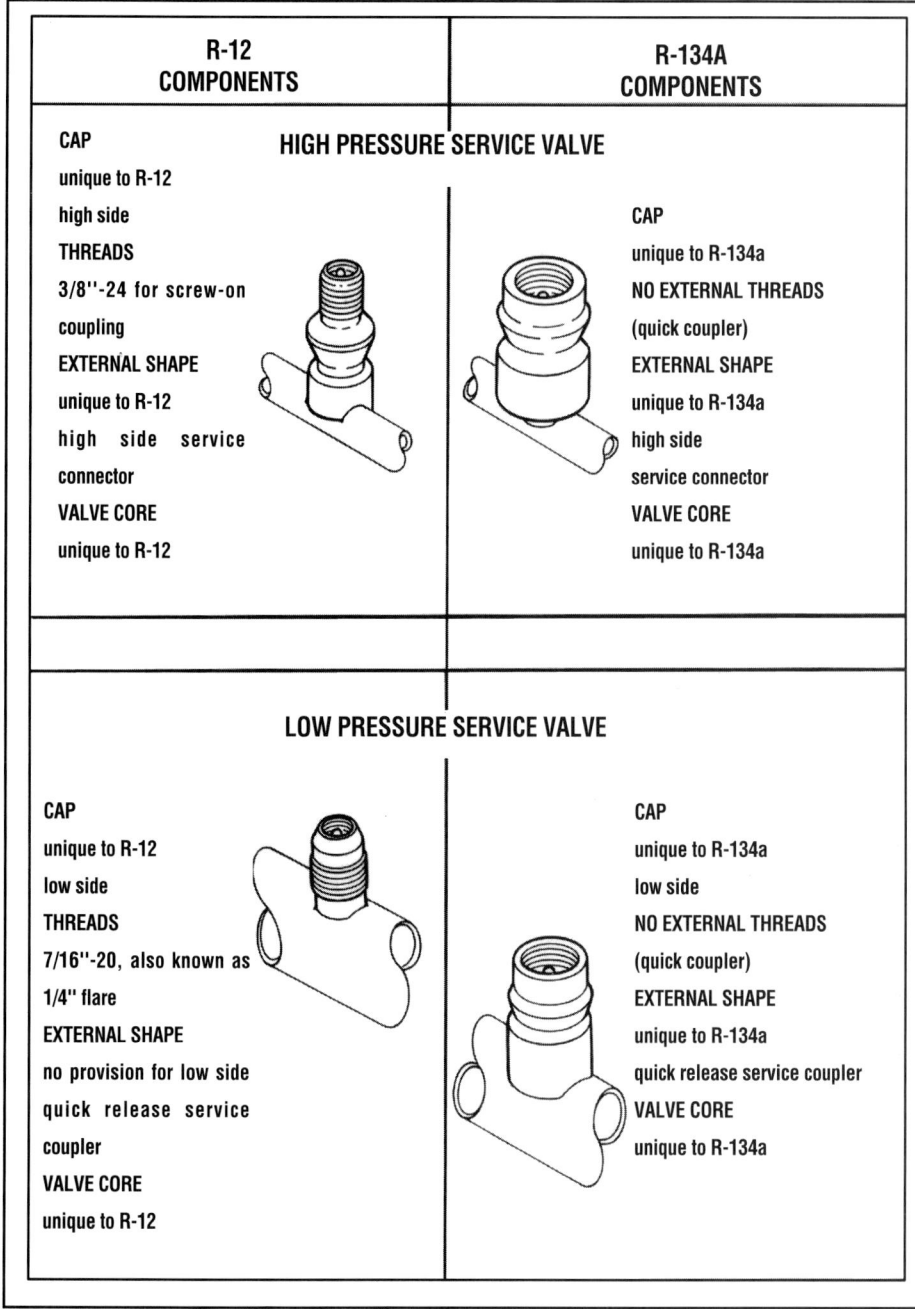

A comparison of R-12 and R-134a service valve fittings.

at the time). So in the late 1970s, CFC propellants in aerosol cans were banned in the U.S. as well as a number of European countries. Unfortunately, the rest of the world didn't follow the initiative and did nothing. Even so, since we were the leading user of CFCs at the time, it was thought the aerosol ban would make a difference. It didn't. The ozone hole kept reappearing over the South Pole. What's more, the growth in use of CFCs worldwide and for many other purposes than aerosols and automotive air conditioning more than offset any reduction in emissions achieved by the aerosol ban. Use of CFCs to produce plastic foam packaging and insulation in particular had grown significantly.

Environmentalists were convinced that some kind of coordinated international effort was necessary to curtail the uncontrolled growth in CFC usage. It was also apparent that substitutes for CFCs would have to be developed, a problem that would require considerable research and development by CFC producers and have a major impact on CFC users.

The result was the Montreal Protocol. Signed by 24 nations on September 16, 1987 in Montreal, Canada, the historic agreement established a plan to cut the release of CFCs into the atmosphere 50% by 1999. The plan called for freezing CFC production at 1986 levels, reducing CFC consumption 20% by July 1993, and an additional 30% by July 1998. Allowances would be made for 10% to 15% growth in consumption among developing nations.

The timetable was devised to allow sufficient time for substitute products to be developed and for an orderly reduction in CFC usage. As more scientific data came in, however, the consensus shifted toward even more stringent reductions in CFC usage with the goal being a total phase out by the turn of the century. Then the environmental bandwagon really got rolling and the ban was moved up to the end of 1995.

The R-134a Alternative

The new "ozone-friendly" CFC-free refrigerant that's replacing R-12 is DuPont's "SUVA," also known as "HFC134a" or "R-134a." Unfortunately, R-134a has a few drawbacks. For one, it has smaller molecules, which means it will leak out of older A/C systems unless all the hoses are replaced with nylon barrier style hoses.

The A/C systems in 1993 and newer vehicles that are charged with R-134a can be distinguished from R-12 systems by the type of service fittings used and by labels located on the firewall. Unfortunately, some opportunistic entrepreneurs have been selling adapters that allow R-134a to be put into R-12 systems and vice versa. If the two are intermixed, you can end up with evaporator core icing problems as well as desiccant problems (R-134a requires either XH-7 or XH-9 desiccant).

R-134a systems currently use polyalkylene glycol (PAG) oil rather than mineral oil (the two don't intermix). There

Many 1993 and most 1994 vehicles including this Dodge Ram pickup now have R-134a refrigerant in their A/C systems. All Chrysler Corp. vehicles built after January 1, 1994 will carry R-134a. All other car manufacturers have acted in a similar fashion.

are different formulations of PAG oil and currently no such thing as a "universal" PAG oil. You must use the type of PAG oil specified by the vehicle manufacturer. The wrong type may not be compatible with the seals or desiccant that's used in the system. Retrofit conversions must use an oil that's compatible with R-12 mineral oil unless the system can be thoroughly cleaned and all traces of the old compressor oil removed (a difficult if not impossible task).

Converting older vehicle with R-12 A/C systems to R-134a is going to be a booming business in the latter half of the 1990s, but it's also going to be expensive for motorists. A typical conversion may cost upwards of several hundred dollars, depending on which hoses, seals and other parts have to be replaced. On some, it may even be necessary to replace the compressor, because the old compressor's seals may not be compatible with the new refrigerant and lubricant.

Replacing R-12—As R-12 production comes to an end, the existing supplies of R-12 refrigerant will gradually

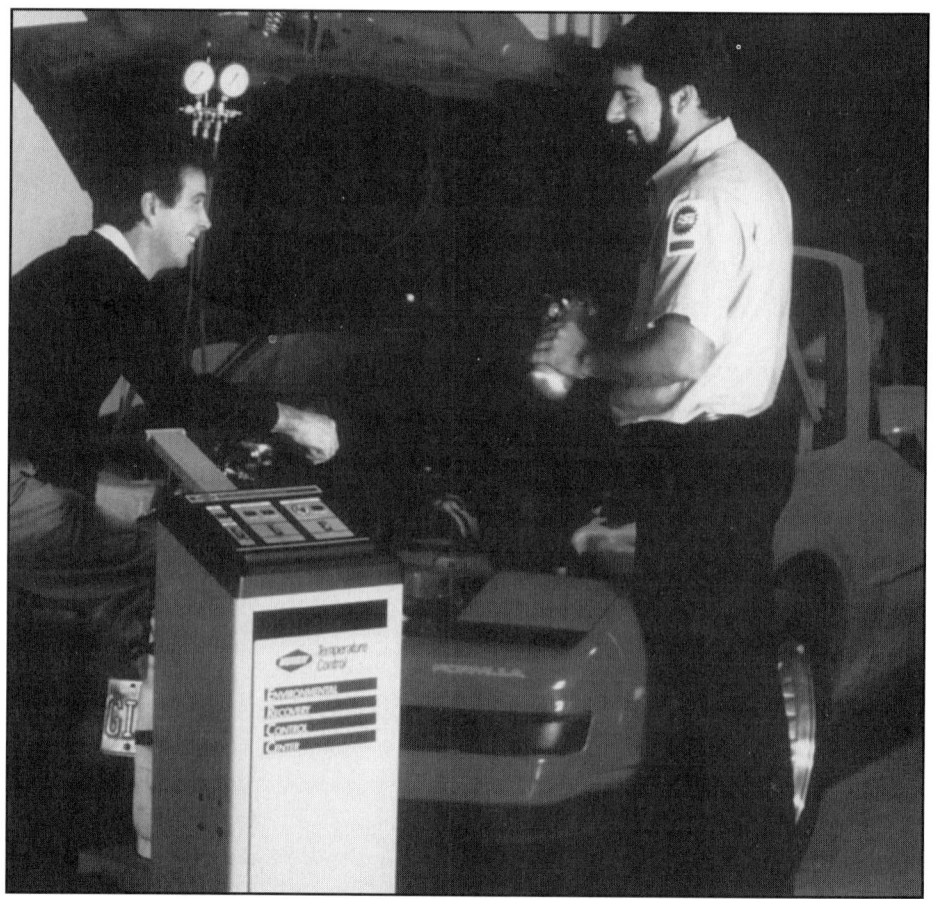

Additional training is generally required to learn how to recover R-12 and convert over to R-134a. Most shops won't touch a system that has another refrigerant in it. Don't add any bootleg stuff.

disappear, creating shortages and a rapid increase in price. So if you have an R-12 air conditioning system in your vehicle (which almost all 1992 and earlier vehicles do), you'll find yourself paying more and more for refrigerant as the years go by. Eventually you'll reach a point where you'll have to decide between converting your A/C system to R-134a or doing without your air conditioner.

There is a third option, but one that is fraught with danger. Some people will try to get around the rising price of R-12 and dwindling supply by using some type of blend or bootleg refrigerant in their A/C system. A number of chemicals will work as refrigerants, but some are corrosive while others are flammable. One thing you don't want is a pressurized flammable gas that could leak into the passenger compartment in the event of an accident!

The EPA has moved to ban flammable substitutes on the grounds of safety. Blended refrigerants made from R-22 (which is also a CFC) or other chemicals are also being banned to prevent cross-contamination of refrigerant recovery and recycling equipment. What's more, it may get to the point where shops will refuse to service vehicles that contain anything other than R-12 or R-134a refrigerant. So don't be tempted to put anything other than the correct type of refrigerant in your vehicle's A/C system!

CFC Refrigerant Laws

Since January 1, 1993, any shop that performs A/C work (even if it's only one job a year) must comply with the following regulations:

• The shop must have technicians trained and certified by an approved testing program such as those offered through the International Mobile Air Conditioning Society (IMACA), the Mobile Air Conditioning Society (MACS), the National Institute for Automotive Service

Taboo! It's now illegal to discharge R-12 Freon into the atmosphere. If you're caught doing this, you're setting yourself up for a hefty fine.

R-12 vs. R-134a SYSTEMS

	R-12	R-134a
Trademark: Mobile	Freon-12	SUVA Trans A/C (1/2" ACME)
Stationary	Freon-12	SUVA Cold-MP (1/4" Flare)
Boiling Point (F):	-15.7	-21.6
Vapor Pressure @ 70° F:	70.2	70.7
Compressor:	existing	similar
Discharge Pressure:	existing	10-20% higher
Lubricant:	mineral oil	Polyalkylene Glycol (PAG)
Hoses:	nylon lined (some)	nylon lined (all)
Desiccants:	4A-XH-E, XH-7, XH-9	XH-7, XH-9
Fittings: Low Side	1/4" Flare Threaded	16mm quick disconnect
High Side	3/8" Flare Threaded	13mm quick disconnect
Recyclable:	Yes (required now)	Yes (required 1/1/95)
Service Hoses:	Blue/Red/Yellow 1/4" Flare (female)	Blue/Red/Yellow w/Black Stripe 1/2" ACME (male & female)
Visible Indication of Refrigerant Type:	None, except on can or cylinder	System tags on hoses, marked on gauges, hoses, recycling and recharging equipment, cans, cylinders

A comparison of R-12 and R-134a refrigerants for automotive air conditioning systems.

New legislation means the do-it-yourselfer won't be topping up his A/C system with those 12 oz. Freon "grenades" anymore. The only people who can legally work on A/C are certified professionals using proper recovery/recycling equipment.

Excellence (ASE), or others that have received EPA approval.

• Have an EPA-approved and Underwriters Laboratories (UL) certified R-12 recovery and recycling unit. This type of equipment costs upwards of $3000 per machine. A separate machine is required for recovering and recycling R-134a.

The mandatory certification, recovery and recycling requirements do not apply to do-it-yourselfers. But there's no need to since you can no longer purchase the small 12- or 14-oz cans of refrigerant in retail stores—unless you're certified. Section 609 of the Clean Air Act revisions stipulate that only "certified" technicians (those who have taken and passed an EPA-approved refrigerant recovery and recycling test) may purchase refrigerant. Any store that sells R-12 refrigerant to a non-certified individual today is risking a fine of up to several thousand dollars! In states that have not banned the small cans altogether, certified professionals can still legally purchase the 12- or 14-oz cans of refrigerant for small jobs (or friends, perhaps?). Otherwise, R-12 can only be sold in large "bulk" containers.

The purpose of the law is to discourage do-it-yourselfers from recharging leaky A/C systems. Every A/C system leaks a little refrigerant over time (usually past the compressor shaft seal, but sometimes through leaky hose connections or pinholes in the evaporator). But a loss of more than 1/2 an ounce of refrigerant a year is considered major enough to require attention. By restricting the availability of R-12 refrigerant to certified professionals, the EPA is effectively forcing motorists to take their vehicle to a shop when the A/C system needs topping off or service. The shop, in turn, should thoroughly check your A/C system for leaks and encourage you to have any leaks repaired before recharging your system. In some states (Florida being one), it is illegal for a shop to recharge an A/C system that is known to be leaking. The shop must be authorized to repair the leaks before it can recharge the A/C system.

Cost—At the same time, additional federal excise taxes on R-12 have pushed its market price up to $5 or more per pound as of this writing, increasing its recycling value and giving shops a faster payback on their mandatory recovery and recycling equipment.

The law doesn't prevent you from recharging your own A/C system, providing you can get your hands on some refrigerant. But recharging a leaker wouldn't be very environmentally responsible, would it? Besides that, the soaring price of R-12 makes the stuff a fairly pricey commodity these days. So why would you put it into a system if you know it's going to leak right back out again? Unless you enjoy throwing away money, the sensible (and environmentally responsible) thing to do is to (1) fix the leaks, and then (2) recharge the system.

Finding and Stopping Leaks

You could once buy small cans of R-12 refrigerant for "diagnostic" purposes, but no more. Such products contained special dyes that allowed you to pinpoint leaks when the refrigerant was put into your A/C system. As the refrigerant leaked out, it left a telltale stain that showed you where the leak was occurring. These were a good do-it-yourself product but are now banned because they contain R-12.

Some of these products were also claimed to contain "stop leak" additives to plug leaks. Don't believe it. Most A/C leaks are too large for a stop leak to have any measurable effect. If the product were effective enough to stop the average leak, it would also plug up the orifice tube. Professional mechanics now use an electronic leak detector to find refrigerant leaks. An electronic leak detector emits a beep when it sniffs refrigerant. The detectors are very sensitive but also expensive, so they're generally beyond the budgets of most do-it-yourselfers. Even so, you may find a store that rents them or a professional friend who would loan you his detector.

Another important point to keep in mind about refrigerant leaks is that leaks are a two-way path. As refrigerant leaks out, air and moisture can enter the system. Air reduces the efficiency of your A/C system and reduces cooling output. Moisture reacts with the refrigerant to form corrosive acids and sludge that can damage the compressor and plug up metering valves and the orifice tube.

If an A/C system has lost all its refrigerant either due to leakage, intentional venting, an accident or because you had to replace a major component such as a hose, compressor or whatever, two things must be done before the system is recharged. The moisture-absorbing desiccant in the receiver-drier or accumulator should be replaced to restore the system's ability to protect itself against acid and sludge formation. Once that's been done, the system should be connected to a vacuum pump and purged to remove air and moisture. Since few do-it-yourselfers have an A/C vacuum pump sitting around in their garage, the only way to pump out the air is to take your vehicle to a shop that has the proper service equipment.

Another thing that's lost when there's a leak in an A/C system is the vital compressor oil. Tiny droplets of oil circulate throughout the system with the refrigerant to keep the compressor from seizing up and self-destructing. Loss of compressor oil or a blockage that prevents the oil from reaching the compressor are the leading causes of A/C compressor failures. So if you're recharging a leaker, it's also important to replenish the vital oil supply that may have been lost. The total amount of oil in the system isn't very much (usually a few ounces), so it's important not to overfill it. ∎

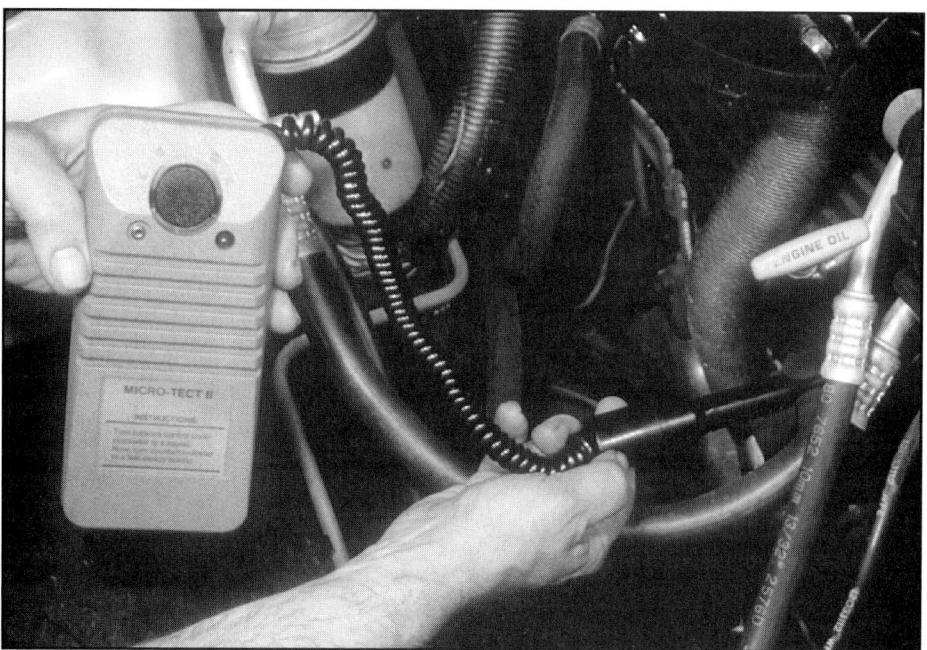

Even a professional technician isn't supposed to recharge until he's fixed the leak. The only approved way to locate it is with an electronic leak detector.

Reclaiming R-12 and converting old A/C systems over to the new R-134a is going to be big business in the last part of the '90s and beyond. Professional equipment such as this IG-LO 1400 are mandatory. The units must be approved by the EPA and Underwriters Laboratories (UL).

RULES, REGULATIONS & LAWS

3

Like it or not, federal clean air legislation has restricted your freedom to tinker under the hood of your vehicle—unless you drive a pre-emissions era vehicle (generally 1967 or earlier). Since November 15, 1990, when the latest revisions to the Clean Air Act became law, it has been and will continue to be illegal to remove, disconnect, defeat or otherwise render inoperative any emissions control or emissions-related device on a street-driven vehicle. A violation of this law may make you liable for a $2,500 fine!

The law does not, however, apply to race cars, other full-time off-road vehicles, show cars that are not driven on the street, or vehicles that were not factory-equipped with emissions controls.

The new law has broadened the scope of emissions control "tampering" to include virtually any type of engine or exhaust system modification that alters what comes out the tailpipe. In an open letter addressed to all automotive parts manufacturers, distributors, retailers, professional installers and do-it-yourselfers from the EPA's Field Operations and Support Division, the agency warned that the anti-tampering provisions in the revised act now apply to everyone:

"The Clean Air Act now prohibits any person from manufacturing, selling, offering for sale or installing any part or component intended for use with or as part of any motor vehicle, where a principal effect of the part or component

Is this an emissions-legal engine? Only if it's in a pre-emissions vintage vehicle or race car. There's no PCV system, no EGR system, no heated air intake, no spark controls or other pollution controls.

An annual or biennial vehicle emissions test is something that fewer and fewer vehicles are able to avoid.

is to bypass, defeat or render inoperative any [emissions control] device or element of design, and where the person knows or should know the part or component is being put to such use. A civil penalty of up to $2,500 may be imposed for each violation of this defeat device prohibition."

Under the old law, the EPA was primarily concerned with aftermarket suppliers who made or installed emissions bypass or defect devices, such as catalytic converter "test pipes," which everyone knew was a ruse for replacing catalytic converters so people could burn less expensive leaded gasoline and/or improve engine performance by reducing exhaust system backpressure (which really made no difference unless the converter was clogged). They were called "test pipes" or "test tubes" because they supposedly allowed you to "test" how your car ran with the converter removed. Removing the converter presumably allowed you to determine whether or not the converter was plugged. The trouble was that once somebody installed a test tube to supposedly "check" his converter, he usually left it in place and never reinstalled the converter. (A simple check of intake vacuum with a hand-held vacuum gauge is a much easier way to

EMISSIONS RESTRICTIONS

The following restrictions apply to any street-driven vehicle with an emissions-controlled engine:

- Removing the EGR valve.
- Disconnecting or plugging the EGR valve vacuum hose or ported vacuum switch.
- Removing or disconnecting the PCV valve.
- Removing the stock air cleaner and heat riser duct plumbing.
- Removing the catalytic converter.
- Removing or disconnecting the air pump.
- Removing or modifying the stock distributor vacuum advance/retard.
- Altering the stock ignition advance mechanism or timing curve.
- Replacing the stock distributor with an aftermarket unit that is not emissions-certified.
- Modifying, removing or replacing the stock computer or PROM chip with a non-certified component.
- Blocking the heat-riser duct under the intake manifold.
- Knocking out the filler restrictor on the fuel tank inlet pipe.
- Replacing the stock non-vented gas cap with a vented cap.
- Removing or disconnecting the fuel vapor recovery canister.
- Changing the idle mixture or stock carburetor jetting.
- Removing or modifying the carburetor choke.
- Modifying or replacing the carburetor accelerator pump with non-certified components.
- Installing an intake manifold that lacks provisions for the stock EGR valve and/or a heat riser duct.
- Installing a tunnel ram or dual-quad manifold.
- Installing a carburetor that lacks the stock emissions hookups.
- Installing non-certified fuel injectors.
- Installing a non-certified nitrous oxide injection system.
- Installing a long duration cam that changes the idle quality.
- Installing exhaust headers that lack provisions for a heat riser valve, an air cleaner preheat stove or fittings for an oxygen sensor (if required).
- Installing valve covers with open breathers or no fittings for a PCV valve.
- Installing any induction, fuel or ignition system component that isn't emissions-legal.

test a converter for a possible restriction. A low vacuum reading or one that progressively drops would indicate a buildup of backpressure—see Chapter 9.)

The old law also only barred professional mechanics (not vehicle owners) from removing or disconnecting emissions-control devices. This included the installation of aftermarket engine performance products which did not include provisions for retaining the stock emissions-control hardware (i.e. headers without air pump or oxygen sensor fittings, intake manifolds without EGR, air cleaners without heat riser or PCV hose connections, etc.). Violators could (and were) fined up to $10,000 per offense!

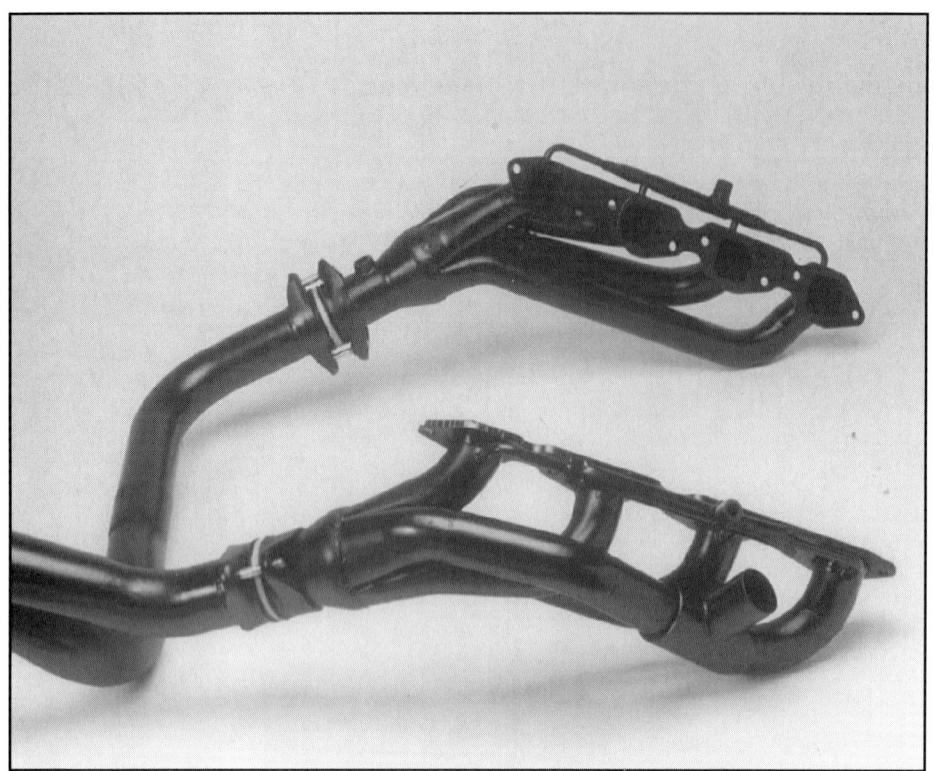

These exhaust headers are an example of "emissions-legal" aftermarket parts because they have all the required hookups for the air pump, EGR and heated air intake plumbing. They also have an "EO" number and CARB approval.

LOOPHOLES PLUGGED

The old law had two loopholes, however. One was that motorists could still modify their own vehicles by removing emissions devices or installing parts that didn't have the required hookups (as long as any such modifications weren't noticed during an emissions inspection, if the state or municipality involved even had one). The other loophole was that manufacturers could still sell parts that did not contain the required emissions hookups as long as the parts were labeled "for off-road use only"—a tactic that many used to simply get around the language in the law. But all that has changed now.

The new law makes it illegal for anyone to disconnect or modify any emissions-control device, or to manufacture, sell or install any part that defeats or alters the performance of an engine's emissions-control system. The EPA is primarily interested in eliminating obvious defeat products such as catalytic converter replacement pipes, and engine computer PROM (Program Read Only Memory) chips that alter emissions. But the EPA is also keeping a watchful eye on the entire spectrum of aftermarket performance products to determine which ones have a potentially adverse impact on emissions—particularly the so-called off-road products that are ending up on the street. The agency says that when it finds significant numbers of such off-road products being used on the street, it will hold the manufacturers of such products liable.

The new law does not define "tampering," which complicates the whole issue of certification and compliance for parts manufacturers. But thus far, the EPA says parts that comply with California's certification program or an EPA document called "Memorandum 1-A" are acceptable. Consequently, the aftermarket suppliers of engine and exhaust system performance products have been certifying their products (where possible), re-cataloging other products that can't be certified for pre-emissions applications only, or discontinuing certain products altogether.

PARTS AFFECTED

The type of aftermarket performance products affected by the new law include:

• Performance computer chips or modules that alter the air/fuel ratio and/or ignition timing at throttle positions other than wide open throttle

• Exhaust headers that lack emissions hookups

• Exhaust systems that require the removal or relocation of the catalytic converter

• Intake manifolds that don't have EGR or preheating provisions

• Performance camshafts with longer durations and overlap that alter engine breathing and increase exhaust emissions (most radical street cams and race cams)

• Performance carburetors with non-stock metering, jetting or emissions hookups

• Nitrous oxide injection systems

• Aftermarket or modified fuel injection systems that alter the stock air/fuel feedback system or are user adjustable

• Air cleaners that do not have provisions for a preheated air snorkel, PCV connections or other emissions hookups

• Non-stock ignition systems that alter stock timing or spark advance. This includes mechanical and electronic distributors, spark advance control modules, knock retard modules, etc.

• Turbochargers, superchargers and related

components. Included here are turbo and blower kits, wastegates, boost control modules, intercoolers, and turbo, blower and intercooler plumbing.

• In short, any fuel, ignition, exhaust or engine component that could potentially alter engine breathing, timing, fuel delivery, fuel mixture or emissions.

WHAT IS EMISSIONS CERTIFICATION?

When a manufacturer submits a component for emissions certification to California's Air Resources Board (CARB), they must submit detailed proof in the form of laboratory dyno test results that shows their part does not have an adverse effect on exhaust emissions. These tests are very expensive and must conform with specified test procedures. CARB then reviews the data and may or may not ask for additional information and/or testing. If the product meets CARB's criteria, CARB issues an "executive order" (EO) number (also called an "exemption" number) certifying that the part is in compliance with the applicable clean air rules.

An EO number means the component can be legally manufactured, distributed, sold and installed on a street-driven vehicle in the state of California. It also means the component is legally acceptable in all 50 states because the EPA recognizes the CARB exemption program as meeting their Memorandum 1A requirements for certifying emissions-legal parts.

It's important to note, however, that EO exemption numbers are granted for specific applications only. They are not a blanket approval that allows a given component to be installed on any engine it might fit. Each application that is significantly different must be individually approved for a CARB exemption number. In other words, a component that happens to fit a

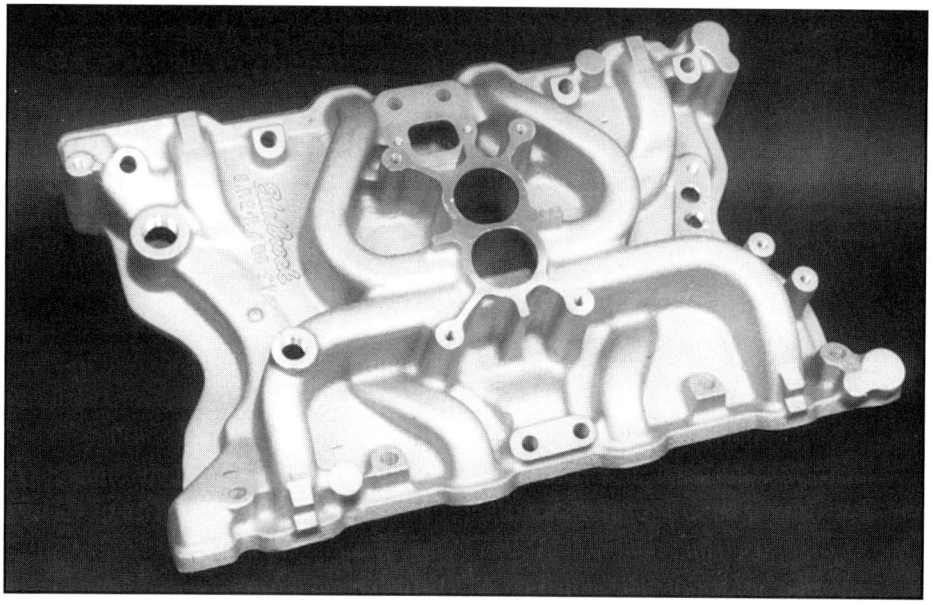

Aftermarket replacement manifolds such as this can usually meet CARB's requirements for emissions legality as long as they have all the required emissions hookups and do not cause the engine to produce increased emissions.

small-block Chevy engine may or may not be certified for all small-block Chevy applications, depending on what it is and how it might affect emissions in various applications. The best advice is don't try to second guess the component manufacturer. If it doesn't say a particular part is certified for your specific application in their catalog, then you'd better think twice about installing it on a street-driven vehicle.

Installing Certified Performance Parts

Though it hasn't gotten to the point yet where you need a lawyer and a second opinion before buying or installing aftermarket performance products on your vehicle, you should protect yourself from potential legal problems by adhering to the rules. If you're replacing a stock engine component like a carburetor,

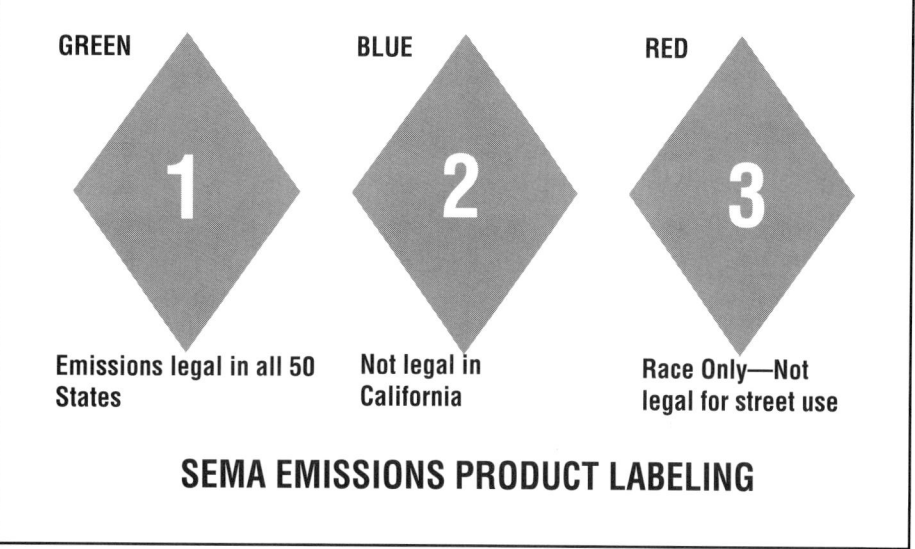

This is the Specialty Equipment Market Association's (SEMA) emissions-product labeling key. If you're purchasing specialty performance equipment, look for one of these labels on the box. However, beware that not all manufacturers follow SEMA's guidelines.

23

This type of aftermarket performance manifold, however, would be for "off-road use only"—unless it was being installed on a pre-emissions vintage vehicle.

distributor, intake or exhaust manifolds, cam, etc., with a non-stock aftermarket performance product that falls into one of the categories which require an exemption—and you plan to drive your vehicle on the street—make sure the product you're buying is emissions certified and has an EO number.

• Look for wording on the box that says the product is emissions-legal or emissions certified for street use in compliance with the EPA and/or CARB rules.

• Look for the EO (executive order) exemption number issued by the California Air Resources Board on the box, product or in the catalog. Remember, the product must have an EO number to be street-legal.

• If there is no EO number and one is required to be street-legal, it cannot be legally installed on a street-driven vehicle—period.

Installing Non-Certified Parts

It's important to keep in mind that there are a lot of aftermarket performance products that do not have to be emissions certified to be street-legal. As long as a part has no direct or indirect effect on emissions, it does not require an exemption number. Such parts include:

• Custom wheels and tires

• Suspension components such as shocks, lift kits, lowering kits, springs, sway bars, bushings, etc.

• Brake parts, such as linings, rotors, drums, hydraulic components

• Aerodynamic add-on appearance packages, spoilers and air dams

• Many engine dress-up accessories such as oil pans, timing covers, fasteners, hoses, tubing, etc. Valve covers for emissions-controlled engines must have provisions for a stock PVC valve and no external venting. Air cleaners must retain all emissions hookups as well as have a heated air snorkel. An open-style chrome air cleaner would not be street-legal on an emissions-equipped vehicle.

• Most internal engine parts, such as lifters, pushrods, rocker arms, valves, valve springs, bearings, connecting rods, pistons, rings, crankshafts, timing chains and gears (except those that allow for cam advance/retard, which can affect emissions), windage trays, oil pumps, bolts and gaskets.

• Drivetrain components such as

MOTOR VEHICLE REQUIREMENTS STACKING UP IN THE 1990S

MODEL YEAR PHASE-IN

	'92	'93	'94	'95	'96
Tailpipe Emissions Controls			■	■	■
Onboard Vapor Controls					■
Evaporative-Running Loss Controls				■	■
Emissions Diagnostic Systems			■	■	■
Car Full Passive Restraints			■		
Car Side Impact			■		
Truck Passive Restraints				■	
Truck Side Impact			■		
Truck Roof Crush			■		
CFC Product Phase-Out	■	■	■	■	■

Source: Motor Vehicle Manufacturers Association, January, 1992.

Government is forcing the automakers to re-engineer vehicles to make them cleaner and safer. The chart above is an example of the type of vehicle safety and emissions regulations set forth by the federal government.

flywheels, clutches, torque converters, transmissions, differential gears, axles, U-joints and CV joints.

• Lighting products such as driving lights, fog lights, etc.

• Fuel pumps, fuel lines and filters.

COMPLIANCE & ENFORCEMENT

Though the EPA doesn't have the manpower to actually go out and hunt down motorists who are running off-road performance equipment on the street, the mere threat of prosecution and a $2,500 fine is considered sufficient to discourage such activities—that, and the threat of an annual emissions inspection in areas that require it. The EPA's primary enforcement efforts will continue to be directed at manufacturers, distributors, retailers and professional installers who either ignore the law or try to get around it.

Thirty-five states currently have some type of annual emissions testing program in place. Of these, 27 include various anti-tampering checks (looking under the hood to see if emissions components have been removed or disconnected, checking to see if the catalytic converter has been removed, seeing if "off-road" performance parts have been installed, etc.).

In 1993, roughly 60 million vehicles were subjected to such emissions

Classic vehicles built prior to 1967 are generally exempt from annual emissions inspections, although check your local state laws to be sure. Photo by Michael Lutfy.

However, if you have a vintage vehicle that has been fitted with a late-model engine (such as installing a TPI-equipped 350 small-block from a 1989 Corvette into a street rod), then the vehicle must conform to the guidelines required of the engine for that particular model year (at least in California). Again, check local state laws regarding engine swapping. Photo by Michael Lutfy.

inspections. By 1996, the number of vehicles being subjected to emissions tests is expected to increase to about 100 million because the Clean Air Act also contains provisions to force areas that don't comply with federal clean air standards to add or expand vehicle emissions testing programs. So if you've been able to avoid emissions testing thus far, your days may be numbered.

As we mentioned earlier, the EPA isn't out to hunt down and prosecute individuals who violate the law. Its focus is primarily on manufacturers, distributors and retailers who aren't playing by the rules. Even so, no government agency at either the state or federal level is going to turn down a chance for "revenue enhancement" if the opportunity presents itself. In other words, if you happen to be the unlucky one who gets caught and are fined, don't say we didn't warn you. You're better off following the rules and saving yourself the headache and hassle of getting caught.

WARRANTIES

Everybody is for clean air but few want to shoulder the burden of paying for it. Thus CARB and EPA came up with the concept of a government-mandated emissions warranty. To reduce the out-of-pocket expenses of maintaining the pollution control equipment on new cars, the government required the automakers to provide an extended emissions warranty.

This is a complicated subject, but we will try to make is as clear as possible. First off, the whole thing started with this passage from the 1970 Clean Air Act:

"The Administrator shall prescribe regulations under which the useful life of vehicles and engines shall be determined for the purposes of (this Act). Such regulations shall provide that useful life shall, in the case of light-duty vehicles and light-duty vehicle engines, be a period of use of five years or of 50,000 miles, whichever comes first. Any cost obligation of any dealer incurred as a result of any requirement imposed by (this Act) shall be borne by the manufacturer."

In other words, Congress, in the heady atmosphere of the Earth Day era, ordained that the automakers had to guarantee that their products wouldn't befoul our atmosphere for a specified period after being put into service. But the legislative branch doesn't implement laws; the executive branch does, and in this case, that's the Environmental Protection

Agency (the "Administrator" in the quote). A fair interpretation of such a broadly stated law naturally required detailed regulations, so what sounded pretty simple at the beginning became pretty complex.

There are actually two separate federally mandated emissions warranties, one for performance (how far does an actual specimen in the real world push those CO and HC needles?), and the other for defects.

Performance Warranty

The Performance Warranty kicks in when ALL of the following stipulations are met:

• The vehicle is an '81 or later model ('82 for high-altitude)

• The vehicle has failed an EPA-approved state or local pollution test

• The state or local government requires that repairs be made to enable the vehicle to pass the test. To put it another way, the owner will have to bear a penalty (a fine, the cost of repairs, the loss of the right to legally use the car) because of the test failure.

• The vehicle has been operated and maintained in accordance with the instructions in the owner's manual (that includes using unleaded only in catalyst-equipped conveyances)

Given this set of circumstances, for the first two years or 24,000 miles of service, the automaker is obliged to pay an authorized dealer to make ANY repairs necessary to get the car through smog inspection. Our reading of the regulation made us assume that this includes taking care of burned valves, leaky valve seals, worn-out or broken rings, jumped time, etc., not just components designed to reduce emissions. However, warranty liaison people at various car companies

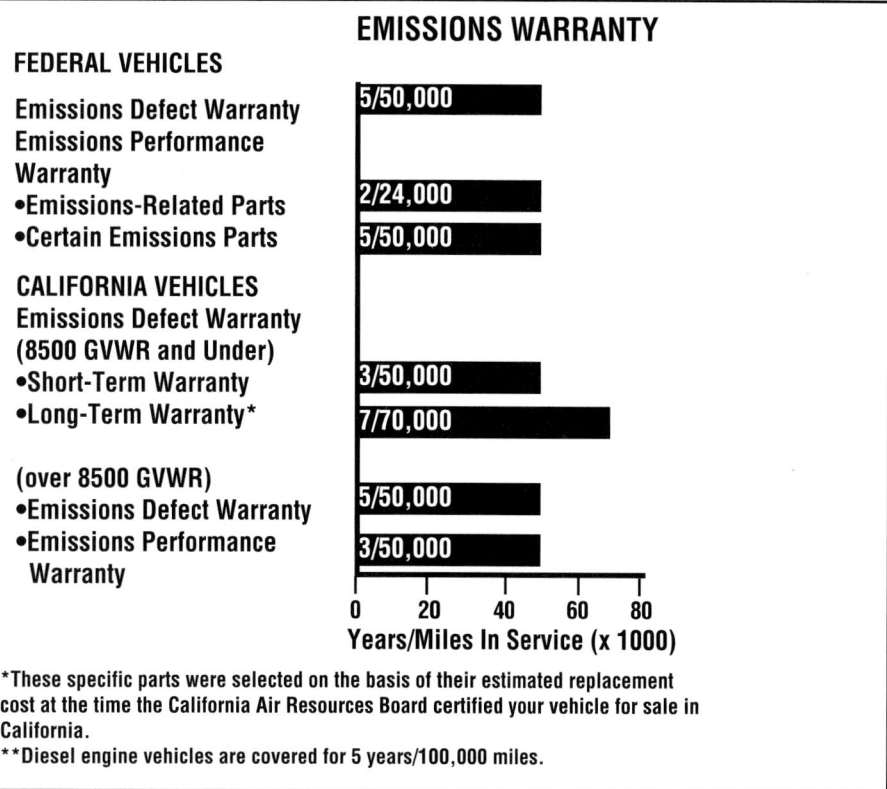

This is a summary of emissions warranty information, both federal and California guidelines.

have told us that internal engine problems almost never fall under this stipulation. Instead, they're usually covered by the regular powertrain guarantee.

Beyond that time and distance of use up to five years and 50,000 miles, the manufacturer is liable for the cost of fixing any emissions-control devices (things not in general use prior to 1968, such as air injection pumps) that must be working properly to allow the vehicle to pass the pollution test. Also covered is any repair or adjustment needed to assure the proper operation of a part that is specifically designed to fight smog. In other words, if a component is not directly covered after 24 months/24,000 miles it might be included because it is keeping some anti-pollution device from doing its job—for instance, a blown head gasket that has poisoned an oxygen sensor.

You can see that if your state or local government does not have an I/M program, you can forget about the Performance Warranty.

Defect Warranty

Then there is the 5/50 Defect Warranty, which applies to the repair of emissions-control-related parts that have gone bad (the vehicle does not have to fail a test), and this is where arguments often start. For instance, suppose the car has a driveability problem that is caused by a faulty thermostatic bleed valve in the heated air intake system. According to GM's list of eligible components, this is covered only up to 24/24 under the Performance Warranty. Ditto for diesel injection pumps, nozzles, and high pressure lines and seals. But the Defect Warranty says these items must keep performing up to 5/50. The two warranties overlap in some cases, and the one with the broadest coverage applies. We have seen more than one dealer go into an elaborate song and dance about what is in force in order to confuse the issue and avoid making a repair (you would think the manufacturer was not paying for it).

It is always best to refer to the list in the

Buying a new car or truck? All cars are protected by a federally mandated warranty that covers emissions-control systems and all related parts. If you have trouble with a catalytic converter, for instance, and it falls within the time frame specified in the text or chart on page 27, then the factory must replace it free of charge. Don't let a shady muffler repair shop sell you a new converter for your new or slightly used car.

warranty booklet for the car in question, but the EPA has its own list, which may leave out a few things, but is very specific:

•Exhaust Gas Recirculation: valve, spacer plate, solenoid, thermal vacuum switch, backpressure transducer, sensors and switches used to control flow.

•Early Fuel Evaporation: valve, thermal vacuum switch.

•Air Injection System: air pump, anti-backfire valve or deceleration valve, diverter, bypass, or gulp valve, reed valve.

•Exhaust Gas Conversion System: catalytic converter, thermal reactor, oxygen sensor, dual-walled exhaust pipe.

•Positive Crankcase Ventilation System: the PCV valve and solenoid.

•Evaporative Emission Control System: purge valve, purge solenoid, fuel filler cap, vapor storage canister and filter.
•Fuel Metering System: electronic control module, deceleration controls, fuel injectors, fuel injection units and fuel rails developed for feedback EFI or TBI, air flow meter, module, or mixture control unit, mixture control solenoid, diaphragm or other fuel metering components that achieve closed-loop operation, electric choke, altitude compensator sensor, other feedback control sensors, switches, valves, thermostatic air cleaner.

•Ignition systems: electronic ignition, electronic spark advance, timing advance/retard systems.

•Miscellaneous Components: hoses, gaskets, brackets, clamps, and other accessories used in the above systems.

•Cold start enrichment system

•Cold start fuel injector

•Electronic engine control computer, wiring harness & sensors

•Electronic fuel injection system, including fuel injectors, fuel pressure regulator and fuel rail assembly (but not the pump)

•Exhaust head pipe from the manifold to converter

•Exhaust manifolds

•Fuel filler cap and neck restrictor

•Fuel vapor storage canister (charcoal canister), purge valve & related plumbing

•Fuel tank

•Fuel tank pressure control valve

•Intake airflow meter or sensor, manifold & gaskets

•Malfunction Indicator Light (check engine/sensor warning light)

•Oxygen sensor

•PCV valve & related plumbing

•All spark control components

•Spark plugs and wires (plugs must be replaced at specified mileage interval)

- Supercharger & related hardware

- Throttle body assembly

- Turbocharger, intercooler, wastegate & related plumbing

Qualifying—To qualify for free repairs under the emissions defect warranty, you have to maintain your vehicle properly (this means according to the vehicle manufacturer's recommended maintenance schedule as described in your owner's manual). It's a good idea, therefore, to keep a log as well as receipts for any maintenance you or anyone else performs on your vehicle in case you're asked to prove the vehicle has been properly maintained.

Most new car dealers are anxious to please their service customers and will do everything they can to correct an emissions problem. But if you think your dealer is not honoring the terms of the federal emissions defect warranty, you should first try to resolve the matter by calling the car company customer service hotline. They'll sometimes bend the rules a bit to keep a customer happy. For example, if your car has just turned 5 years old but has only 20,000 miles on the odometer, they may decide to make the emissions repairs for free even though your warranty has technically expired. In some cases, they may also "extend" the warranty on a certain component if there's been a history of failures.

If you can't get an emissions warranty issue resolved with the dealer or car maker, contact the EPA:

Field Operations & Support Division
Environmental Protection Agency
401 M Street S.W.
Washington, D.C. 20460

California Warranty Rules

California has always played by its own rules when it comes to clean air, and the state's warranty coverage provisions are no different:

- 1989 and older vehicles are covered by a 5 year/50,000 mile emissions warranty.

- 1990 and newer cars and light trucks (under 8500 GVW) are covered by a 3 year/30,000 mile warranty. Emissions related parts costing more than $300 are covered by a 7 year/70,000 mile warranty.

- 1990 and newer gasoline engine-powered trucks over 8500 GVW are covered by a 5 year/50,000 mile emissions warranty. Diesel-powered trucks over 8500 GVW are covered by a 10 year/100,000 mile warranty.

Anyone having questions about the California emissions warranty program or warranty coverage should contact:

California Air Resources Board
Mobile Source Division
9528 Telstar Ave.
El Monte, CA 91731

Note that no matter whose list you use,

There are strict rules for replacing catalytic converters. Most vehicles under 5 years old or with less than 50,000 miles are covered by federal warranty. See sidebar on next page for more rules and regulations regarding converters.

items such as spark plugs and oxygen sensors having a recommended replacement interval of less than 5/50 are only warranted up to that interval. Also, the list for California is greatly expanded over that of the other 49 states, and diesels in that state get 10 year/100,000 mile coverage.

A few miscellaneous notes on the emissions warranty situation:

- No matter how many owners a car may have had, these warranties still apply up to their time and mileage limits.

- Proper maintenance records and receipts must be kept.

- A claim can't be disallowed because a non-OEM part was installed providing it is an approved equivalent labeled "Certified to EPA Standards."

- The vehicle manufacturer has 30 days to approve or reject a claim. If that time limit is exceeded, the company loses the right to deny coverage, except if you request a delay or the missed deadline was caused by factors outside the dealer's control.

SO YOU NEED A NEW CONVERTER, DO YOU?

Federal law prohibits the replacing of a catalytic converter unless (1) the vehicle is more than 5 years old or has more than 50,000 miles on the odometer and a legitimate need for replacement has been established and documented, or (2) the vehicle has flunked a state or local emissions test because of a bad converter, or (3) the converter is missing. The converter is covered by a 5 year/50,000 mile emissions warranty, which extends to 8 years/80,000 miles on 1995 and later vehicles. A violation of the above restrictions will make a shop eligible for a $2,500 fine—per incident. Also, if the converter is replaced, the shop must:

•Certify in writing why the converter needs to be replaced (lead fouled, clogged, rusted out, missing, failed emissions test, etc.). In addition, the repair invoice must include your name, complete address, and the make, model, year and mileage of the vehicle. The shop must retain this paperwork for six months.

•The shop must receive your written permission prior to replacing the converter if the reason for replacement is not because of an emissions violation. Catalytic converters are emissions-control devices which are designed to last the life of the vehicle and do not normally require replacement. Furthermore, if the vehicle is properly used and maintained, original converters are covered by the emissions-control warranty for 5 years or 50,000 miles. Federal law prohibits repair businesses from replacing these devices except under certain limited circumstances. In order to verify that the proper circumstances exist, the owner of the vehicle on which such repairs are made and a facility representative must sign the following statement:

"The vehicle is over 5 years old or has more than 50,000 miles on it and the catalytic converter required replacement because_____" or *"The vehicle's catalytic converter was missing when the vehicle was brought in."*

•The shop must keep a copy of this signed statement for six months, and the old converter for 15 days. Replaced converters must be marked in such a way that they can be identified with particular customer invoices and statements, and be available for EPA inspection.

•The shop must only install an EPA "approved" converter (which is identified with a special ID tag). The replacement converter must also be the same type (two-way, three-way or three-way plug oxygen) as the original—and in the exact same location as the original.

•The converter must be hooked up properly (connected to the air pump or air injection plumbing).

•The shop must give you a warranty card for the replacement converter. Aftermarket converters carry a minimum performance warranty of 25,000 miles and a workmanship defect warranty of 5 years or 50,000 miles.

•When a motorist goes directly to the dealer for diagnosis of a problem, he can't legally be charged for this troubleshooting if the cause is found to be something that's covered by the warranties. If, however, the component at fault is not on the list, or is not covered by the 24/24 portion of the Performance Warranty, he will have to pay.

Other Warranty Coverage

In addition to the federal and California emissions warranty rules, emission problems on newer vehicles may also be covered by a powertrain warranty, a bumper-to-bumper warranty or an optional extended warranty program. An oil consumption problem due to a bad valve or broken ring, for example, would be covered under an engine/powertrain warranty. Again, you may have to do some digging to find out the terms of your warranty coverage, what's covered, how much of a deductible you have to pay, and so on.

Clean Air Act—When the Clean Air Act was revised in 1990, Congress and the automotive industry reached another compromise on warranty coverage. This one rolls back the federal warranty on most emissions-related components to two years and 24,000 miles, but extends the warranty on the catalytic converter and engine computer to eight years and 80,000 miles. The change begins in the 1995 model year. But the rollback is really no rollback at all for two reasons. The first is that the automakers are being pushed to build emissions-control components with 10 year/100,000 mile durability. The other is that automakers are offering longer and longer "bumper-to-bumper" and extended powertrain warranties. So even though certain emissions-control components on 1995 or later vehicles may no longer be covered by a federally mandated emissions warranty after two years or 24,000 miles, they may still be covered by the manufacturer. ∎

EXHAUST EMISSIONS: HC, CO & NOx

4

When most people think of automotive related pollution, they think only about what comes out the tailpipe. But cars and trucks can actually emit pollutants three different ways:

- Gasoline vapors from the fuel tank and carburetor. These are called "evaporative emissions" and are covered in Chapter 6.

- Combustion byproducts and vapors from the engine's crankcase. These are "blowby" emissions and are covered in Chapter 5.

- Exhaust gases produced by combustion that come out the tailpipe. These include unburned hydrocarbons (HC), carbon monoxide (CO), carbon dioxide (CO_2), oxides of nitrogen (NOx), water vapor (H_2O), particulates (soot or chunks of carbon), various sulfur compounds and other substances.

EXHAUST EMISSIONS

Exhaust emissions are the most difficult to control because there are so many variables that affect their formation. The most important factors are the air/fuel ratio, ignition timing and advance, design of the combustion chamber, combustion temperature and composition of the fuel. Other factors that influence what comes out the tailpipe include camshaft timing, valve duration and overlap, intake manifold design and temperature, engine compression, piston-to-cylinder wall clearances, the type of valve seals used (and their condition), and external factors such as humidity, temperature and barometric pressure.

Excessive Emissions

Two things typically contribute to excessive exhaust emissions. One is incomplete combustion. The other is the presence of unwanted substances in the combustion process.

When gasoline is burned inside an engine, combustion is never totally complete. There is always a tiny amount of fuel that fails to burn or is only partly burned. To understand why this is so, we

Older cars emit far higher levels of pollution than newer ones. But any car can be a polluter if it isn't tuned properly or has a major emissions problem.

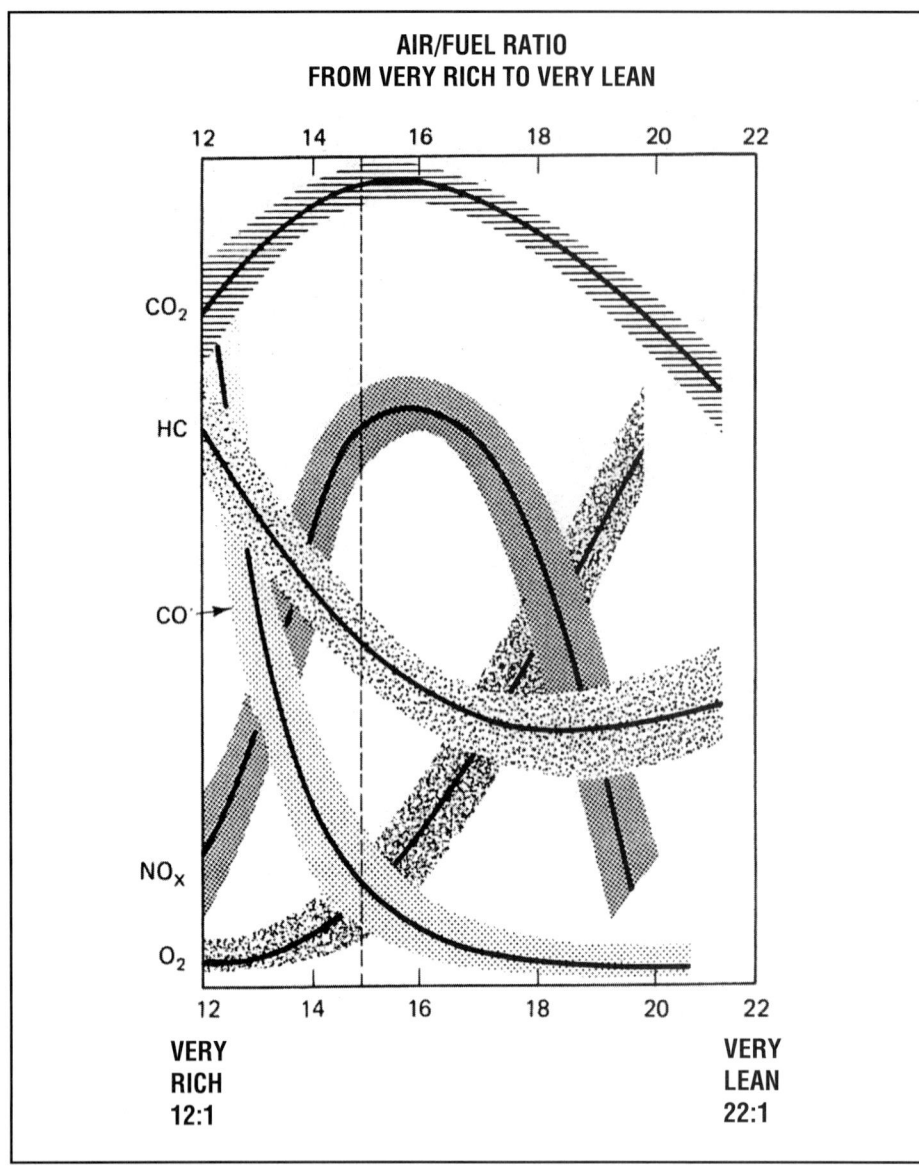

This chart shows how the composition of the various exhaust gases varies with the air/fuel ratio. The dotted vertical line is at the ideal 14.7:1 ratio. Of these five gases, only HC, CO and NOx are pollutants. CO_2 is a greenhouse gas but is otherwise harmless.

Since water vapor and carbon dioxide are harmless byproducts of combustion, we would have no air pollution problems whatsoever if combustion inside a real engine were this simple. But it isn't. For one thing, the relatively cool surfaces of the piston, cylinder walls, and cylinder head have a quenching effect on the tiny droplets of fuel. Some of the fuel will cling to these metal surfaces and not burn. Some fuel will also not burn when it gets trapped down between the piston and cylinder wall just above the top compression ring. This results in unburned hydrocarbons in the exhaust.

Intake manifold design can also interfere with combustion by allowing the tiny droplets of fuel to separate from the airflow. This causes an uneven mixing of the air and fuel, which hinders complete combustion. It can also cause variations in combustion efficiency and power output from one cylinder to another, especially in V6 and V8 engines where the end cylinders often run leaner on carbureted and throttle body-injected applications.

Too Rich—An engine needs a richer (more fuel) than usual air/fuel ratio when it is first started because the fuel does not vaporize very easily in a cold engine. A richer mixture is initially provided by the choke on a carbureted engine, or a "cold start valve" or increased injector duration on engines with electronic fuel injection. As the engine warms up, the mixture is gradually leaned until it is balanced. On late-model engines with oxygen sensors and either feedback carburetion or electronic fuel injection, the engine computer then takes over the task of keeping the fuel mixture balanced.

But sometimes things get out of whack. A choke may be stuck or misadjusted. The fuel level inside the carburetor bowl may be high due to a misadjusted, leaky or fuel-saturated float. There may be excessive fuel pressure or a leaky needle valve that's flooding the carburetor with fuel. Someone may have replaced the main metering jets in the carburetor with

need to look at the basic chemistry of combustion.

Combustion—Gasoline is a blend of various hydrocarbons. A hydrocarbon is any substance whose molecules contain hydrogen and carbon atoms. The molecules in gasoline consist of long chains of hydrogen and carbon atoms. When gasoline is drawn through the carburetor venturis or sprayed out of a fuel injector, it is broken up into tiny droplets. To burn, the tiny droplets of fuel must be mixed with oxygen in the right proportions. Since the air we breathe contains 16% to 20% oxygen, it takes a lot of air to provide enough oxygen for combustion. The ideal or stoichiometric air/fuel ratio for gasoline is 14.7:1. That means 14.7 parts of air for every 1 part of gasoline.

The air/fuel mixture is drawn into the combustion chamber, compressed, and ignited by a spark. If the combustion process were totally complete and the air and fuel were mixed in the correct proportions, all the oxygen would combine with all the gasoline to produce heat energy, water vapor and carbon dioxide:

$$O_2 + HC = H_2O + CO_2$$

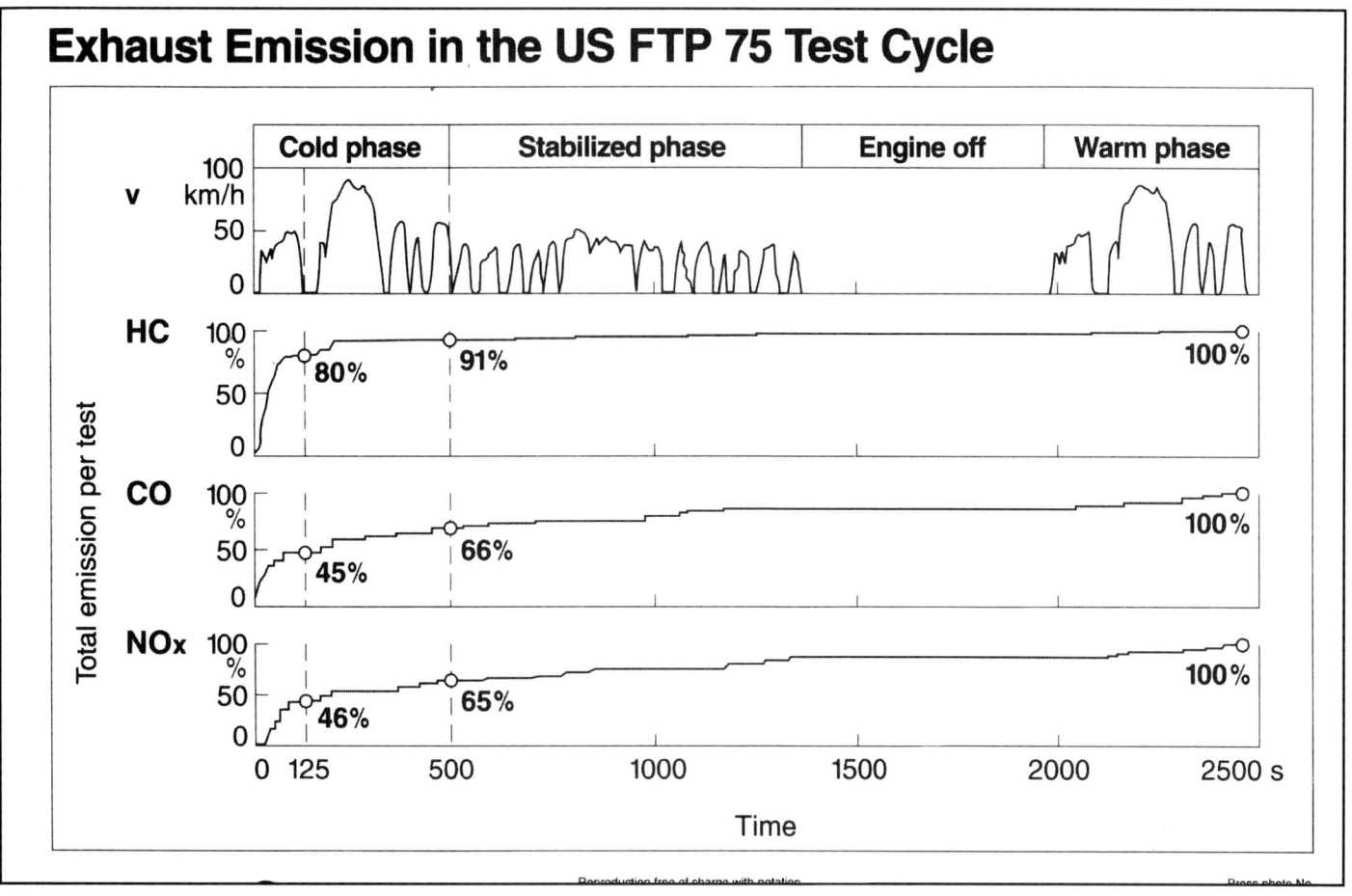

An engine needs a richer air/fuel ratio during cold starting. As the mixture warms up, less fuel is needed. This chart shows how the pollution produced by a typical engine varies with operating temperature, time and speed during the federal FTP 75 test cycle. Courtesy Robert Bosch Corp.

ones that are too large, or screwed up the idle mixture adjustment screws. A plugged PCV valve or hose can also contribute to a rich fuel mixture. On late-model engines with electronic feedback carburetors, the carburetor mixture control solenoid may be defective or misadjusted, or the oxygen sensor may be bad. On a fuel-injected engine, a leaky cold start valve can leak extra fuel into the intake manifold. A bad fuel pressure regulator may be raising the fuel pressure too much. A computer or sensor problem may be throwing the mixture off. Any of these conditions can cause a rich fuel mixture.

When the air/fuel mixture is too rich (too much fuel, not enough oxygen or an air/fuel ratio lower than 14.7:1), there will not be enough oxygen to burn the fuel completely. This results in incomplete combustion and the formation of carbon monoxide (CO) in the exhaust:

$$\text{Insufficient } O_2 + HC = H_2O + C$$

If the air/fuel mixture is excessively rich, it may cause the formation of soot particles in the exhaust:

$$\text{Insufficient } O_2 + HC = H_2O + CO_2$$

Or, it may not ignite at all, causing the entire mixture to simply pass through the combustion chamber unburned and be dumped into the exhaust. This can be especially damaging to a catalytic converter, because raw fuel in the exhaust makes the converter overheat. If it gets too hot, the ceramic or metallic substrate that holds the catalyst can melt, causing an obstruction or blockage in the exhaust.

Too Lean—Sometimes, the air/fuel mixture is too lean (more air than fuel). Any number of things can cause this kind of problem, however a vacuum leak in the engine's vacuum plumbing, carburetor or intake manifold is the most common cause. Others include a float level in the carburetor that's too low, jets that are too small, misadjusted idle mixture screws, low fuel pressure or obstructions in the fuel line, clogged fuel injectors or injectors with insufficient flow capacity for a modified engine, etc.

Lean mixtures (those with too much oxygen and not enough fuel, or an air/fuel mixture ratio greater than 14.7:1), can also create emissions problems. If the air/fuel mixture is leaner than about 18:1, or too lean for what the engine needs to handle its present load, the mixture may fail to ignite. This condition is known as "lean misfire" and may cause a rough idle or

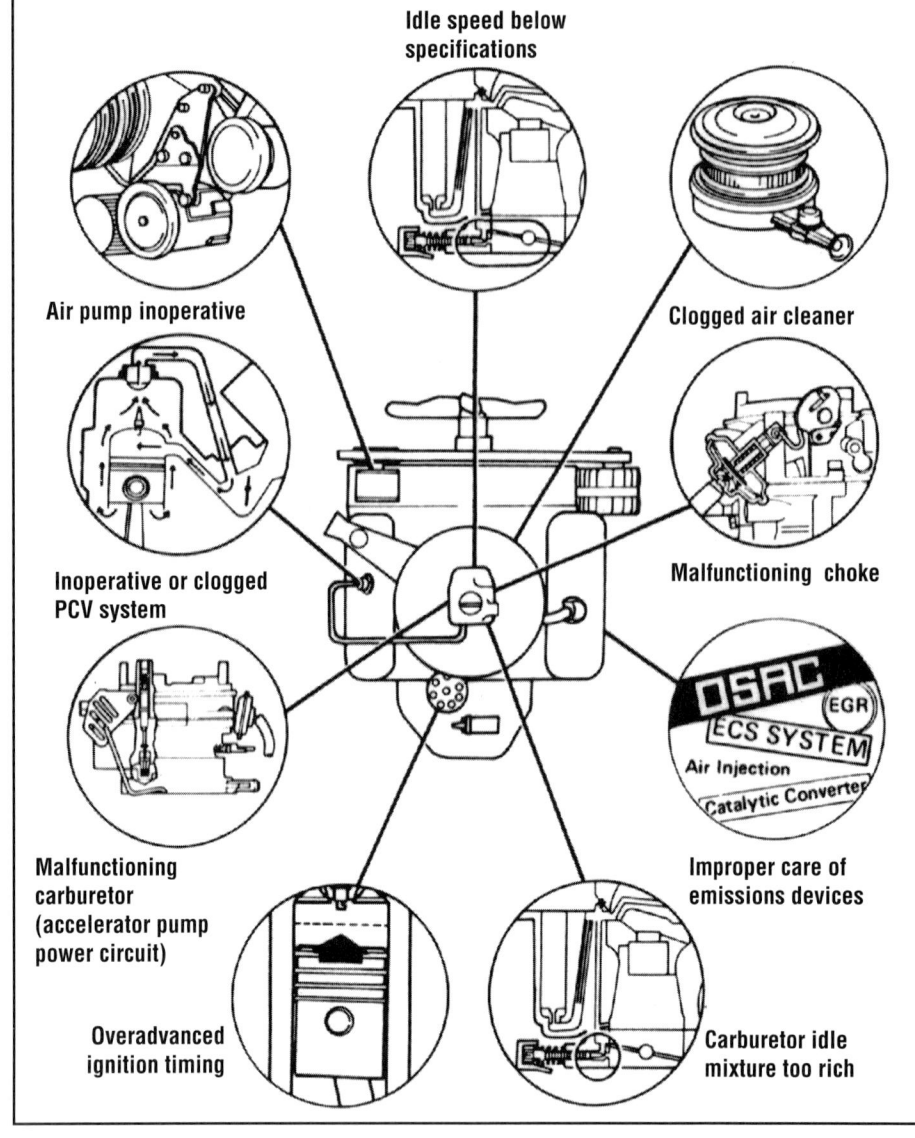

This chart illustrates the primary causes of high CO emissions. These include a rich air/fuel mixture, overadvanced ignition timing and an inoperative AIR system. (Chrysler).

misfiring at high speed. If the condition occurs momentarily while accelerating, it can cause a hesitation stumble or bog when the gas pedal is floored. This can be caused by a weak accelerator pump in a carburetor, dirty fuel injectors or sometimes even an accumulation of heavy carbon deposits on the intake valves.

Lean mixtures can also be harmful in that they elevate combustion temperatures. If things get too hot, the engine may go into preignition and/or detonation and suffer expensive damage (like a burned piston).

If an engine is suffering from a lean misfire condition, it will have excessively high HC emissions because unburned fuel is passing through the combustion chamber and entering the exhaust. And as we said earlier, that may overheat and damage the catalytic converter.

Too Hot—High combustion temperatures inside an engine is also undesirable. Air is almost 80% nitrogen. Normally, nitrogen is inert and does not do much of anything. But at temperatures above 2500 F, nitrogen and oxygen combine to form oxides of nitrogen. The abbreviation "NOx" is used to describe the various nitrogen compounds that are formed.

The higher the combustion temperature, the greater the tendency to form NOx. In an engine without emissions controls, combustion temperatures can easily exceed 2500 F, so some means of lowering the temperatures must be used to minimize the formation of NOx. This system is known as exhaust gas recirculation (EGR), which is covered in Chapter 10.

Other Factors

What else can contribute to exhaust emissions? Oil seeping past worn piston rings or valve guides will increase hydrocarbon emissions. Oil burning will produce a characteristic odor and a bluish colored smoke in the exhaust. The cure for an oil consumption problem is a valve job (new guides & seals) and/or an overhaul.

CARBON MONOXIDE

Of the three main pollutants, carbon monoxide (CO) is the deadliest, because you can't see it or smell it. An exhaust analyzer is used to measure the percent of carbon monoxide in the exhaust in percent (%). A concentration of only half a percent (0.5%) CO in the air can render a person unconscious—and kill within 10 to 15 minutes! Even concentrations as small as four hundredths of a percent (0.04%) can cause headaches and be life-threatening after several hours exposure.

Carbon monoxide is deadly because it displaces the oxygen in your bloodstream. A molecule of carbon monoxide has 210 times the affinity with joining red blood cells as does oxygen. Breathing it in reduces the amount of oxygen that reaches your brain. This starves the brain for oxygen and rapidly leads to unconsciousness and death. When the CO concentration in a person's blood reaches 2 to 5 percent, vision begins to blur and reaction time starts to drop. It can also cause headaches, dizziness, chest pains, and breathing difficulty. Carbon monoxide concentrations of 20 ppm are

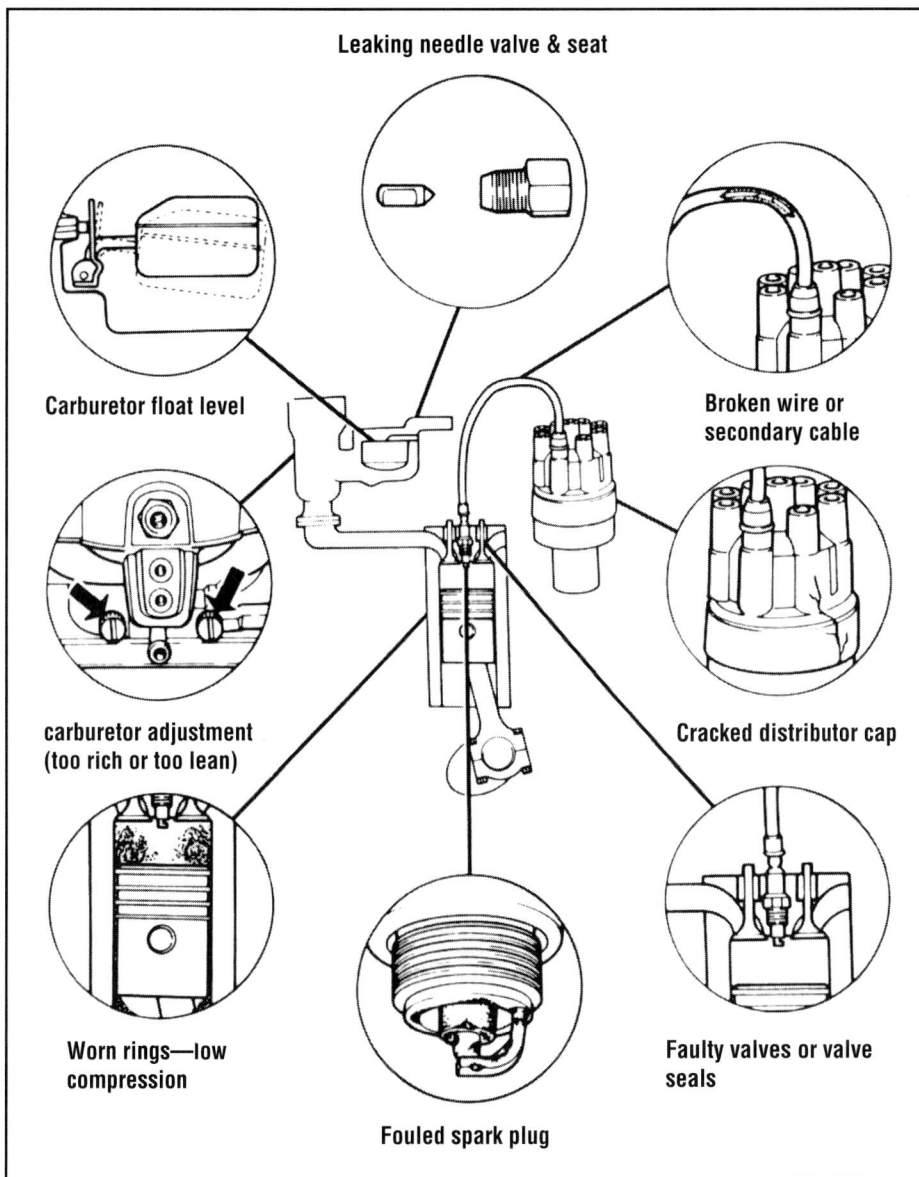

The primary causes of high HC emissions are indicated here. Frequent causes include ignition problems, poor compression, oil consumption and excessively rich fuel mixtures. (Chrysler)

not uncommon in heavy traffic.

Formation

Carbon monoxide is formed when the fuel mixture is rich and there is insufficient oxygen to completely burn all the fuel. The richer the fuel mixture, the greater the quantity of CO produced. That makes CO a diagnostic indicator of incomplete combustion, carburetor maladjustment and similar problems (clogged air filter, sticking choke, defective heated air intake system, leaky cold start injector or fuel injectors, etc.).

Carbon monoxide emissions are highest when an engine is first started because the fuel mixture is richer than normal during this time and the catalytic converter has not yet reached operating temperature. On 1994 and newer cars, the converters are designed to reach operating temperature more quickly to reduce carbon monoxide emissions while the engine is warming up.

Carbon monoxide formation is minimized in the engine by leaning out the mixture as quickly as possible as the engine warms up, and then maintaining a balanced fuel mixture (which depends heavily on a good oxygen sensor). The carbon monoxide that is formed during combustion is changed into carbon dioxide (CO_2) in the catalytic converter. Carbon dioxide is not considered a pollutant, but it is a "greenhouse gas" that may contribute to global warming.

A well-tuned engine with a good converter will produce CO levels in the exhaust that are practically zero—compared to as much as 2% to 3% in an engine without a converter.

HYDROCARBONS

Hydrocarbon emissions are unburned gasoline and oil vapors. Though not directly harmful, hydrocarbons are a major contributor to the formation of atmospheric smog and ozone. Hydrocarbons react with sunlight and break down to form these other chemical compounds that irritate the eyes, nasal passages, throat and lungs.

Ozone

Ozone is probably one of the most toxic and dangerous air pollutants known. Ozone is formed when an extra atom of oxygen attaches itself to the normal oxygen molecule ($O_2 + O = O_3$ = ozone). The extra oxygen atom causes the molecule to be very toxic to other materials. It irritates the eyes and lungs. It causes a variety of symptoms including coughing, headaches, choking, and a feeling of weariness in concentrations as low as 0.5 ppm. Ozone attacks rubber products and is toxic to many types of plants and microbes. In fact, industry sometimes uses ozone under controlled conditions to disinfect certain products. The maximum concentration of exposure over an 8-hour period to ozone is considered by most medical experts to be 0.1 ppm.

Elevated HC emissions, which are measured in "parts per million" (ppm) with an exhaust analyzer, usually result from ignition misfiring (a fouled plug or bad plug wire), lean misfiring, loss of compression (such as a burned exhaust valve), or engine wear that causes the engine to burn oil (worn valve guides, seals and/or rings).

Formation

Hydrocarbon formation is minimized by maintaining a balanced air/fuel ratio, by making sure the engine has steady and reliable ignition (proper plug gap, clean plugs, good wires and distributor cap, etc.), by having close tolerances in the engine (piston rings that are fully seated and seal properly, valve guides and seals that don't leak oil, etc.), and by having good compression (no leaky valves). The hydrocarbons that are formed in the combustion chamber are "reburned" in the catalytic converter and transformed into water vapor and carbon dioxide.

A well-tuned, late-model engine with a good converter should produce HC exhaust readings of less than 50 ppm—compared to several hundred ppm HC for an engine without a converter.

OXIDES OF NITROGEN

As we said earlier, nitrogen makes up almost 80% of the air we breathe. Though normally inert and not directly involved in the combustion process itself, flame temperatures above 2500 degrees F cause nitrogen and oxygen to combine and form various compounds called "oxides of nitrogen" or NOx. This typically occurs when the engine is under load and combustion temperatures soar.

Formation

Most of the NOx that comes out the tailpipe is in the form of nitric oxide (NO), a colorless, poisonous gas. It then combines with oxygen in the atmosphere to form nitrogen dioxide (NO2), which is the brownish version of the gas that creates a brownish haze in badly polluted areas.

NOx is a nasty pollutant both directly and indirectly. In concentrations as small as a few parts per million, it can cause eye, nose and lung irritations, headaches and irritability. It has an odor that becomes noticeable in concentrations as small as 1 to 3 ppm. When levels reach 5 to 10 ppm, NOx causes eye and nose irritation in some people. Higher concentrations can cause bronchitis and aggravate other lung disorders. Prolonged exposure to 10 to 40 ppm can have serious health consequences. Once in the atmosphere, it reacts with oxygen to form ozone (which is also toxic to breathe) and smog.

Reducing NOx—To minimize the formation of NOx in the engine, exhaust gas recirculation (EGR) is used. By recirculating a small amount of exhaust gas back into the intake manifold to dilute the air/fuel mixture, EGR actually has a "cooling" effect on combustion, thus keeping temperatures below the NOx

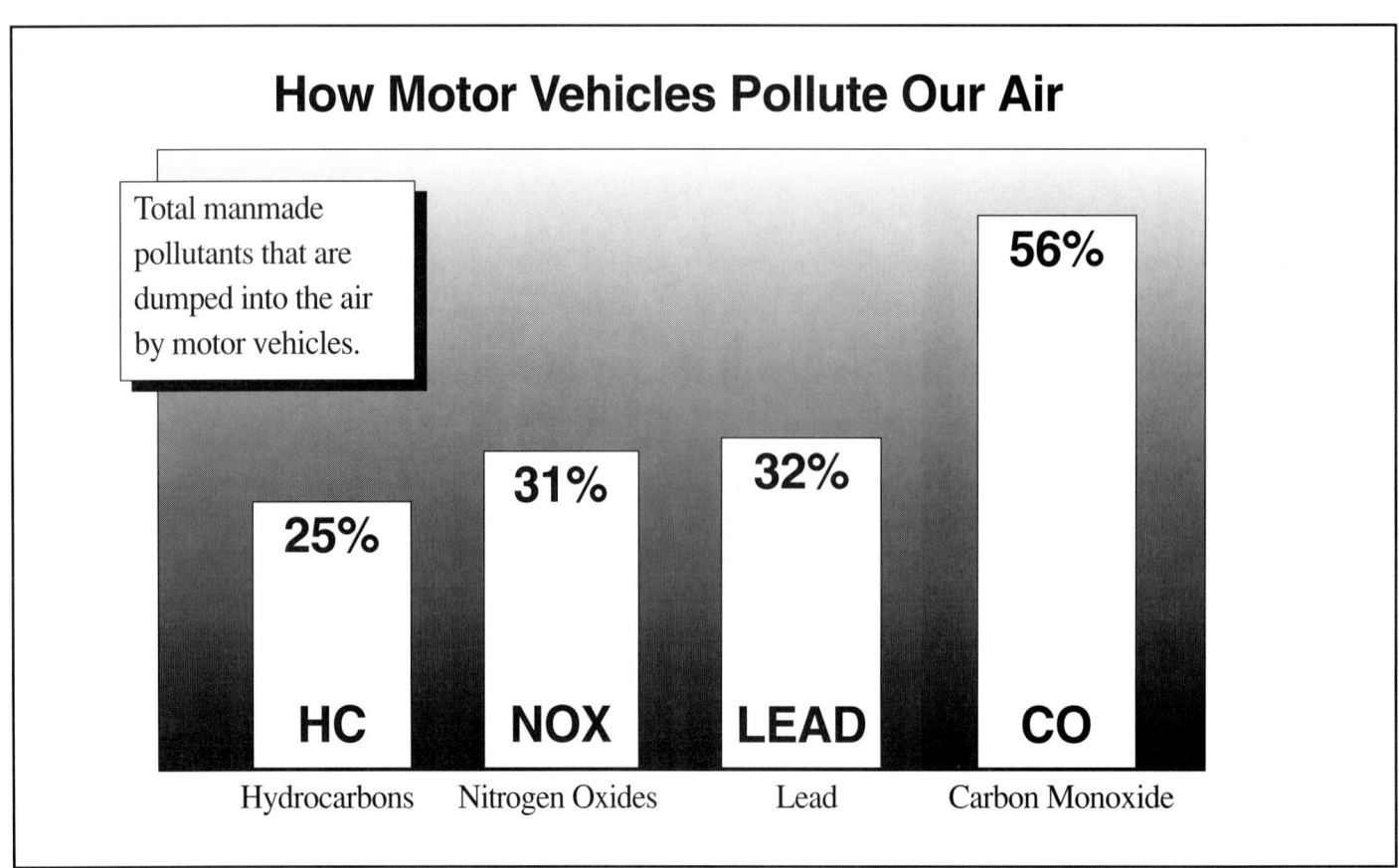

This chart illustrates a percentage breakdown of the pollutants emitted by motor vehicles. As you can see, carbon monoxide, the deadliest gas, has the highest percentage. Lead emissions, however, have been virtually eliminated with the phase-out of leaded gasoline.

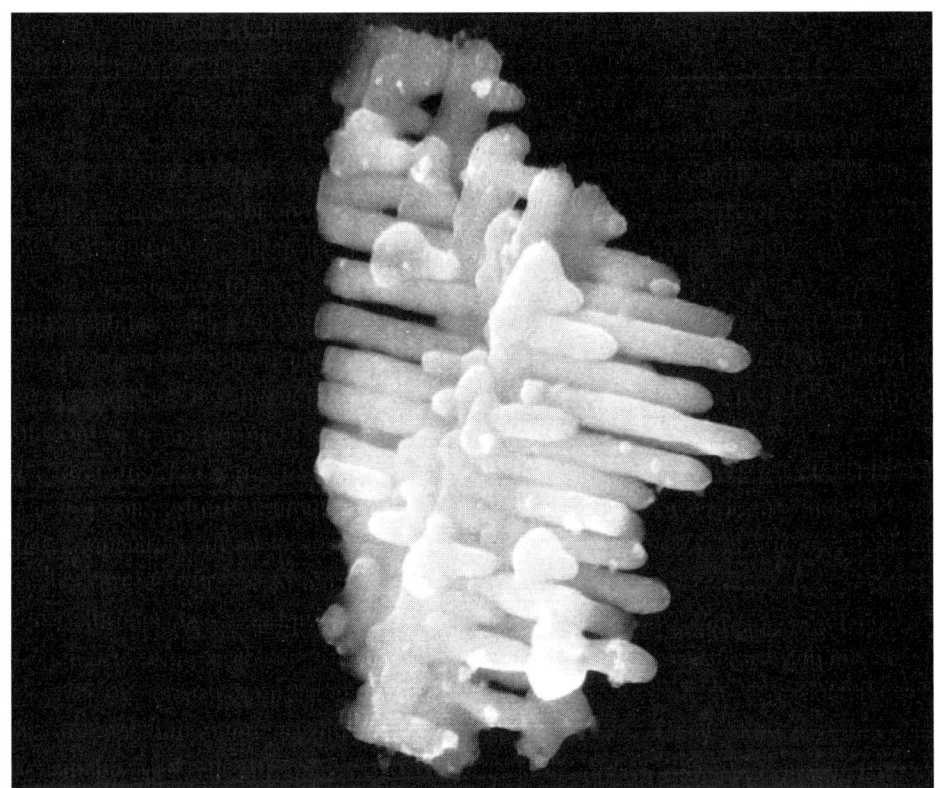

This is what an emissions particulate looks like magnified 5000 times. It measures 5 to 10 microns in diameter and was captured by a scanning electron microscope by Dow Chemical researchers. They are small enough to pass through the protective mucous membranes in our nose and throat, and therefore become lodged in our lungs, increasing our vulnerability to respiratory diseases.

formation threshold.

On 1981 and later engines with computerized engine controls, a special "three-way" catalytic converter is used to further reduce NOx in the exhaust. The first chamber of the converter contains a special "reduction" catalyst that breaks NOx down into oxygen and nitrogen. The second chamber in the converter contains the "oxidation" catalyst that reburns CO and HC.

Until the arrival of the new I/M 240 emissions inspection program, the only way to measure exhaust NOx was in a laboratory. Ordinary emissions test equipment could not measure it, only HC and CO (and CO_2 and O_2 with four-gas analyzers). The only way to tell if an engine was producing excessive NOx emissions, therefore, was to inspect the EGR valve and system to see that it was working (not an easy task on many engines) or to listen for a detonation problem when the engine was under load (a good clue that the EGR system isn't working). But the new I/M 240 test equipment is capable of measuring NOx emissions, so now there's a way to measure it directly.

OTHER POLLUTANTS

In addition to this terrible trio (HC, CO & NOx), other pollutants may also be found in the exhaust.

Particulates—These are tiny particles of carbon soot that result from incomplete combustion and excessively rich air/fuel ratios. Diesel engines are notorious for high-particulate emissions. Particulate emissions are typically measured with an "opacity" test that measures the darkness of the smoke that's coming out the tailpipe. Accurate injection timing plays a critical role in diesel particulate emissions.

Scientists estimate that as much as 100 tons of particulates fall to the ground or are breathed in by the public in most large American cities every day. Most of these come from trucks and buses with diesel engines. Many particulates are smaller than 1 micron in diameter (39 millionths of an inch). Particles this small are called "aerosols" because they tend to remain suspended in air rather than settle to the ground. Such particles pose a significant health hazard because they are too small to be filtered out by the mucous membranes in our nose and windpipe. They penetrate deep into our lungs and remain there, accumulating with each passing year. This can increase the odds of developing emphysema and cancer, especially if a person smokes.

Sulfuric Acid—This is another pollutant that's formed by exhaust pollution in catalytic converter-equipped vehicles. All crude oil contains a certain amount of sulfur. Crude oil from the western United States contains much more sulfur than do eastern crude oils. When the crude is refined to make gasoline, some of the sulfur remains in the fuel. The concentrations are small, but enough to cause pollution problems.

When gasoline containing sulfur is burned, the sulfur combines with oxygen to form sulfur oxides, and hydrogen to form hydrogen sulfides. When sulfur oxide concentrations reach 8 to 12 ppm, most people's eyes will water. Such levels also cause coughing, breathing difficulty, and can increase the likelihood of heart disease over time. Hydrogen sulfide (SO_2) produces the familiar rotten-egg odor.

When exhaust sulfur emissions combine with water, it forms sulfuric acid, which in turn contributes to "acid rain," a corrosive brew that eats away at buildings and statues, ruins paint jobs and kills fish in lakes.

Acid rain is causing long-term changes in our environment. The Adirondack mountain range is a pristine wilderness area in upstate New York. There are no factories and few roads. Yet acid rain caused by pollution sources hundreds of miles upwind is killing the fish in many of the lakes. Over 200 lakes that once provided excellent fishing are now devoid of marine life because acid rain has made the lake water toxic to fish. Hundreds of other lakes in the region are threatened by the same fate. The lakes are dying because they lack any natural means (no limestone) of countering the cumulative effects of acid rain. Automotive exhaust may be a contributing factor, but most of the acid rain causing these environmental problems is coming from stationary utilities that burn sulfur-laden coal.

No More Lead

Lead pollution is not a problem with cars that burn unleaded gasoline. But it was in vehicles that used leaded regular gasoline. Tetraethyl lead was long used in gasoline for two reasons. One was to raise the octane rating of the fuel. This allowed refiners to take a lower grade of gasoline and boost its performance to acceptable levels by adding lead as an octane enhancer. Federal regulations limited the amount of lead that could be added by refiners. But many secondary refiners or blenders added up to five times the maximum permissible amount of lead to marginal gasoline to make the product salable. The other reason why lead was used was because it had a lubricating effect on exhaust valves, which helped prolong valve life.

If leaded gasoline is used in a car with a catalytic converter, the lead fouls the catalyst and renders it useless. This, in turn, causes a significant increase in HC and CO emissions. It can also foul up the oxygen sensor in computer-controlled engines. So many emissions testing programs include a visual check of the restrictor in the fuel tank filler neck to see if it has been punched out to accept a pump nozzle for regular leaded gasoline.

Lead is considered a serious threat to the environment because it literally lasts forever. When ingested into the body, either by breathing in lead-polluted air or by drinking water from lead-polluted wells or by eating plants contaminated by lead dust or lead-polluted soil, the metal accumulates in the body and eventually produces lead poisoning. There is ample evidence to suggest that lead from automobile exhaust in metropolitan areas adversely affects the learning abilities of inner city school children. Lead is absorbed into nerve cells in the brain, where it inhibits normal functions. Tests have shown that school children who live near areas of traffic congestion score lower on intelligence tests and are more prone to behavior problems than those who live away from lead-polluted environments. Fortunately, leaded fuels are now a thing of the past. Lead has been gradually phased out over the years and is no longer available in pump gasoline. ■

PASSENGER CAR EXHAUST EMISSION REDUCTION PROGRESS FEDERAL 49 STATE STANDARDS (Grams Per Mile)

Model Year	Hydrocarbons**		Carbon Monoxide		Nitrogen Oxides	
	Grams	Reduction	Grams	Reduction	Grams	Reduction
Precontrol*	10.6	–	84.0	–	4.1	–
1968-1971**	4.1	62%	34.0	60%	NR	–
1972-1974**	3.0	72%	28.0	67%	3.1	24%
1975-1976	1.5	86%	15.0	82%	3.1	24%
1977-1979	1.5	86%	15.0	82%	2.0	51%
1980	0.41	96%	7.0	92%	2.0	51%
1981-1982	0.41	96%	3.4(1)	96%	1.0	76%
1983-1993	0.41	96%	3.4	96%	1.0(2)	76%

* AAMA 1960 baseline data.
** Pre-1975 standards have been adjusted according to current test procedure, EPA data.
(1) Waivers allowed manufacturers to use 7.0 grams per mile CO for 1981-82 model years.
(2) Waivers allowed 1.5 grams per mile NO_x for light-duty diesel engines only through the 1984 model year.

LIGHT-DUTY TRUCK EXHAUST EMISSION REDUCTION PROGRESS FEDERAL 49 STATE STANDARDS (Grams Per Mile)

Model Year	Hydrocarbons		Carbon Monoxide		Nitrogen Oxides	
	Grams	Reduction	Grams	Reduction	Grams	Reduction
Precontrol*	8.0	–	102	–	3.6	–
1975-1978	2.0	75%	20	80%	3.1	14%
1979-1983	1.7	79%	18	82%	2.3	36%
1984-1987	0.8	90%	10	90%	2.3	36%
1988 and Later:						
Less than 3,750 lbs. GVWR	0.8	90%	10	90%	1.2	67%
3,751 lbs. GVWR & Over	0.8	90%	10	90%	1.7	53%

* EPA 1969 baseline data gasoline engines.

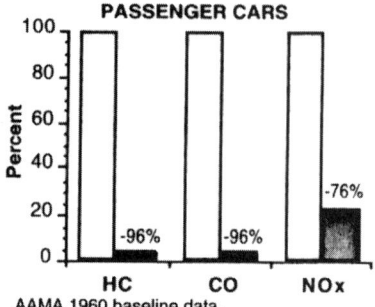

EXHAUST EMISSION PROGRESS FEDERAL 49-STATE STANDARDS

PASSENGER CARS
AAMA 1960 baseline data.

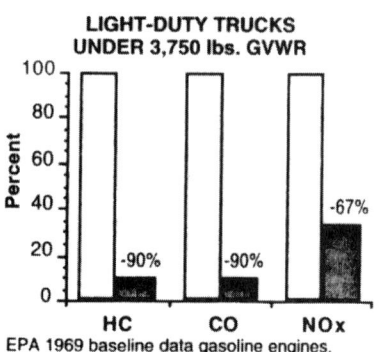

LIGHT-DUTY TRUCKS UNDER 3,750 lbs. GVWR
EPA 1969 baseline data gasoline engines.

☐ PRECONTROL ■ 1993

This chart shows the progress that has been made in reducing the emissions of hydrocarbons, carbon monoxide and oxides of nitrogen. The bar graph at right compares 1960 (1969 for trucks) emissions versus 1993. Other details trace the percentage of reduction by years.

5
POSITIVE CRANKCASE VENTILATION (PCV)

PCV was one of the first emissions-control systems developed. Not only does it help to control harmful emissions, but it does so without robbing horsepower with the use of accessory units. As an added benefit, it also protects the engine. Even the most anti-emissions control hot-rodder would have a hard time disliking this system.

During the 1950s, society became concerned with how our industrial way of life affected the environment. In particular, the automakers privately felt some sense of responsibility, because they knew full well the contribution their products made to atmospheric pollution.

Because of this concern and, more likely, anxiety about the possibility of government intervention, the automakers voluntarily added positive crankcase ventilation (the true definition of PCV, not "pollution control valve") systems to California cars in 1961 and to all cars sold in this country in 1963.

PCV was a good start for automotive emissions controls. This efficient method of airing out the crankcase is necessary because even the most precise, well-engineered set of piston rings can't seat against the cylinder walls perfectly. High cylinder pressures, reciprocal motion, ring end gaps, ring land clearances and piston slap are just a few of the reasons why complete sealing isn't possible. This inability of the piston

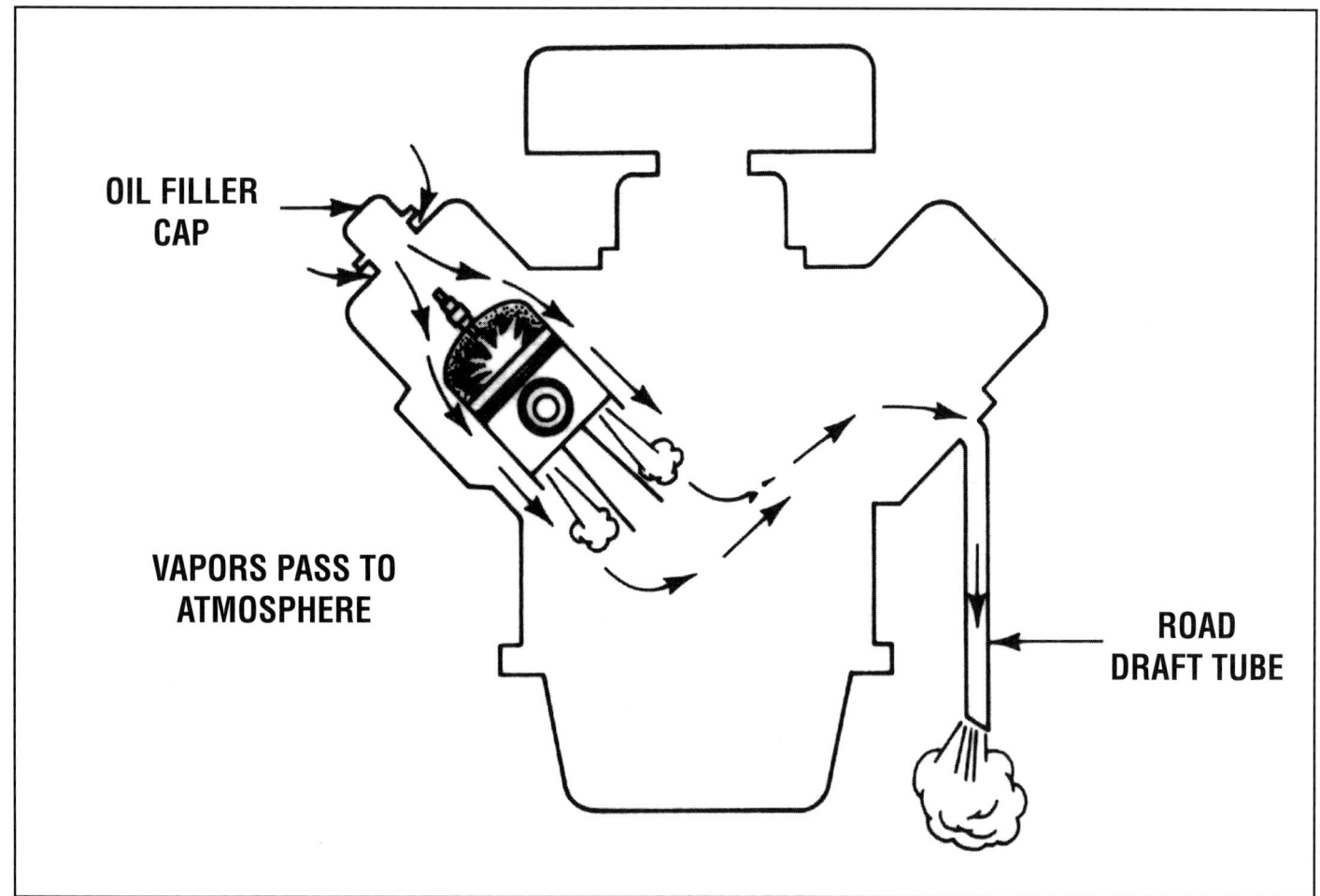

In the days before emissions controls, blowby was allowed to escape through the road draft tube (Chrysler).

39

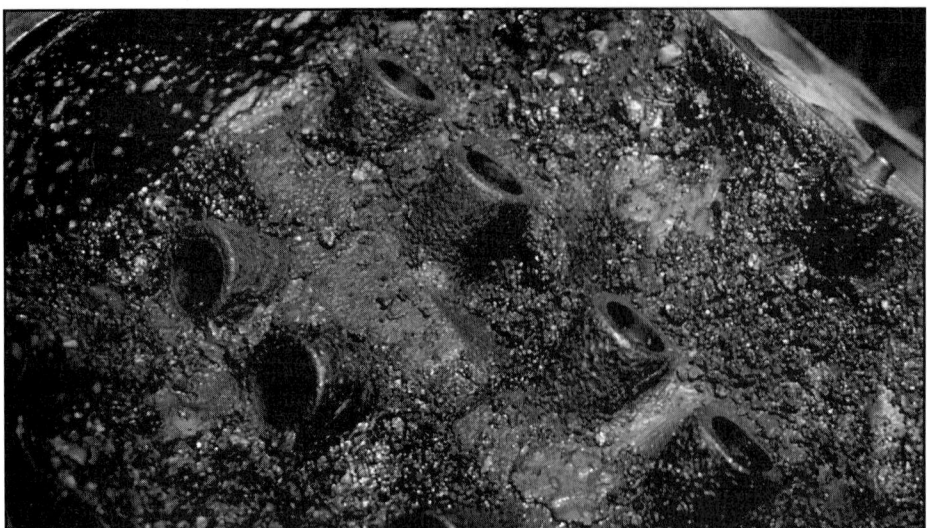

Besides reducing emissions by 20%, PCV kept internal engine deposits like these from forming.

engine to contain all of the fuel and burning gases within the combustion chamber and cylinder results in blowby, the passage of raw fuel and the byproducts of combustion into the crankcase.

THE OLD WAY

In the B.C. (Before Controls) era, the crankcase was simply vented to the atmosphere to allow these gases to escape and to relieve the pressure building up inside the engine. There was a road-draft tube that ran down under the chassis at an angle that produced a small amount of vacuum as the vehicle traveled forward. Fresh air was drawn through a mesh filter in the oil filler cap, circulated around inside the crankcase, and exhausted through the road-draft tube, carrying the blowby with it.

The increased pressure in the engine compartment caused by the air forced through the front grille helped this circulation by adding to the pressure differential between the air inlet and the end of the tube. In many cases, the outlet was positioned in the block at a point that took advantage of the pumping action set up by the rotation of the crankshaft.

Blowby gases are primarily HC from uncombusted gasoline and hot lubricating oil. With a car idling, these fumes could be observed escaping from the oil filler cap; they were also visible at the road-draft tube. If the engine was in poor condition internally, this puffing was very noticeable—and serious.

This was the most obvious place to start a campaign against air pollution, and also the least troublesome. The technology involved was not entirely new. Years before automotive air pollution became a widespread concern, many military and commercial vehicles used engine vacuum to positively ventilate the crankcase in order to reduce the formation of internal engine deposits.

The Open System

The PCV system uses the vacuum an engine naturally produces to draw fresh air through the crankcase, into the intake manifold, and into the cylinders to be burned. Not only does this eliminate about 20% of the total amount of emissions a car generates, it also keeps the inside of the engine much cleaner than was possible with the road-draft tube, allowing extended oil change intervals without the sludge and varnish buildup (providing the system is working properly).

A typical, early PCV system consisted of a hose that routed blowby from a valve cover into a spacer plate between the carburetor and the intake manifold, and an uncomplicated valve that regulated flow. This was known as the "open" system because fresh air was still admitted through a mesh filter in the oil filler cap. If blowby exceeded the capacity of the

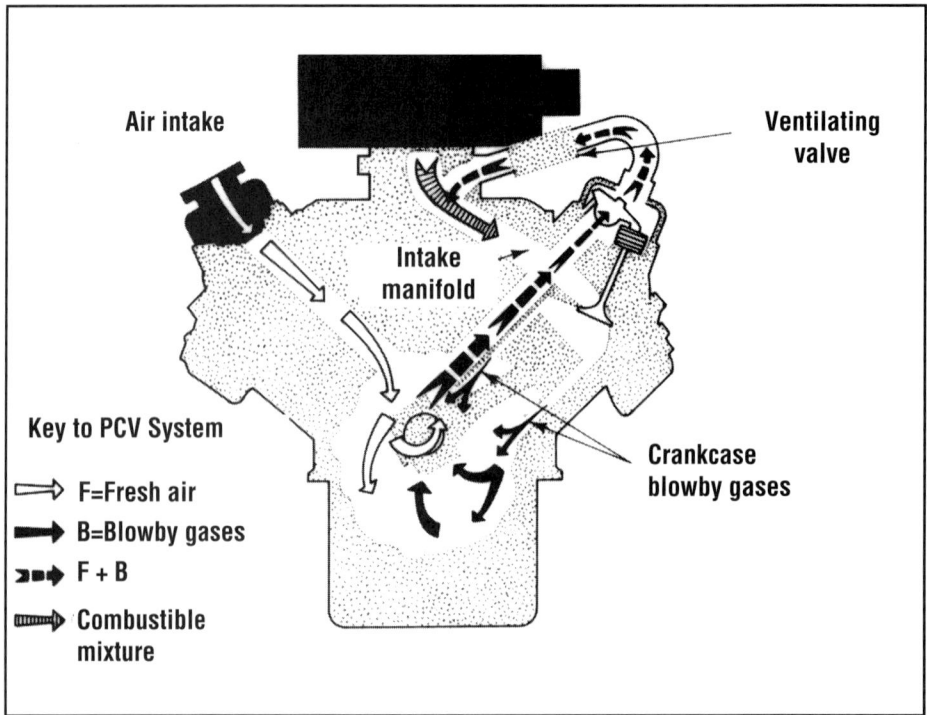

The open PCV system was installed on all cars sold in the U.S. starting in 1963 (Chevrolet).

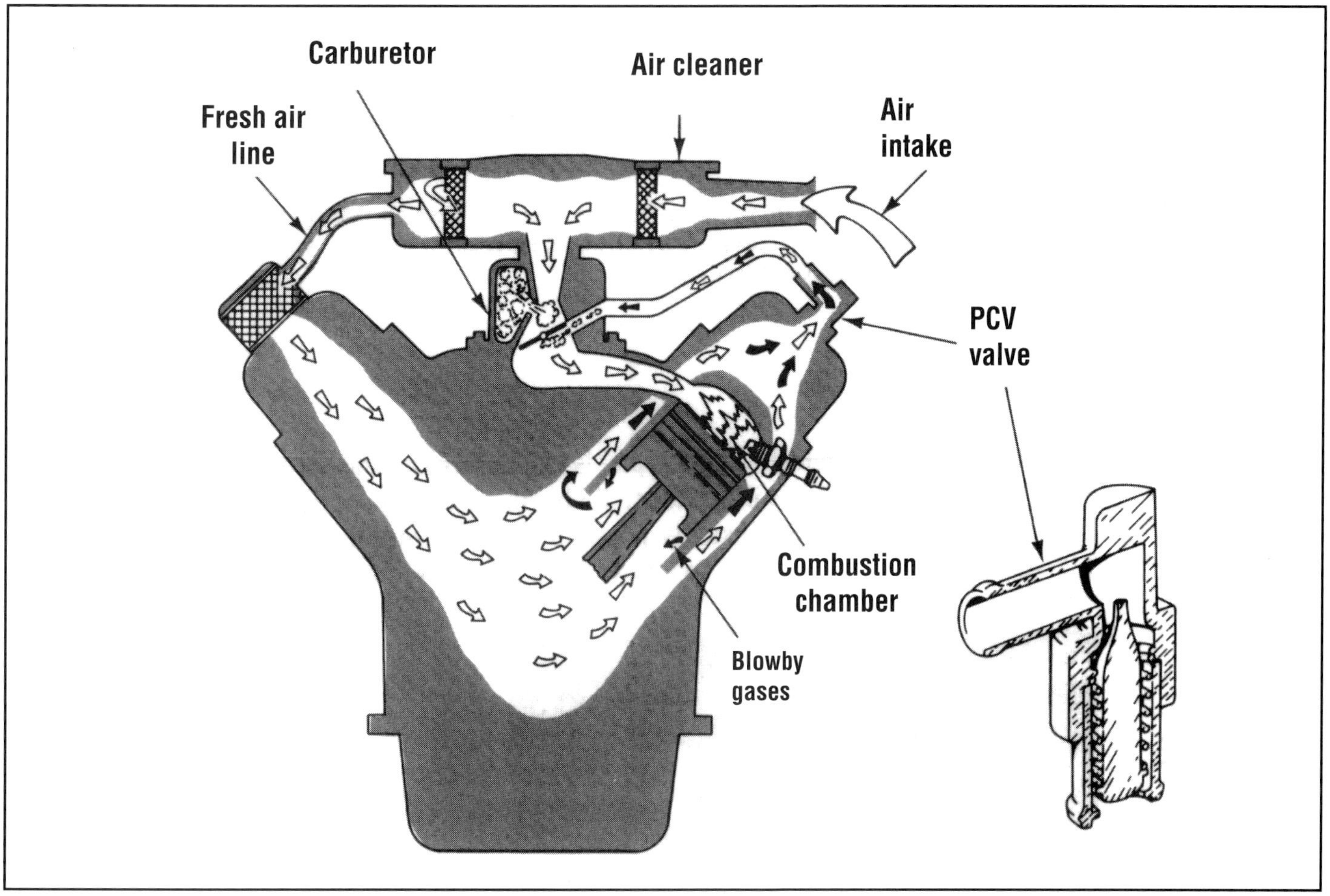

The closed PCV system appeared in 1968. Crankcase fresh air intake was at the air cleaner so that if the valve's capacity was exceeded, blowby backed up into the intake stream to be drawn into the carburetor (Chrysler).

valve, say during full-throttle operation, these gases could still escape into the atmosphere by backing up through the filler cap.

The Closed System

To eliminate this possibility, the closed system was introduced in 1968. This was a very simple modification consisting of a second hose that ran from the valve cover to the air cleaner housing, where it terminated in a mesh filter. The oil filler cap was now sealed instead of vented.

In normal operation, the closed system picks up fresh air from inside the air cleaner in the same way the open system did at the oil filler cap. But whenever there is more blowby than can be handled by the valve, the excess backs up through the hose into the air cleaner, where there is always a partial vacuum if the engine is running anyway, so it's drawn into the

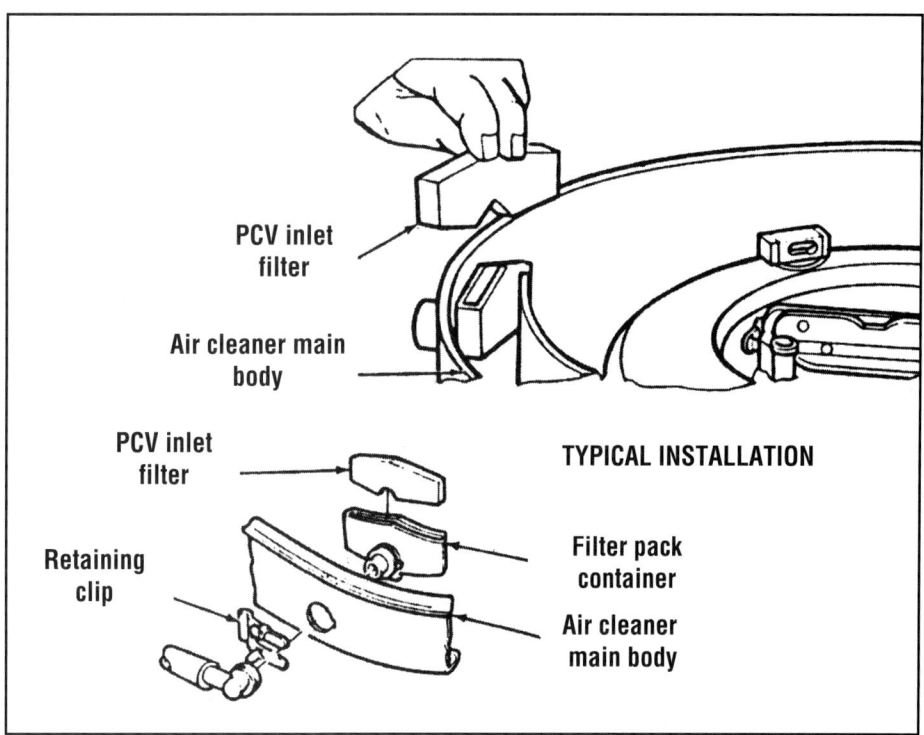

Closed PCV systems get their supply of fresh air through an inlet filter in the air cleaner (Chrysler).

41

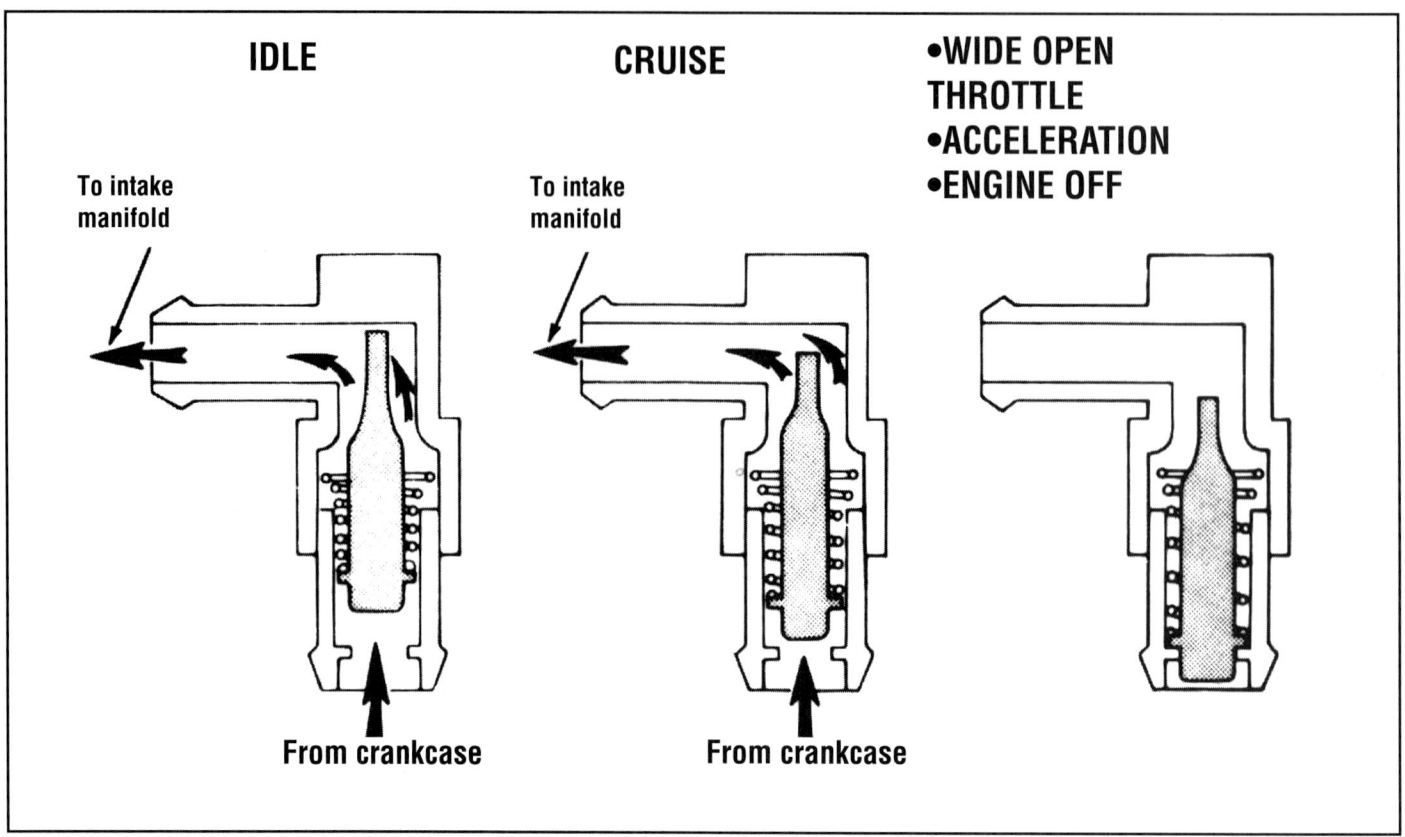

PCV valves are designed to allow less flow during high-vacuum conditions (as at idle) than during lower vacuum conditions. At wide open throttle or during an intake manifold backfire, the valve closes (Chrysler).

carburetor along with intake air and there is no chance of it escaping into the atmosphere.

THE PCV VALVE

The PCV valve itself, while fairly simple, is still widely misunderstood, so a discussion of its working principles is important. To begin with, any garden-variety valve consists of a steel or plastic housing, one end of which usually plugs into a grommet in the valve cover and the other end into a hose that goes to the spacer under the carburetor, or to an intake manifold port with fuel-injected engines. Inside the housing is a moveable plunger and a light coil spring that bears against it.

How It Works

One of the characteristics of the gasoline piston engine is that it produces more intake manifold vacuum when idling than when accelerating or cruising. This is simply a function of the amount the throttle plates are open and hence how much atmospheric pressure is allowed to enter the engine. The PCV valve takes advantage of this situation in order to meter the flow of blowby.

At idle, a strong vacuum forces the plunger against the outlet port, compressing the spring. The end of the plunger and the opening of the port are made so that, when in this position, only a small amount of PCV flow is permitted. This keeps the system from interfering with the idle mixture any more than is necessary. The flow is sufficient to accommodate the relatively small amount of blowby an engine produces at low rpm.

As the throttle plates are opened and rpm increases, vacuum is reduced, and the spring begins to overcome it, forcing the end of the plunger away from its seat and allowing a larger volume of waste gases to pass. This is a progressive action, with flow increasing with rpm, so that the PCV system keeps up as the engine produces more blowby. A typical valve will allow 1 to 3 cfm of flow at idle, and 3 to 6 cfm at 6 in. Hg. (inches of mercury) of vacuum.

If there should be a backfire in the intake manifold, the shock waves slam the plunger momentarily against a seat in the other end of the housing, shutting the valve off completely and eliminating the possibility of flammable fumes igniting in the crankcase.

PCV's Without Valves

Some vehicles, notably imports, don't have PCV valves. Instead, ventilation of the crankcase is accomplished by means of a one- or two-stage orifice arrangement. Honda's Dual Return System is a common example. A liquid-vapor separator is built into the top of the camshaft cover, and connected by a hose to a condensation chamber at the bottom of the air cleaner. Another hose runs from the condensation chamber down to a fixed orifice vapor passage connected to the carburetor insulator that's mounted on the intake manifold.

During idle and part throttle operation when vacuum is high, blowby flows from the liquid-vapor separator to the condensation chamber, then down through the fixed orifice and into the intake stream. The fixed orifice is calibrated so that it doesn't cause idle mixture problems by allowing too much flow.

At throttle positions approaching wide open, vacuum in the air cleaner housing increases and that in the intake manifold decreases. Therefore, blowby leaves the liquid-vapor separator, enters the condensation chamber, and leaves through the condensation chamber's upper passage. This exits inside the circumference of the air filter element itself so that the vapors don't soak and clog the element. Finally, the blowby is drawn into the carburetor's throat. Very little enters the intake stream through the fixed orifice during this low-vacuum mode. Although it may appear that there is nothing to go wrong with this system since no valve is used, there is always the possibility of the orifice becoming plugged, and that would interfere with the idle mixture and cause all the blowby to enter the air cleaner.

As you can see, there's no provision in this type of system for admitting fresh air into the crankcase, so the flow is limited to the amount of blowby the engine produces. It would seem logical to assume that true crankcase ventilation systems—wherein fresh air passes through the inside of the engine and carries blowby through a PCV valve into the intake manifold—would tend to keep the internal parts and surfaces cleaner than those arrangements that don't admit fresh air, but this isn't necessarily the case in the real world.

Some imports use a more basic system with simply one hose that runs from the top of the camshaft or rocker cover to the air cleaner housing, with no valve or calibrated orifice in between. The fact that there is pressure inside the engine and a partial vacuum in the air cleaner makes the vapors move into the intake stream. This has the disadvantage of wetting the filter element. Also, when the weather is cold, the water vapor present in the blowby can cause icing.

If the PCV Fails

A malfunctioning crankcase ventilation system can cause several problems, the most serious being the accumulation of sludge and varnish inside the engine. Furthermore, if there is a pressure build-up, gaskets and seals will leak. Other troubles include rough idling or stalling if the PCV valve is stuck open, and the soaking of the air filter element if there is a blockage. Therefore, it's extremely important that the system be kept in good working order. You'll also have trouble passing an emissions test if it isn't working properly or if the hose is leaking or disconnected. ■

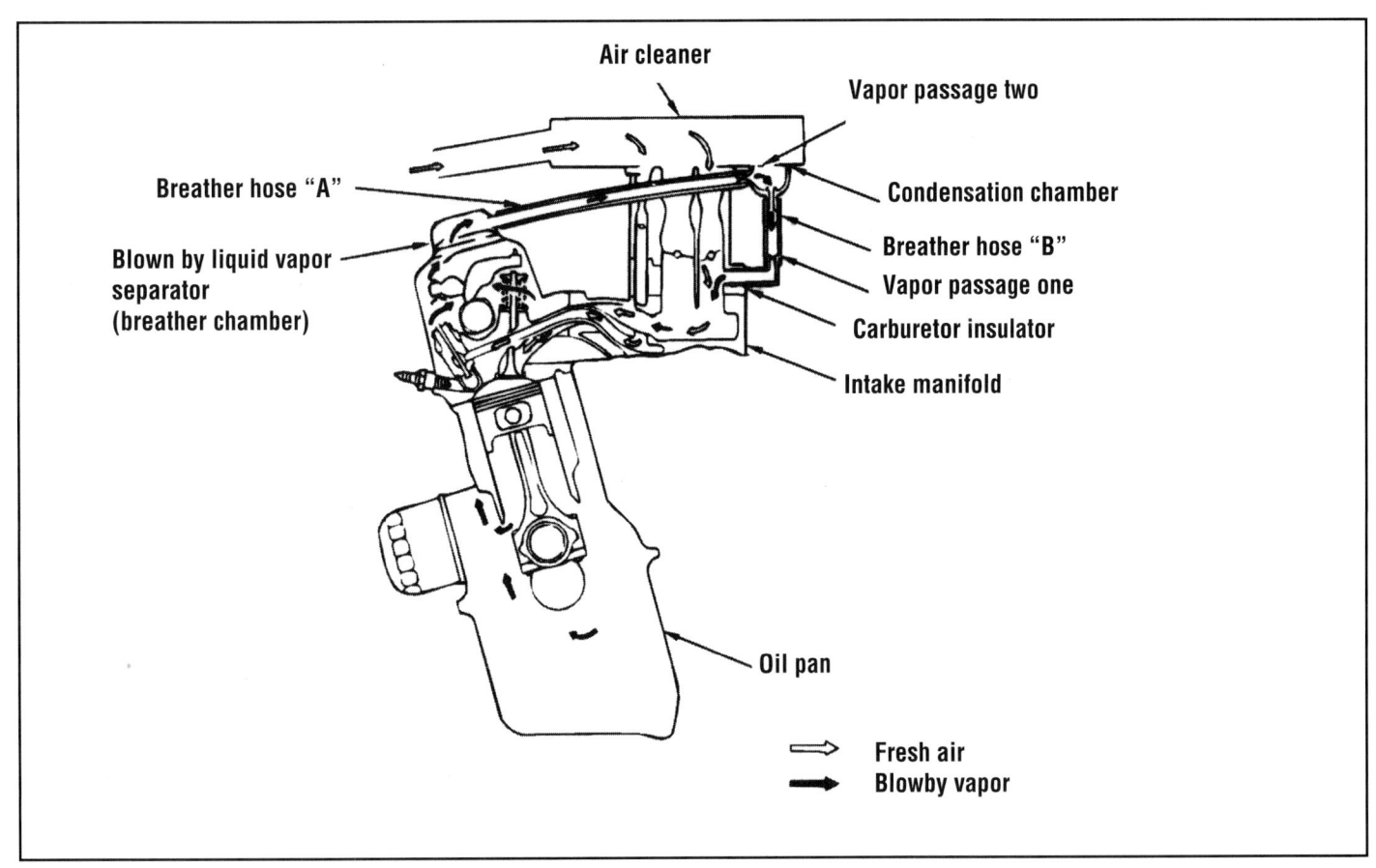

This Honda system uses two calibrated orifices instead of a PCV valve to regulate the flow of blowby, and no fresh air is admitted (Honda).

EVAPORATIVE EMISSIONS CONTROLS (EEC)

Gasoline fuel vapors, or evaporative emissions, as they are called, contain a variety of hydrocarbons (HC). This is because gasoline itself is a blend of many different hydrocarbons. The makeup of any given gallon of gasoline depends on the grade of crude oil from which it was refined, the refining process it underwent, and the additives that were put in to improve the performance characteristics of the fuel. Because of such factors, the nature of gasoline can vary considerably from one area of the country to another, or even from one brand to another in the same town.

DEFINING EVAPORATIVE EMISSIONS

The lighter elements in gasoline evaporate easily, especially in warm weather. These include aldehydes, aromatics, olefins and paraffins. These substances can react with air and sunlight (a photochemical reaction) to form *smog*. Aldehydes are often called *instant smog*, because they can form smog without undergoing photochemical changes.

Evaporative emissions can account for about 20% of a vehicle's total emissions. When you consider the fact that a parked car can pollute the air with hydrocarbon emissions, even though the engine is not running, controlling evaporative emissions becomes an important consideration. The problem becomes especially bad on hot summer days when a vehicle is parked in the sun on an open parking lot. Under such circumstances, gasoline vapors can literally spew out of the vehicle's fuel system unless they are prevented from doing so by an evaporative emissions control system.

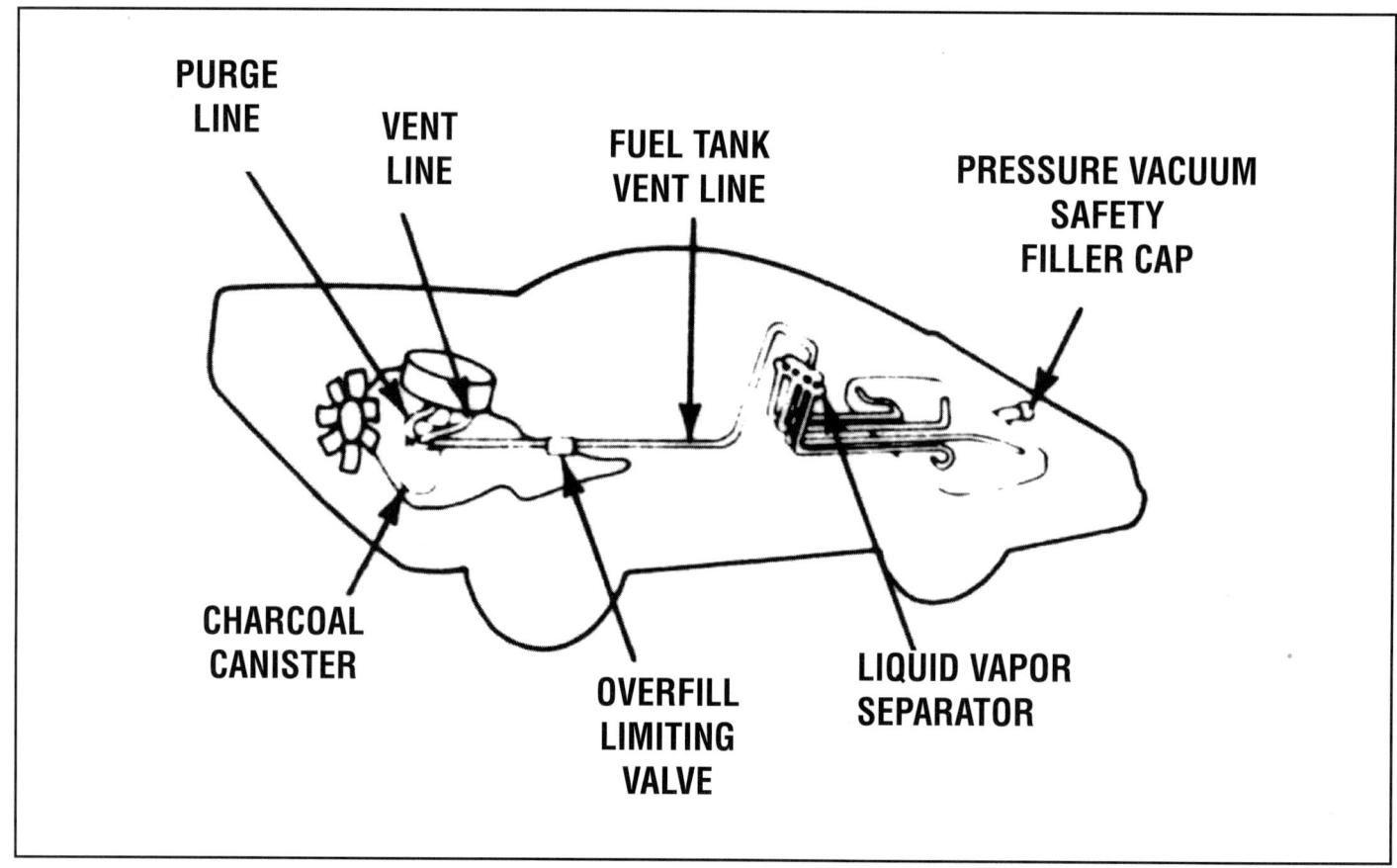

These are the basic components in the evaporative emissions-control system.

Evaporative emissions controls were required on cars sold in California in 1970, but they were federally mandated on all cars sold in all states beginning in 1971.

Controlling Evaporative Emissions

Evaporative emissions are eliminated by sealing off the fuel system from the atmosphere. This prevents the gasoline vapors from escaping from the fuel tank or carburetor bowl. Vent lines from the fuel tank and carburetor bowl route vapors to a charcoal canister, where they are trapped and stored until the engine is started. The vapors are then drawn into the intake manifold and burned.

With fuel injection, the problem isn't as bad because a fuel injection delivery system is a pressurized sealed system. Unlike a carburetor, there is no vented fuel bowl to leak vapors. The fuel is all contained within the pressurized fuel rail and injectors. But the fuel tank must still be sealed to prevent vapors from escaping out the filler pipe.

Evaporative emissions controls were first required on cars sold in California in 1970, and on all other cars since 1971.

SEALED SYSTEM

Sealing the fuel tank and carburetor is a simple way to prevent the escape of fuel vapors into the atmosphere. But doing so is not as simple as it sounds. For one thing, a fuel tank must be vented so that air can enter to replace fuel as the fuel is used up. If a tank were sealed tight, the fuel pump would soon create enough negative suction pressure inside the tank to cause the tank to collapse or to restrict the flow of fuel to the engine. It is something like trying to pour oil out of a can with a single small hole in it. The vacuum created inside the can slows the flow of oil to a trickle. Punch a vent hole in the can and the oil gushes out. A fuel tank needs the same kind of ventilation so that the fuel pump can draw the fuel out. The job of venting the gas tank is usually performed by the gas cap.

A fuel tank must also allow for a certain amount of fuel expansion. Gasoline, like other liquids, expands as it gets warmer. If you fill a tank on a cool morning, rising temperatures later in the day can cause the fuel to expand. If there is not sufficient reserve capacity built into the tank to handle the added fuel volume, the tank will overflow.

Like the fuel tank, the carburetor bowl must also be vented to function properly. If the bowl were sealed tight, negative pressure inside the bowl would decrease the flow of fuel through the metering circuits and venturis, causing a leaning of the air/fuel mixture or possibly fuel starvation. So some means of venting the bowl must be provided.

Evaporation of fuel from a carburetor increases with temperature. The hotter the bowl, the faster the rate of evaporation. While the engine is running, fresh fuel entering the bowl has a cooling effect. This helps to minimize evaporation somewhat. But as soon as the engine is turned off, the carburetor begins to soak up heat like a sponge. Evaporation increases dramatically, and on especially hot days the fuel can literally boil in the bowl.

To keep the gasoline in the fuel system and out of the atmosphere, the evaporative emissions-control system must allow for fuel expansion, tank venting, carburetor bowl venting, and be able to store gasoline fumes for extended periods of time.

The Name Game

A variety of evaporative emissions-control systems have been created by the automobile manufacturers. Although these systems go by a number of different names, most share common elements and operate in the same manner. Some past and present acronyms are:

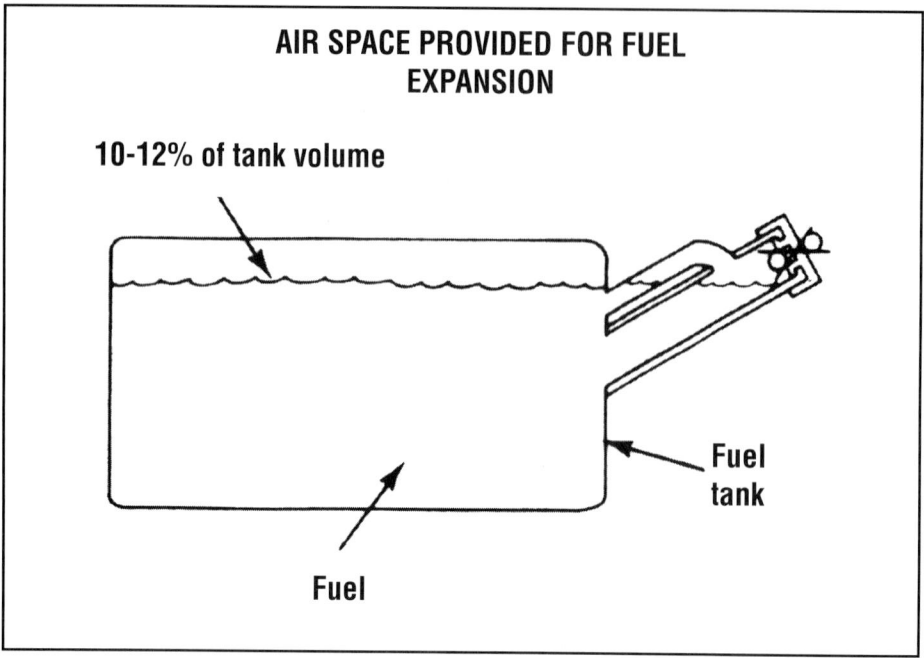

The fill control tube between the fuel tank and filler neck prevents overfilling of the tank. An air space is needed at the top to allow for fuel expansion.

FTVC (Fuel tank vapor control): AMC
ECS (Evaporation control system): Chrysler
EEC (Evaporative emissions control): Ford
EFE (Early fuel evaporation control): GM & Nissan
EECS (Evaporative emissions control system): GM
FVCR (Fuel vapor control recovery): Most Imports

Knowing the system names is not important, because the manufacturers will call different systems by the same name, or the same system by different names. Terminology and consistency have always been a problem in the auto industry, especially with respect to emissions controls. For example, GM's EECS system is not the same for all models. A Chevy may have a somewhat different configuration than that on a Cadillac. Each auto manufacturer usually has a number of variations on the basic theme of evaporative emissions control to suit the needs of their different engine/fuel system/body styles within their product line.

The best method to determine the system specifics for your particular automobile is to refer to a service manual. The official factory shop service manuals are the best guides because they contain more detailed information than the large general repair manuals such as those of Motor, Chilton, and Mitchell (though special emissions manuals are available from these sources).

EECS COMPONENTS

Let's now look at some of the components of a typical evaporative emissions control system.

Fuel Tank

All fuel tanks in today's cars are designed to allow for fuel expansion. The expansion space is usually 10% to 12% of the total tank volume. For example, a tank designed to hold 12 gallons of fuel when filled would need at least an additional one gallon capacity for expansion. There are several ways expansion space can be designed into a fuel tank. The easiest way is to locate the filler neck so that an air space is created at the top of the tank when it is filled. Designing a bulge or dome on the top of the tank serves the same purpose. The air pocket absorbs the increase in volume as the fuel expands.

Another way to create an air space in the top of the tank is to connect a fill control tube to the filler neck. When the tank reaches a certain level as it is being filled, gasoline begins to flow back through the fill control tube into the filler neck. This causes the gas nozzle to kick off and prevents overfilling the fuel tank. The remaining air space at the top of the tank then serves as the expansion reserve.

Some vehicle manufacturers solve the expansion problem by using a small external expansion tank on top of the main fuel tank. The expansion tank has a capacity of about 1 to 2 gallons, and is connected to the main tank with vent lines and a fill control tube. With this approach, the main tank can be filled to capacity. The expansion tank then handles any resulting fuel expansion. It is something akin to the expansion tank on a radiator.

Yet another approach to controlling fuel expansion is to use the tank-within-a-tank method. A little fuel tank with several small orifices punched in the sides is located inside the main fuel tank. The orifices limit the speed with which the smaller tank can fill with fuel. When the main tank is being filled, it will reach capacity long before the smaller tank. So when the gas nozzle kicks off, indicating a "full" tank, gasoline will continue to seep into the smaller tank from the big one. This creates an air space in the top of the main tank for fuel expansion.

The only problems any of these expansion control techniques can create are complaints about slow filling. Many motorists quickly discover that such fuel tanks fill slowly or that they never seem to be quite full. That is because the tanks are designed that way. Overfilling by continually squeezing in a few more cents' worth of gasoline after the nozzle has kicked off defeats the design purpose of expansion control.

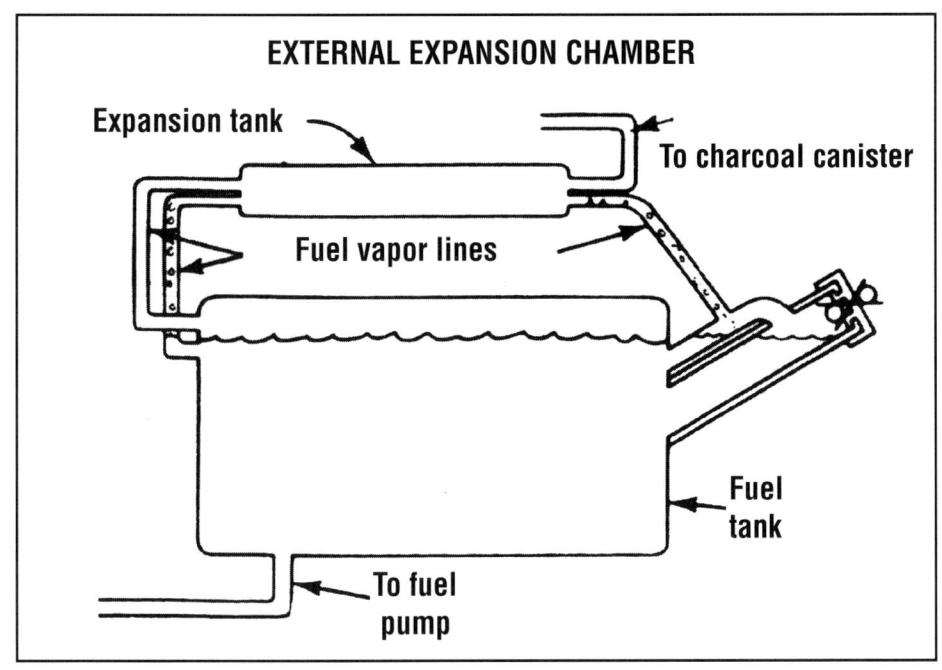

Some vehicles have an externally mounted expansion tank to prevent overfilling.

Some Ford vehicles do not have a pressure-vacuum relief gas cap. Instead they use a three-way valve in the fuel tank vent line to control internal tank pressure. The valve vents tank pressure to the charcoal canister. When there's a vacuum in the tank, the upper diaphragm will allow air to be drawn into the vent line to the tank. The lower diaphragm serves as a safety vent for excessive tank pressure in case the main vent line becomes clogged. *Note:* If you have to replace your gas cap because it was lost, be sure to use the same type of cap (sealed or vented) as the original.

Liquid-Vapor Separator

On top of the fuel tank or as part of the expansion tank is a device known as a liquid vapor separator. The purpose of this unit is to prevent liquid gasoline from entering the vent line to the charcoal canister (located in the engine compartment). You do not want liquid gasoline going directly to the charcoal canister, because it would quickly

Gas Cap

Few people think of a gas cap as being an emissions-control device, but it is. In pre-control days, the gas cap's main job was to keep gasoline from sloshing out of the tank. It was also equipped with a small vent hole so that the tank could breathe. Air could enter through the cap to make up for fuel as it was used, and fuel vapors could exit through the cap as internal pressure built up on warm days.

Types of Caps—Today's emissions-control gas caps are considerably different. They are either of solid construction (venting is provided by other means) or they contain a pressure/vacuum valve. The valve-type cap will vent tank pressure if it exceeds 1 psi. It will also allow air to enter the tank if a vacuum exists within the tank. In other words, the valve-type cap can vent pressure or relieve vacuum as the situation warrants without allowing gasoline vapors to pollute the environment.

The valve itself is a simple double-spring arrangement similar to a radiator cap. One spring reacts to internal pressure while the other reacts to external pressure. A plate or diaphragm between the two springs opens and closes to allow air to pass through the valve in the direction needed.

Internal fuel tank pressure can also be vented by means of a three-way valve in the vapor line to the charcoal canister.

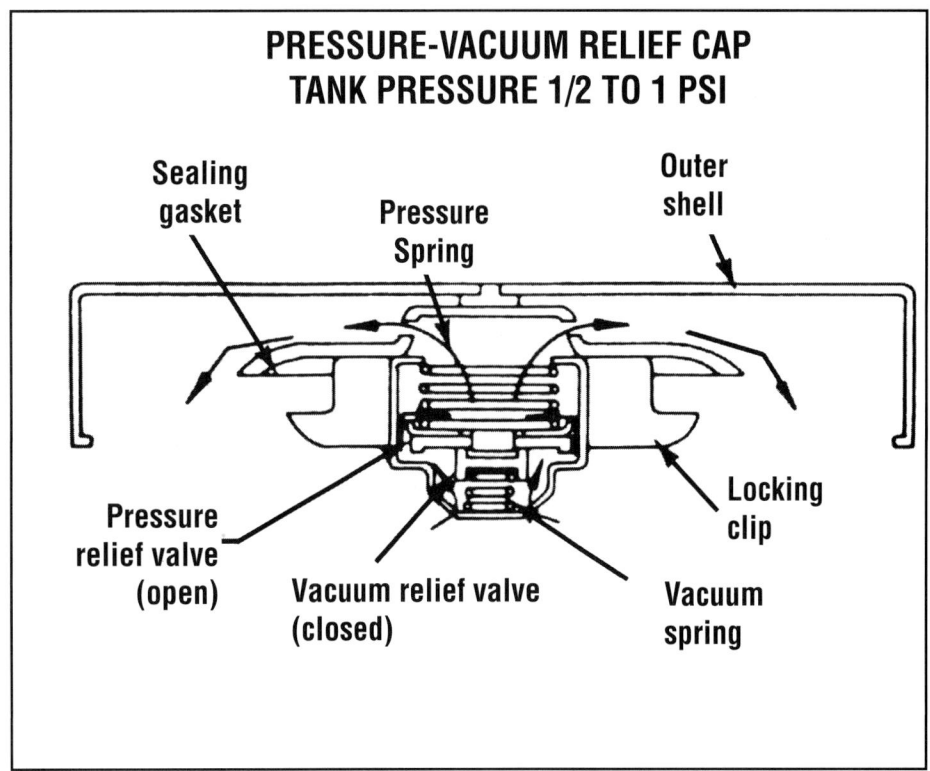

To prevent fuel vapors from escaping from the fuel tank, gas caps are sealed. But many have an internal pressure relief valve to compensate for thermal expansion.

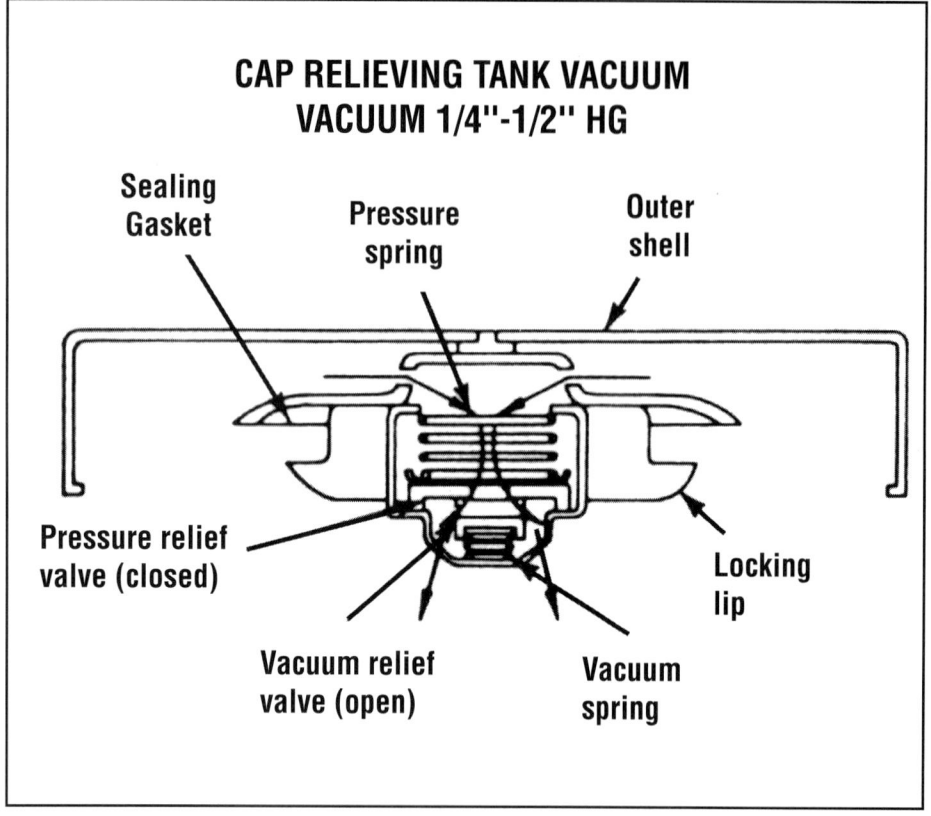

The gas cap relief valve must also allow make-up air to enter the tank as fuel is drawn out of the tank into the fuel line.

becomes blocked or crimped; or if a vent line develops an external leak due to rust, corrosion or metal fatigue from vibration.

Some liquid-vapor separators use a slightly different approach to keeping liquid fuel out of the canister vent line. A float-and-needle assembly is mounted inside the separator. If liquid enters the unit, the float rises and seats the needle valve to close the tank vent.

Another approach sometimes used is a foam-filled dome in the top of the fuel tank. Vapor will pass through the foam but liquid will cling to the foam and drip.

Checking Venting—If a blockage occurs in the liquid-vapor separator or in the vent line between it and the charcoal canister, the fuel tank will not be able to breathe properly. Symptoms include fuel starvation or a collapsed fuel tank on vehicles with solid-type gas caps. If you notice a whoosh of pressure in or out of the tank when the gas gap is removed, suspect poor venting.

You can check tank venting by removing the gas cap and then disconnecting the gas tank vent line from the charcoal canister. If the system is free and clear, you should be able to blow overload the canister's ability to store fuel vapors.

How It Works—The liquid-vapor separator works on the principle that vapors rise and liquids sink. The vapor vent lines from the fuel tank that go to the separator are positioned vertically inside the unit with the open ends near the top. This allows the vapors to rise to the top of the separator. Any liquid that enters the separator through the vent lines (from fuel sloshing around inside the fuel tank as a result of hard driving, parking on a steep hill, excessive fuel expansion, etc.) dribbles down the sides of the vent tubes and collects in the bottom of the separator. A return line allows the liquid gasoline to dribble back into the fuel tank. The vapors then exit through an opening in the top of the separator, which usually has an orifice restriction to help prevent any liquid from getting into the canister vent line.

The liquid-vapor separator is a simple device that is relatively trouble-free. The only problems that can develop are if the liquid return becomes plugged with debris, such as: rust or scale from inside the fuel tank; if the main vent line

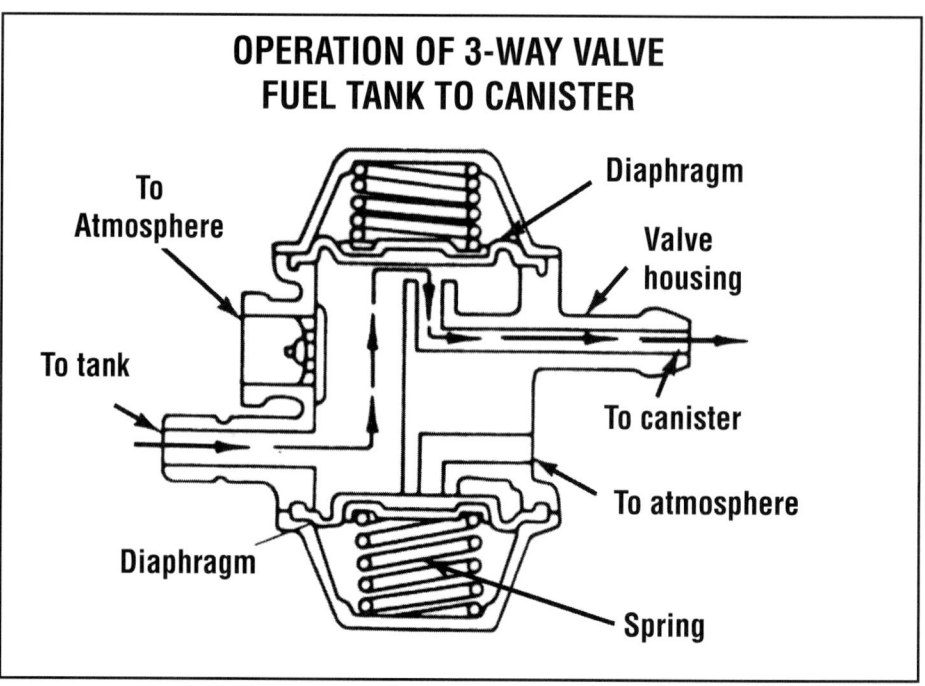

Some vehicles do not use a pressure relief gas cap. They have a three-way valve in the charcoal canister plumbing that serves the same purpose.

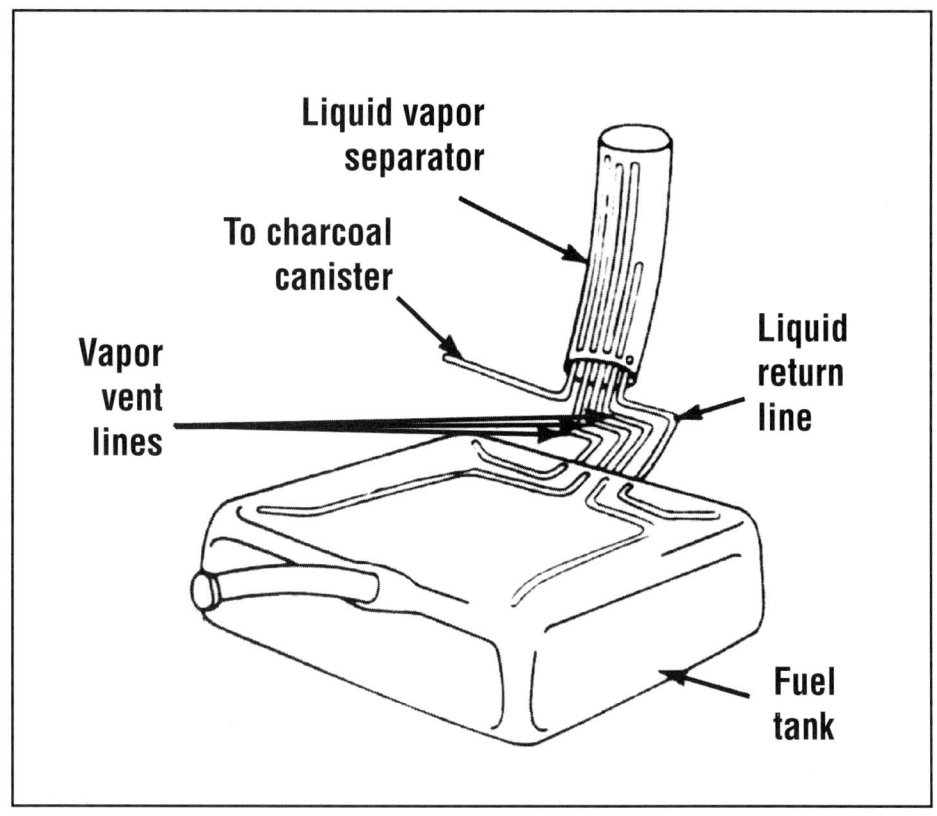

A typical liquid vapor separator.

(about 1-1/2 lbs.). Charcoal acts like a sponge to soak up the gasoline vapors, holding up to twice its weight in fuel.

The vapors are stored in the canister until the engine is started. The vapors are then drawn into the air cleaner, through a vacuum port in the carburetor or intake manifold, or are siphoned into the engine through the PCV plumbing.

Some early Chrysler evaporative control systems did not use a charcoal canister. Instead, the fuel vapors were routed into the engine's crankcase for storage. When the engine was started, the PCV system would draw the fumes out of the crankcase and into the intake manifold. This approach had its drawbacks, though. For one thing, the gasoline vapors tended to dilute the crankcase oil. The vapors also formed an explosive mixture that could literally blow the valve covers off the engine. Because of such problems, the approach was dropped in favor of the charcoal canister method of storing fuel vapors.

The charcoal canister is connected to the fuel tank via the tank vent line, and to the carburetor bowl with another vent through the vent line into the fuel tank. Blowing with compressed air can sometimes free a blockage. If not, you will have to inspect the vent line and possibly remove the fuel tank to diagnose the problem.

Some Chrysler and Ford cars have an overfill limiting valve in the vent line between the fuel tank and charcoal canister. The valve's purpose is to prevent any liquid that might have passed through the separator from reaching the canister. It consists of a simple float valve that closes if liquid fills the small chamber around it.

Charcoal Canister

The charcoal canister is a small round or rectangular plastic or steel container mounted somewhere in the engine compartment. On some late-model vehicles where underhood space is at a premium, the canister may be hidden behind a fender splash panel.

The canister's job is to store gasoline vapors from the fuel tank so that the fumes do not pollute the atmosphere. The canister contains activated charcoal

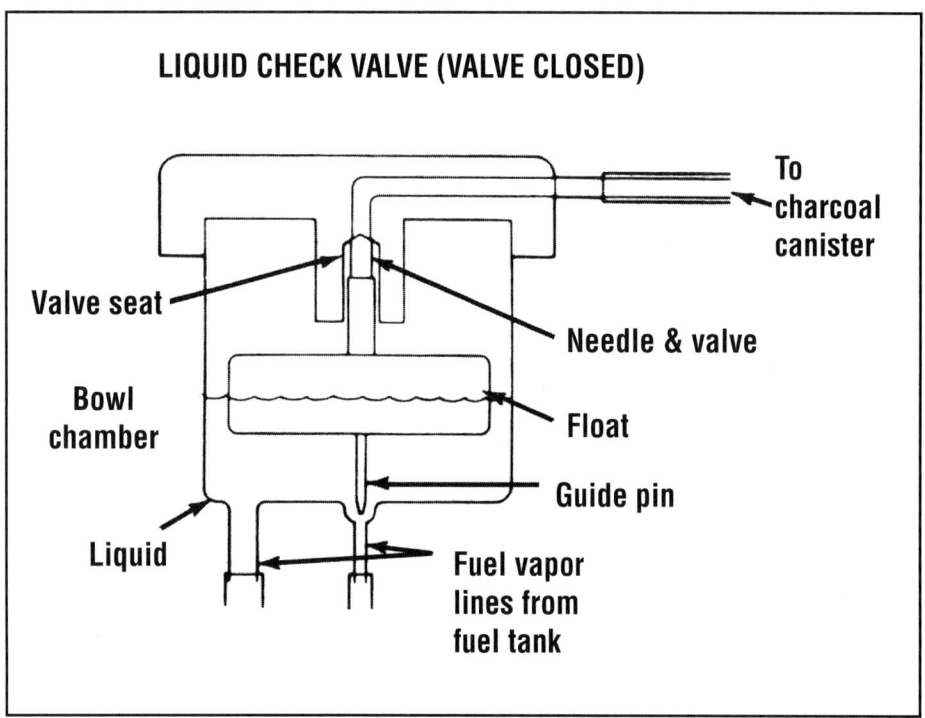

The liquid check valve in the tank vent line closes to prevent liquid from reaching the charcoal canister.

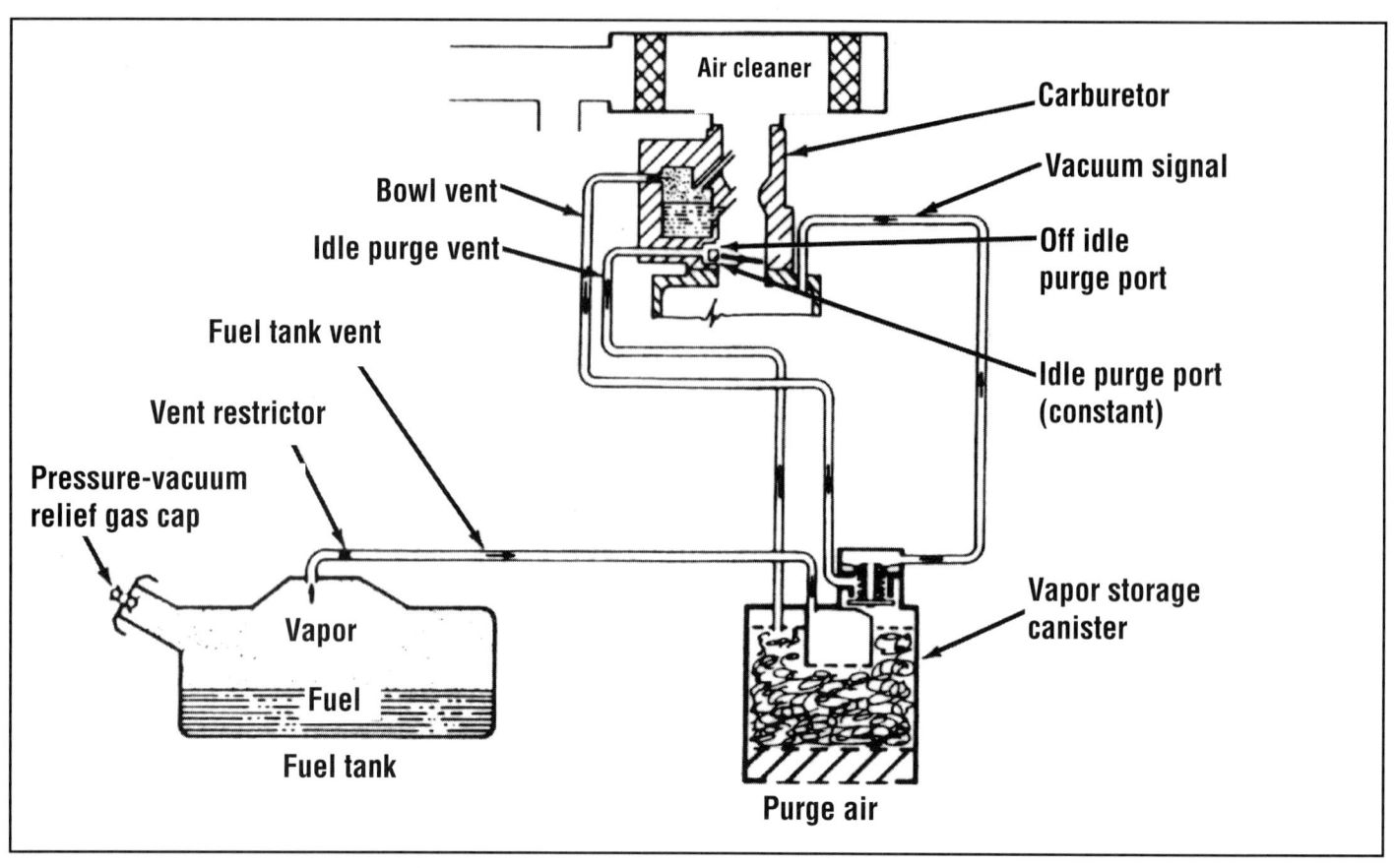

Here's an early charcoal canister evaporative emissions control system. Air to purge the vapors from the canister is drawn up through an open filter in the bottom of the canister (Chevy).

hose. The pressure created by evaporating fuel drives the vapors through the lines and into the canister. Here they stay until the canister is purged by starting the engine.

The auto manufacturers have been quite clever in coming up with various ways to purge the charcoal canister of its contents. Some canisters in older vehicles have an open bottom with a small, flat air filter across the opening. The filter is there to keep dirt out of the canister while it is being purged. On applications that use such a filter, the filter should be inspected periodically and replaced according to the manufacturer's recommendations. A good rule of thumb is to replace the filter every two years.

Open Bottom—On canisters with the open bottom, fresh air is drawn in through the filter by connecting a purge line from the top of the canister to the air cleaner, a carburetor vacuum port, or the PCV plumbing. Airflow through the canister ("purging") is regulated by a purge control valve on the canister. The valve opens in response to a ported vacuum signal. Others use an electric solenoid purge control valve. On these systems, the solenoid is regulated by the engine control computer.

Sealed Bottom—On canisters with sealed bottoms, fresh air is circulated through the canister via a center tube. Air is sucked down the tube, up through the charcoal, and out the purge line. Depending on the design of the canister, purging may be controlled in different ways. Those that use ported vacuum use a technique called constant and demand purge. The purge valve on the canister allows constant purging at a restricted rate through an orifice until a certain level of vacuum exists at the canister outlet. When ported vacuum is applied to the purge control valve, it allows a higher purging rate. The reason for having constant and demand purging is because the engine cannot handle a large flow of air through the canister at idle or slow speeds. The additional air and fuel vapor would upset the air/fuel ratio, causing a rough idle and increased tailpipe emissions. At higher rpm rates, however, the engine can digest the canister's contents without a problem. So purging is calibrated to match the engine's ability to handle it. A vapor feed rate of around 12 cubic feet per hour might be average for a small V-8 cruising down the highway.

Under normal circumstances, the charcoal canister causes few problems. Since the charcoal does not wear out, it should last the life of the vehicle. Problems can result, though, when the canister filter is neglected, when a purge control valve malfunctions, or if someone gets the vent, purge, and control vacuum lines mixed up (the connections are usually labeled to avoid such mistakes).

If vapor is not being purged from the canister, the purge valve may be defective

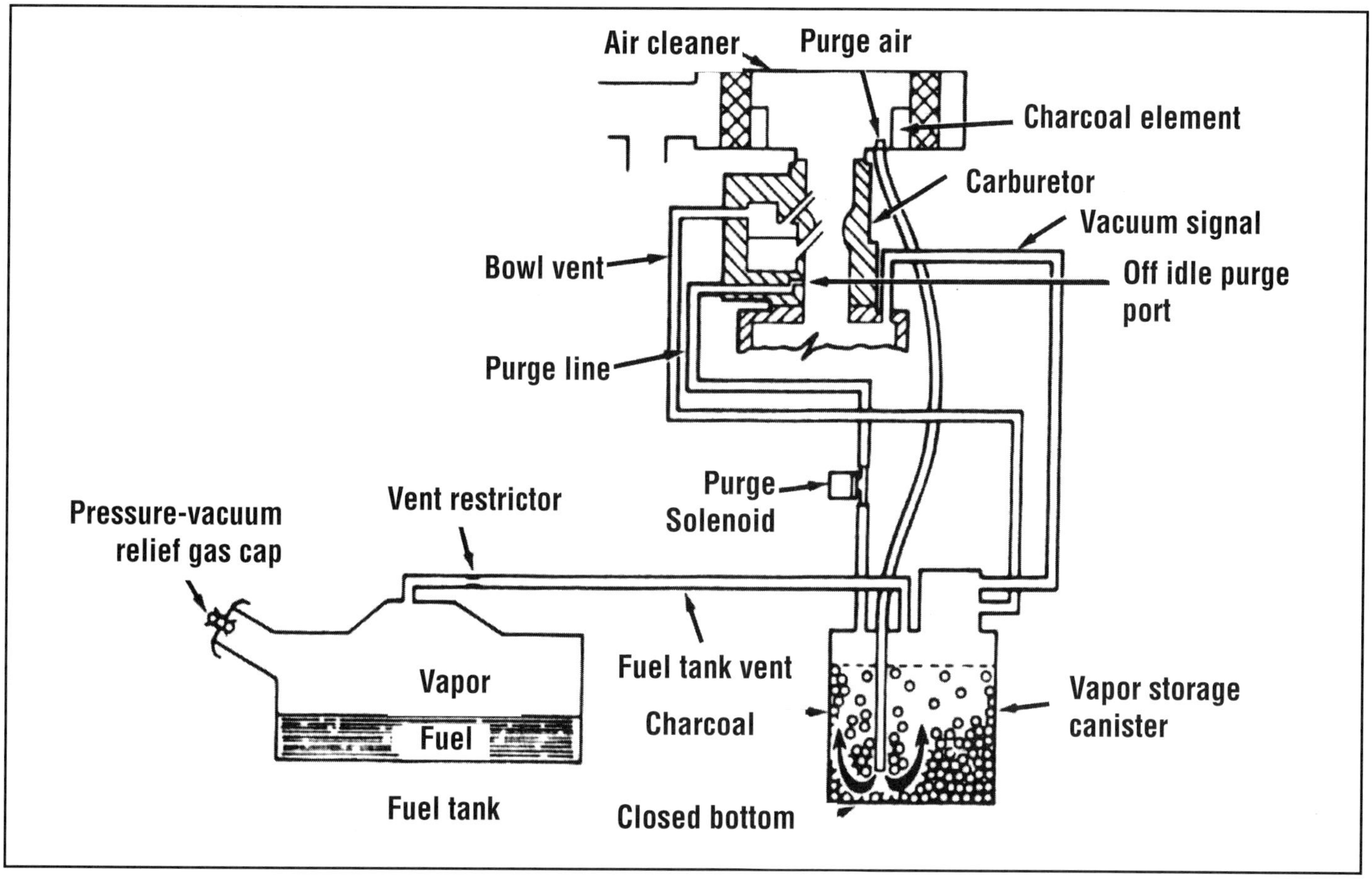

Later systems use a sealed canister that draws air through a vent tube to purge stored fuel vapors from the canister (Chevy).

or the canister filter may be plugged. The purge valve can be tested with a hand vacuum pump. It should hold vacuum for at least 15 to 20 seconds without leaking down. Vacuum connections should be inspected to make sure that they are tight and properly routed.

Computer-Controlled Units—On units equipped with computer-controlled solenoid purge valves, you will have to refer to the manufacturer's shop manual for the specs on when and under what conditions the solenoid is supposed to open. Generally speaking, such systems will not purge the canister until the engine reaches operating temperature. The coolant sensor monitors temperature and when the computer reads the appropriate temperature, it sends a command to open the canister purge solenoid.

On General Motors products with the Computer Command Control system, for example, the computer energizes the canister solenoid when the engine is operating in the "open loop" mode. Open loop means that there is no feedback computer control over the air/fuel mixture. It is set at a fixed value until the engine warms up to improve cold idle. This prevents vacuum from reaching the purge valve on the canister. When the engine enters the "closed loop" mode of operation (when the oxygen sensor is hot enough to produce a signal and the engine is at operating temperature), the solenoid is de-energized and vacuum is allowed to open the canister purge valve and purge the fuel vapors.

Anti-Percolator Valve

To prevent fuel evaporation from the fuel bowl during engine operation, some carburetors have an anti-percolator valve, where the canister vent line connects to the fuel bowl. The valve seals off the vent line while the engine is running. The valve is connected to the throttle linkage so that it will be closed when the throttle is open, and open when the throttle is closed. The valve opens when the engine is off (the throttle closed) so that the hot fuel vapors can boil out through the vent line and into the canister. ■

The charcoal canister may be hidden anywhere, and it's often neglected.

HEATED AIR INTAKE & EARLY FUEL EVAPORATION (EFE)

Everybody knows why a hot-air balloon rises, right? The sea of air we live in consists of a combination of gases, each of which has weight, as little as that might be. If the air is cold, the molecules of these gases are crowded together and a given volume of air will weigh more and contain more molecules than if the air is hot. That's the basic principle behind the heated air intake systems (also known as TAC—Thermostatic Air Cleaner—systems) on emissions-controlled engines.

HEATED AIR INTAKE

Early in their battle against automotive pollution, the engineers realized that the carburetor was crucial to their efforts, so they clamped down on it hard. Before controls, carburetors were calibrated to be rich enough to accommodate wide variations in the temperature of the air that entered them, so they provided a mixture that would fire dependably regardless of the time of year and variations in the weather.

The Problem

But anti-pollution carburetors were built to meter less gasoline into each cubic foot of air in order to reduce HC and CO emissions. The mixture became lean, and, if the density of the incoming air was such that it made the blend even leaner, it would go over the edge into a ratio that simply contained too little gasoline to burn, resulting in hesitation, stalling, and generally poor driveability.

If we go back to our initial illustration on the relatively small number of molecules in warm air when compared to cold air, the problem is easy to understand: Once the choke is open, a carburetor supplies the same amount of gasoline regardless of the temperature of the air that's rushing through it. But the air/fuel ratio can still be affected by the density of the air. That is, the number of oxygen molecules (the only component of air that we're concerned with here) in a cubic foot or a gallon of cold air is greater than the number in that same volume of hot air.

GM calls this system TAC (Thermostatic Air Cleaner), which is essentially an intake system designed to supply the engine with air that is of constant temperature (which also means constant density), by mixing heated and unheated air.

52

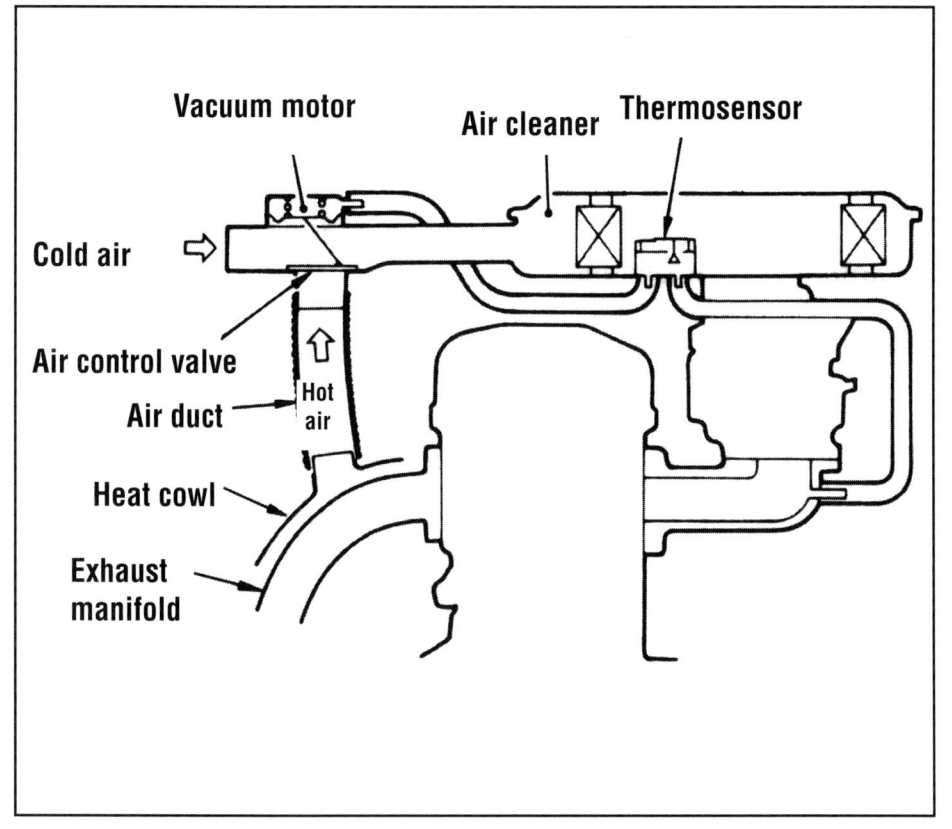

The heated air intake system makes sure the engine always gets air of the proper temperature (Hyundai).

Therefore, introducing cold air into the engine has the same effect as providing a larger amount of air (even though the volume is the same). This raises the ratio of air to fuel, making the blend leaner.

The Solution

So what was needed with tightly controlled carburetors was a way of supplying the engine with air of nearly constant temperature in both summer and winter. The lean calibration of carburetors from this era of pollution fighting simply can't tolerate wide variations in the warmth of the intake air without leaning the blend so much that it won't burn dependably.

Enter the heated air intake system. Its purpose is to duct a combination of heated and unheated air to the carburetor to make sure that the engine is provided with air of the proper density for smooth, responsive running regardless of ambient or engine compartment temperatures. It eliminates the tendency of lean mixture calibrations to cause roughness, stumbling, and missing in cool weather or before the engine has warmed up.

Of course, the choke enriches the mixture for a short time after start-up, but if it were designed to stay on all the time on cold days, emissions would soar while mileage and power would plummet. So the heated air intake system takes over where the choke leaves off. An ancillary benefit is the virtual elimination of carburetor icing.

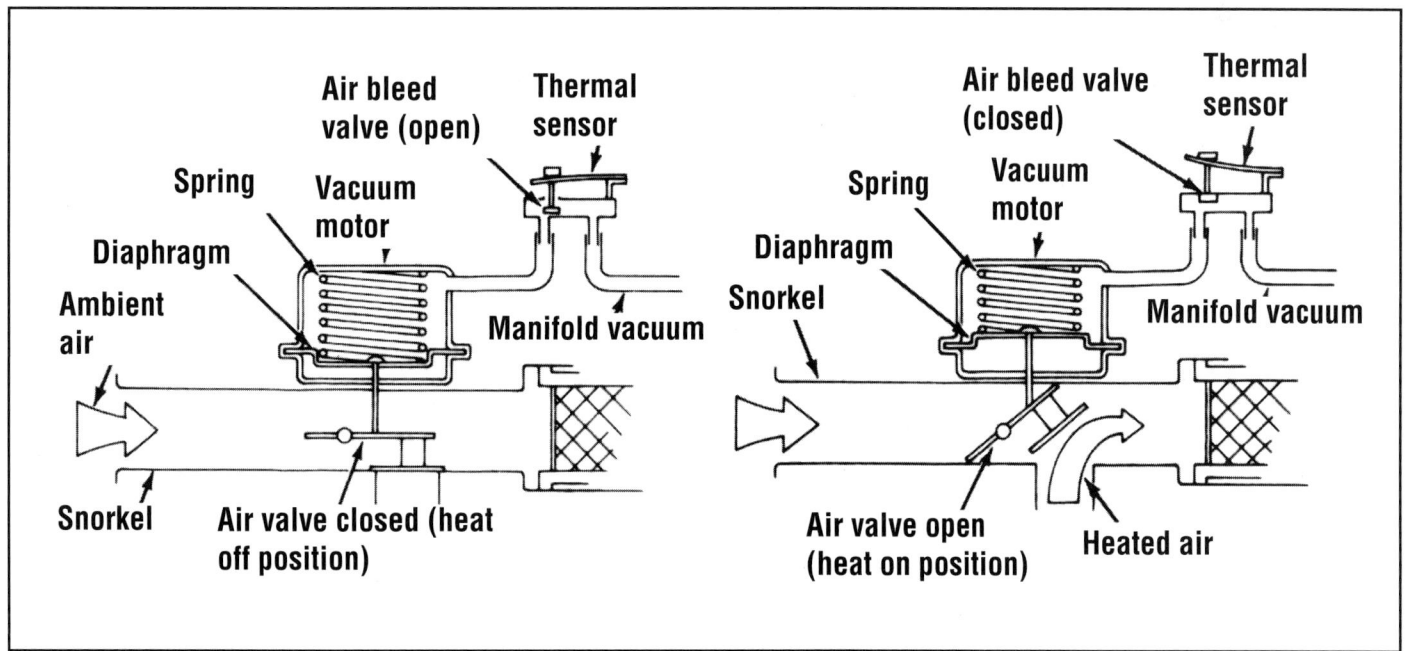

A thermostatically controlled bleed valve controls the vacuum to the vacuum motor. When the temperature is low, the valve is closed so that full vacuum operates on the diaphragm, closing the snorkel to unheated air (American Motors).

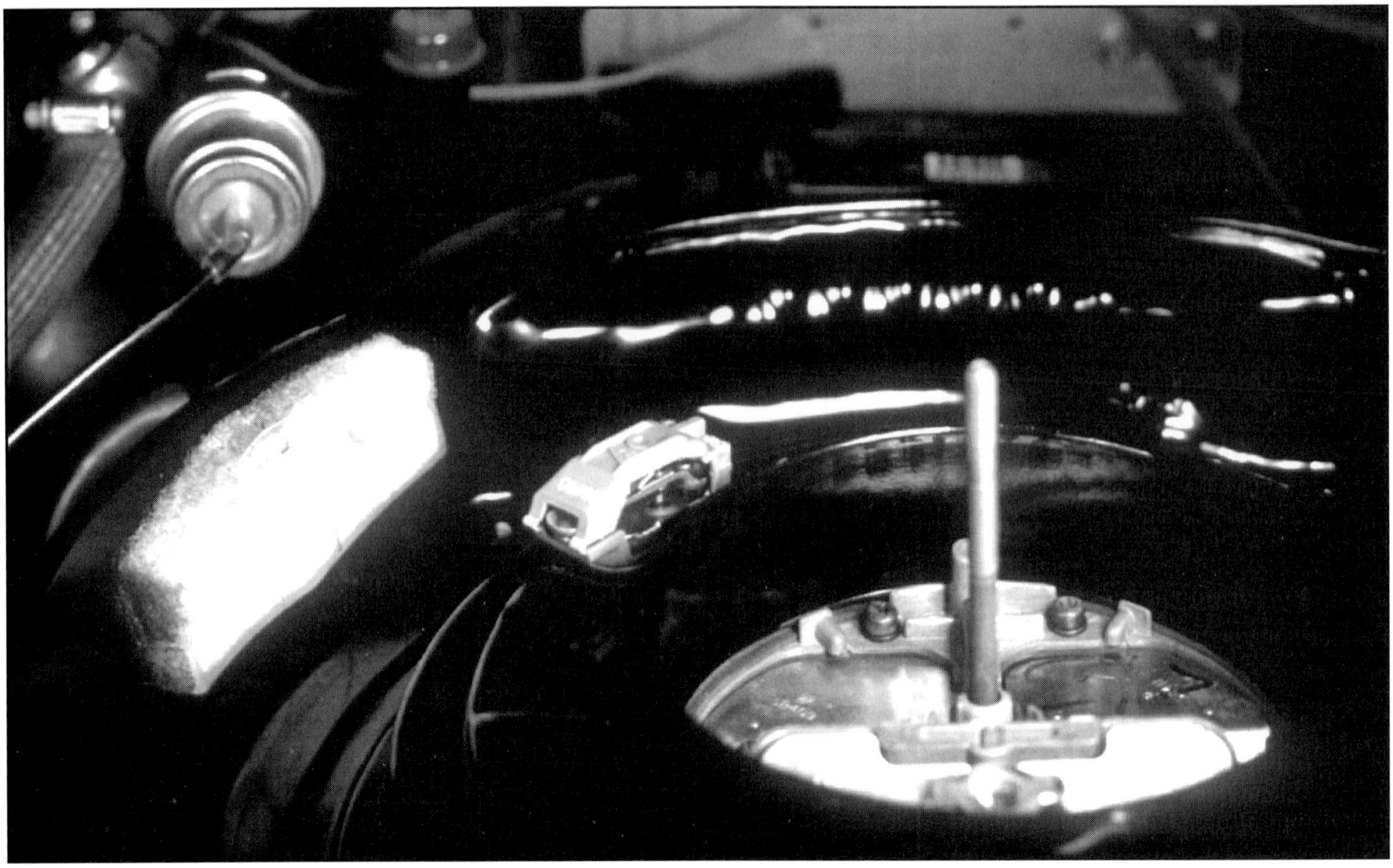

This GM-style vacuum bleed valve is next to the PCV inlet filter.

Components

Intake temperature is stabilized by a system comprising a sealed chamber with a diaphragm and spring inside (called a vacuum motor), linkage that connects the diaphragm to a trap door or flapper valve in the air cleaner snorkel, a metal shroud or stove mounted over the exhaust manifold, a heat tube that runs from the shroud to the bottom of the air cleaner snorkel where the flapper valve is, a thermostatic vacuum bleed valve inside the air cleaner, a vacuum line that connects the bleed valve to the vacuum motor, and another vacuum line that goes to the intake manifold.

How It Works

In a typical system, manifold vacuum is routed to a thermostatic bleed valve inside the air cleaner. When incoming air is cold, the valve closes, allowing full vacuum to reach a diaphragm chamber mounted on the snorkel. This causes the

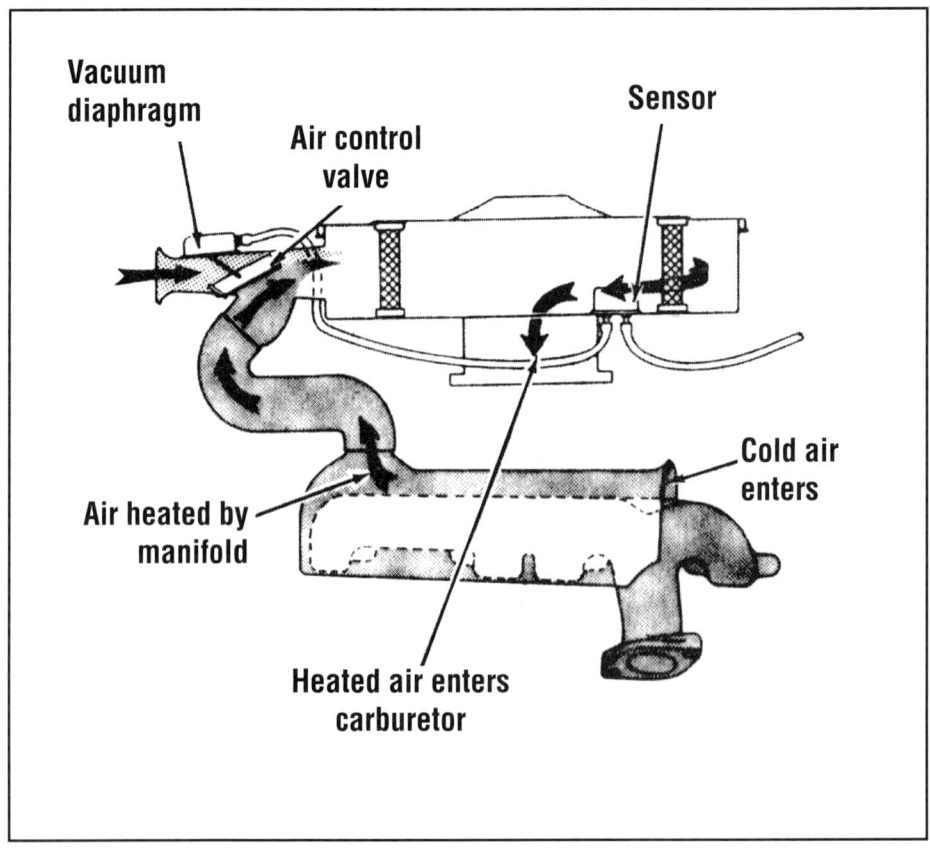

In this Chrysler system, heated air is picked up from a shroud or "stove" over the exhaust manifold.

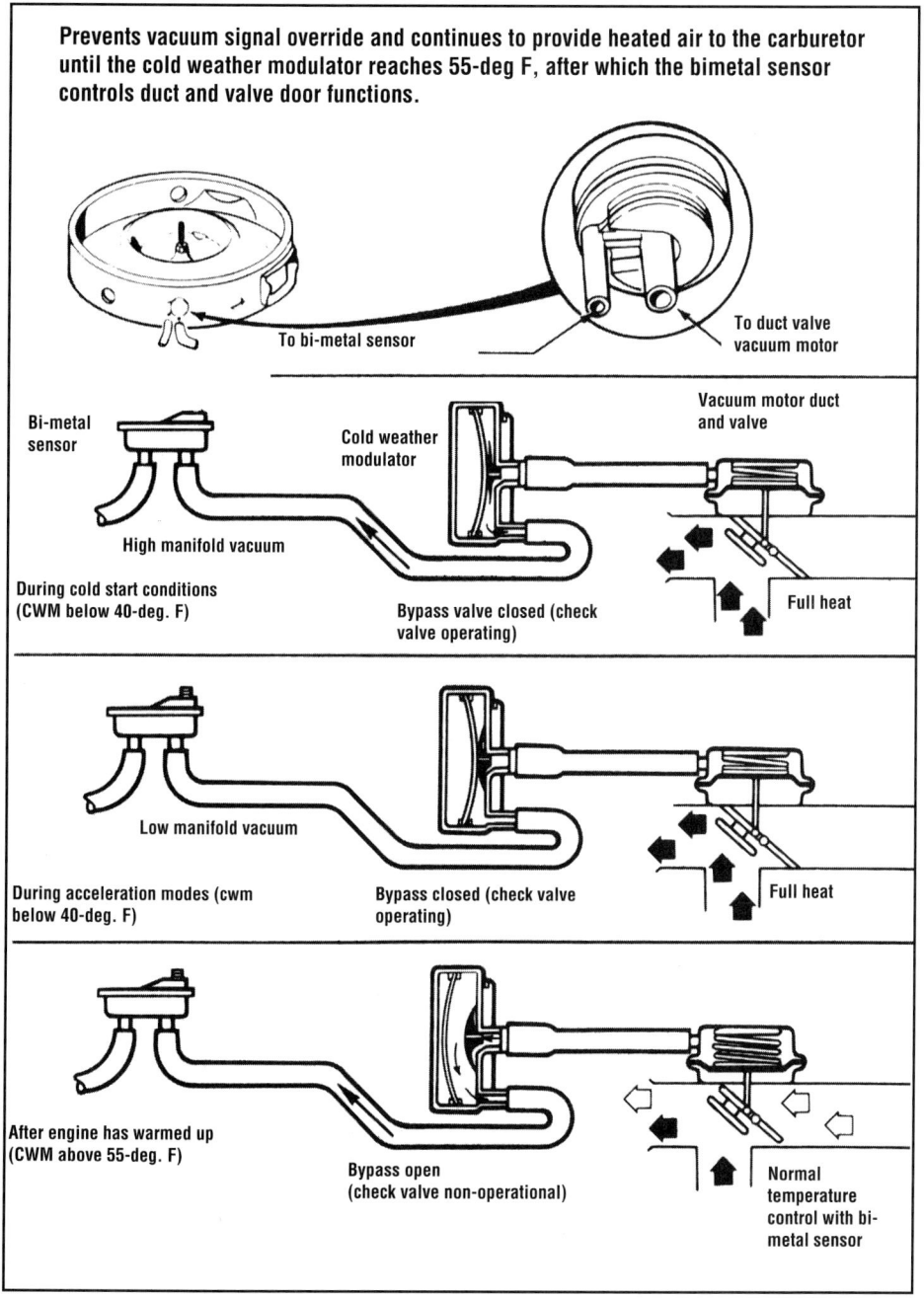

Prevents vacuum signal override and continues to provide heated air to the carburetor until the cold weather modulator reaches 55-deg F, after which the bimetal sensor controls duct and valve door functions.

The Ford cold weather modulator valve is spliced into the hose between the bleed valve and the vacuum motor. It traps vacuum at low temperatures so that even during wide open throttle operation when manifold vacuum is low, the flapper valve stays in the heat-on position (Ford).

diaphragm to rise, lifting the blend door and shutting off the flow from the outside. In this position, the only air the engine gets must pass over the exhaust manifold (which gets hot very quickly) and so is warmed above ambient temperature.

As the engine or the atmosphere warms up, the bi-metallic strip inside the thermostatic bleed valve starts to respond. It moves the valve off its seat gradually, opening the circuit between the intake manifold and the vacuum motor. In other words, it allows the vacuum to bleed off, lessening the strength of the signal to the vacuum motor. The spring inside the diaphragm chamber pushes the door down, closing the hot air duct.

It's important to note that this is not an on/off process. The diaphragm spring and vacuum balance each other so the door stays in whatever intermediate position is necessary for maintaining the proper intake temperature.

W.O.T.—When you floor the accelerator (known as W.O.T. or wide open throttle) the throttle plates are opened all the way and vacuum in the intake manifold drops. So even if the engine is cold, too little vacuum acts on the diaphragm to overcome the spring, and the flapper valve falls, admitting only dense, unheated air into the engine to blend with the rich mixture present during acceleration, and maximum potential power is produced. Variations on this theme include a separate cold-air passage into the air cleaner that's kept closed by vacuum.

Another type of full-throttle control is exemplified by Ford's Cold Weather Modulator. This is a thermostatically controlled check valve that traps the vacuum in the vacuum motor when the car is accelerated hard at temperatures below 55 deg. F. This eliminates hesitation and flat spots by allowing heated air to enter the engine in spite of the drop in vacuum that naturally occurs when the throttle is opened wide. It is the opposite of most systems that allow only unheated air to flow to the carburetor when the accelerator is floored.

Without Vacuum

Some cars don't use vacuum to control air blending. Instead, they employ a direct-acting thermostatic arrangement. A typical version will have a thermostatic bulb that's connected to the rod that operates the flapper valve in the air cleaner snorkel. As the thermostat warms up, it pushes on the rod, closing the door to the heated air from the exhaust manifold shroud and allowing unheated air to enter the air cleaner. The operation of this setup can't be as precisely controlled as that of a vacuum-activated system, but it's simpler.

EARLY FUEL EVAPORATION

Another system for helping lean blends burn has been around for decades: the heat riser. Its purpose has always been to vaporize fuel droplets in a cold intake manifold so that they won't condense on their way to the combustion chamber, thus keeping the mixture that actually reaches the cylinders rich enough to ignite.

But with tightly controlled engines, more precise management of intake manifold heating than could be provided by an ordinary thermostatic coil was needed, so systems such as GM's EFE (Early Fuel Evaporation) were introduced. In the vacuum version, a vacuum motor closes the exhaust outlet by means of a rod, causing hot gases to flow through the heat riser passages in the intake manifold, thus heating it. The signal that acts on the diaphragm of the motor is controlled by a thermostatic vacuum valve similar to those mentioned in the discussion of spark controls (see Chapter 12). When the coolant reaches a certain temperature, the valve cuts vacuum to the motor, allowing the heat riser to open.

The electrical type consists of a heating element that resides in the intake manifold under the carburetor or throttle body. It gets current through a thermostatic switch or from the engine management computer when coolant is below normal operating temperature. ■

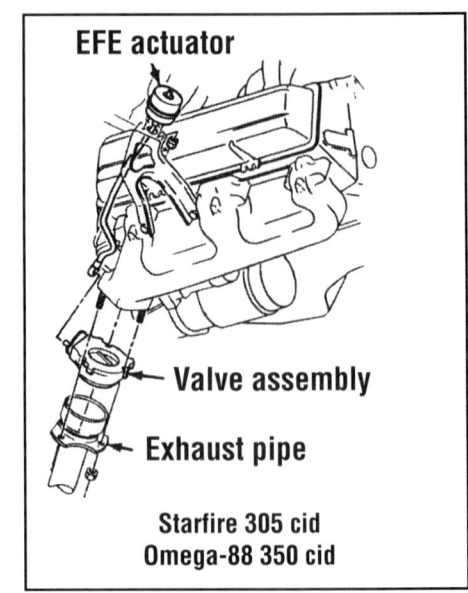

The basic EFE system uses vacuum to control the exhaust manifold valve. In the closed position, exhaust passes through the heat riser passage, warming up the intake manifold (Oldsmobile).

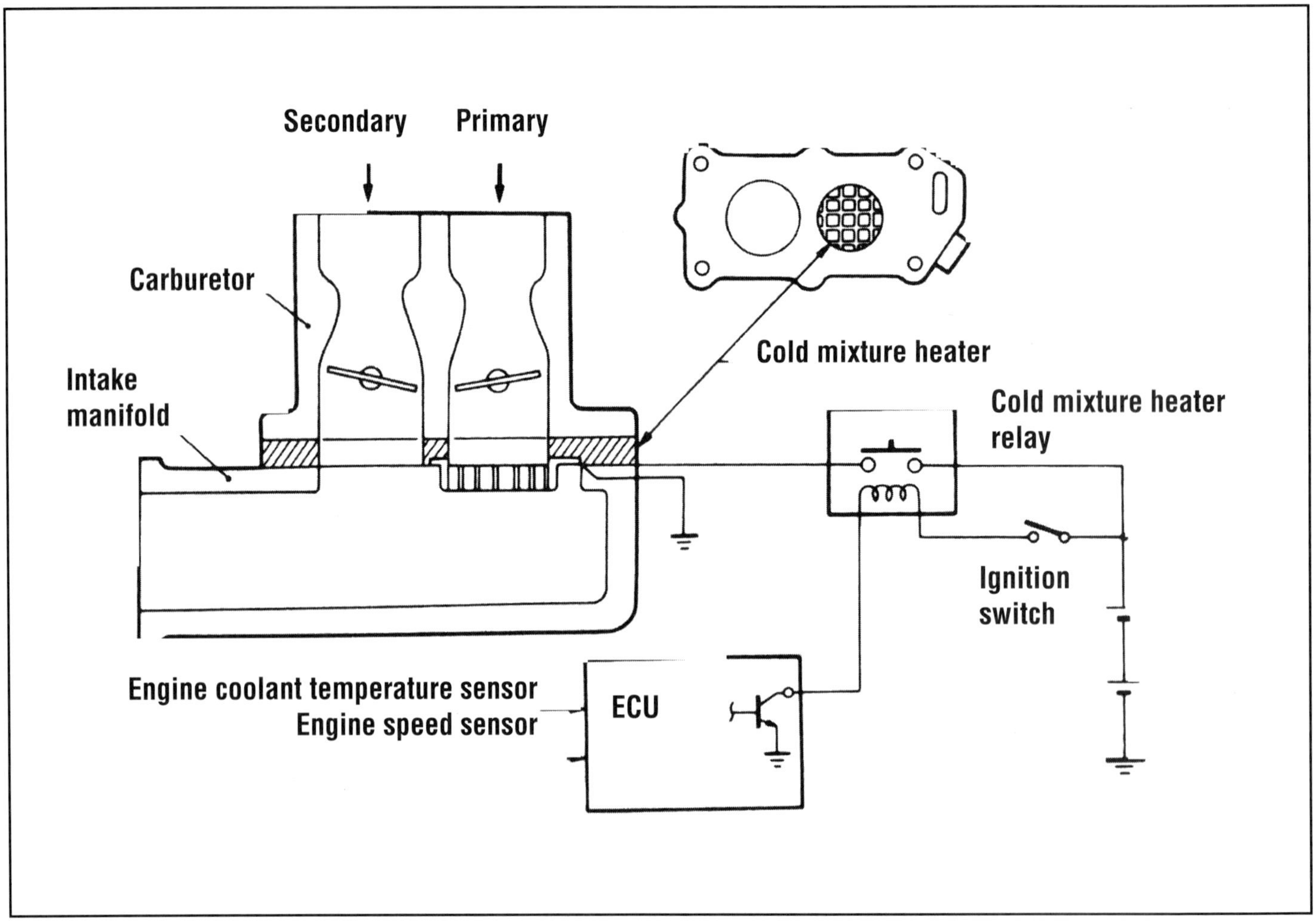

Another EFE approach is to use an electrical heating element under the carburetor (Hyundai).

AIR INJECTION SYSTEMS

It wasn't long after the campaign against air pollution began that a hard fact became painfully obvious: Designing a gasoline piston engine that burns up every trace of HC and CO inside its cylinders is as impossible as achieving a scientist's absolute zero. You can get close, but that's about it.

No matter how cleverly the combustion chambers are shaped to eliminate quench areas, or how accurately the fuel mixture calibration or the spark advance curve is controlled, or how drastic the changes in valve timing, temperature range, and bore-to-stroke ratio are, an engine will still pump a considerable amount of pollutants into the atmosphere. It's simply in the nature of the familiar, dependable gasoline piston engine, mainly because some fuel vapor will always condense back into a liquid on the cylinder walls and the surfaces of the combustion chamber, and only the vapor will burn. This results in the production of HC, which is basically unburned gasoline, and CO, which is a product of incomplete combustion.

Even very efficient designs that allowed an overall lean mixture, such as Honda's CVCC (Controlled Vortex Combustion Chamber) system, could not circumvent the laws of physics, although they went a long way. Clearly, there was, and is, no chance that any gasoline-burning engine as we know it can exhaust only C02 and water vapor, the by-products of perfect combustion (and, incidentally, the same ingredients that,

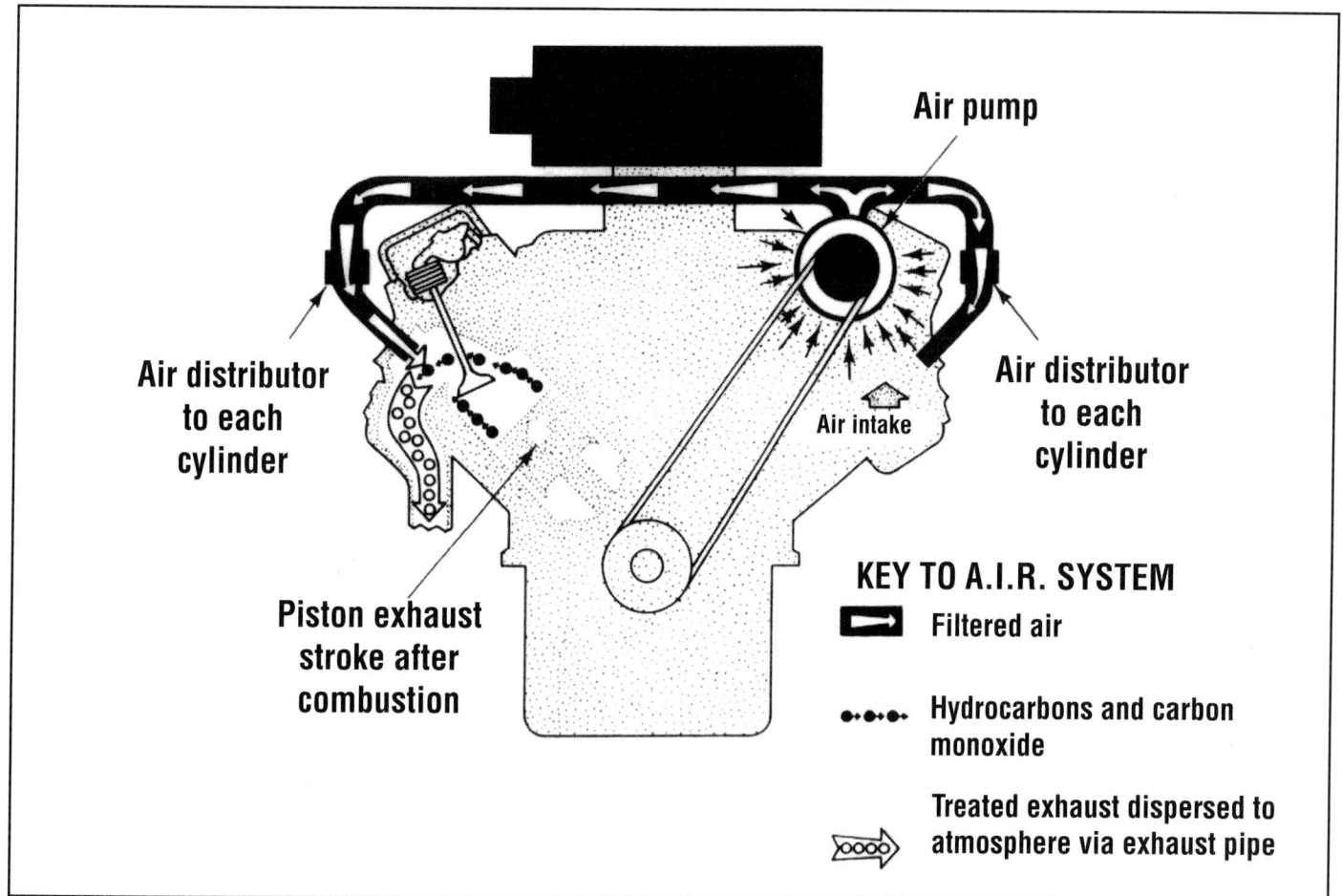

This schematic of a basic air injection system shows how the pump directs air into the exhaust ports (Buick).

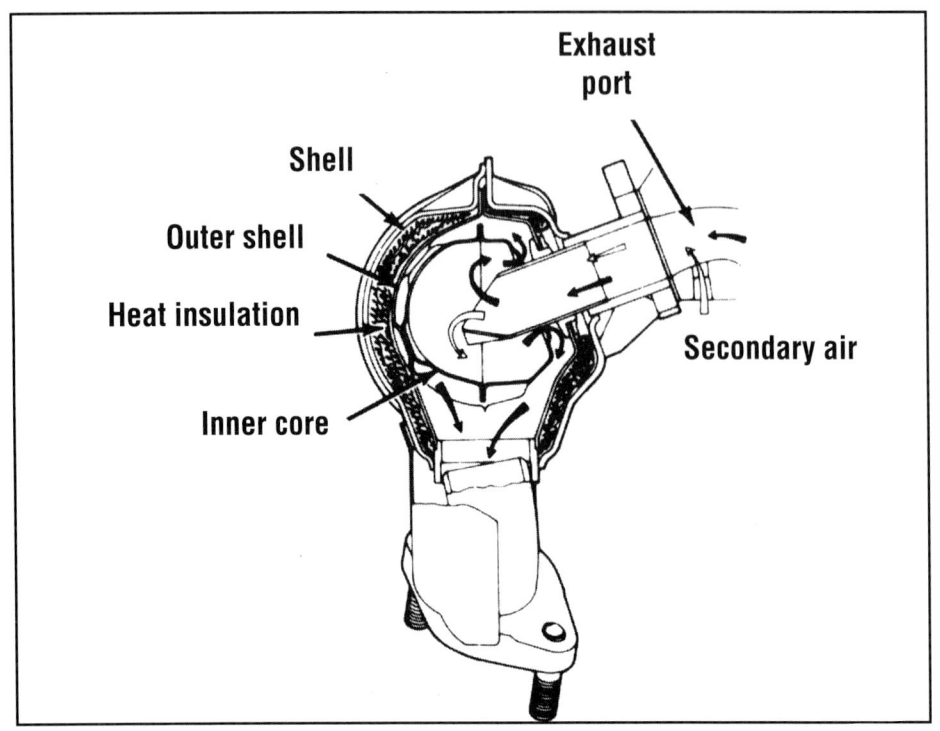

The thermal reactor, really a furnace, represented the furthest the idea of air injection was taken (Dodge).

with sunlight and soil, allow plants to grow).

So, the engineers had to create some crutches to hold up this basic inadequacy of an otherwise very practical engine design. These would have to be add-on systems that treated the exhaust after it left the engine.

THE SMOG PUMP

One of the earliest HC and CO controllers was air injection, which appeared in 1968. Called AIR (Air Injection Reaction) by GM and Thermactor by Ford, the name that stuck was "smog pump." As we said, it's a crutch that has nothing to do with increasing an engine's efficiency, but it helps clean up what actually comes out of the tailpipe.

The basic idea is to provide those hot gases with plenty of fresh air to combine with on their way through the exhaust system, burning a significant amount of them up before they reach the atmosphere.

If you're one of those automotive enthusiasts who's of the opinion that the smog pump is just a performance drag, we should mention that many cases of failed state emissions inspections have been cured simply by replacing a broken pump belt. And, it only requires a fraction of a single horsepower to run.

The Thermal Reactor

The furthest this principle was taken was something called the thermal reactor. Actually, it was just a very large and well-insulated exhaust manifold used in conjunction with a high-capacity air pump. Basically, it provided a furnace-like environment in which the HC and CO was given a good opportunity to combine with oxygen to complete the combustion process.

In the early 1970s, before the catalytic converter came into general use, it looked like this might be the answer to the emissions problem. It had the advantage over the catalyst of not requiring expensive metals or unleaded gasoline. But it didn't do the job as well as the catalytic converter did, and it raised underhood temperatures drastically, so the catalyst won out. However, it was used for years on some cars up until 1980, notably the Mazda RX-7.

Feeding the Cats

The wholesale adoption of the two-way catalytic converter by U.S. automakers in 1975 changed the purpose of air injection. It's now used primarily to give the pollutants in the exhaust stream

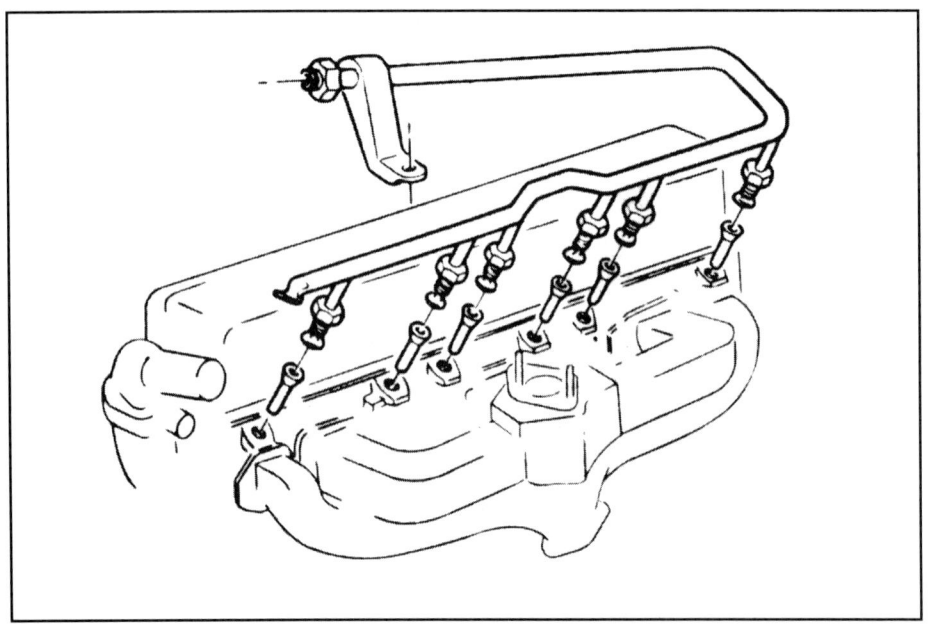

Air flow is routed through a manifold and nozzles that point into the exhaust ports (Chevrolet).

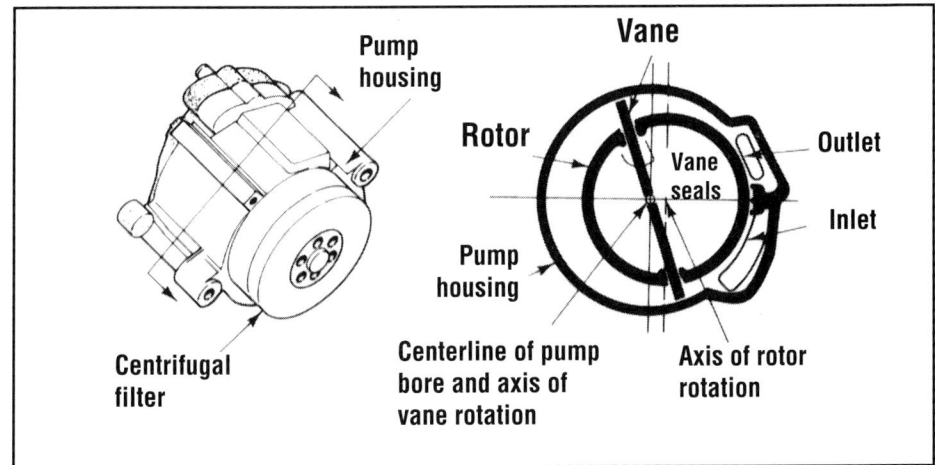

The pump itself is of the sliding vane variety . . .

. . . and is driven off the crankshaft by a belt.

sufficient fresh air to combine with inside the converter. The mixture of oxygen from the air pump and HC and CO flares up in the presence of the catalyst. Temperatures as high as 1600 degrees F are generated, eliminating most of the harmful pollutants by oxidizing them into CO_2 and water. So, the smog pump got a new lease on life, and it is still extremely common.

Some catalyst-equipped cars have been built that don't need air injection. These have engines calibrated in such a way that there's enough O_2 in the exhaust to support the reaction in the converter. But these are exceptions. Most vehicles today have either an air injection system or an aspirator-valve setup (explained later in this chapter) to provide the extra fresh air the catalyst needs to do its job.

Components

You'll find the following components in a typical air injection system:

• A belt-driven vane pump

• A vacuum-operated diverter valve that vents pump output to the atmosphere during deceleration so the combination of a rich mixture and extra oxygen doesn't cause backfiring

• A pressure relief valve that allows excess pump output to escape

• A one-way check valve that allows air flow into the exhaust manifold but keeps exhaust out of the pump if the belt breaks

• The plumbing and nozzles necessary to distribute and inject the air.

Pump—We might as well look at the pump first. When the engine is started, a belt causes the pump pulley to rotate. Inside the pump, vanes riding against the walls of a cylindrical chamber start moving air. The rotor turns on an axis that's different from that of the pump bore, and the vanes slide in and out of slots in the rotor. This causes volume variations that result in the pumping action. Usually, intake is through a centrifugal filter mounted behind the drive pulley, but separate intake filters have been used on some applications.

Most systems use a relief valve that allows excess pump pressure to escape. It may be in the pump itself, or incorporated into the diverter valve. The pressurized air exits into a large-diameter hose that routes it to this valve.

Diverter Valve—The only part of this system that requires any effort to understand is the diverter valve (and related components that do a similar job). The most common type receives a vacuum signal through a hose that runs to the intake manifold or the base of the

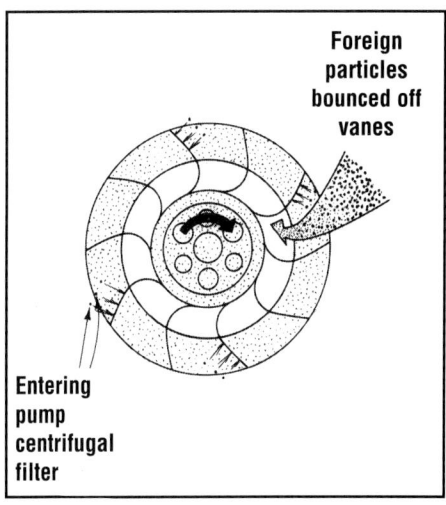

A centrifugal filter behind the pulley keeps dirt out of the pump (Chevrolet).

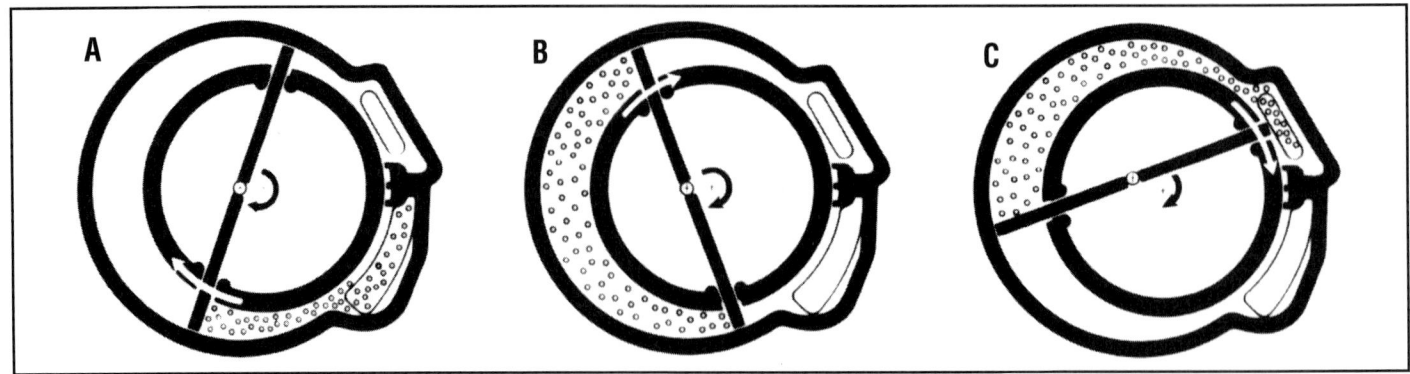

A sliding vane pump moves air as shown here. In figure (A), the vane is traveling from a small area into a larger area—consequently a vacuum is formed that draws fresh air into the pump. As the vane continues to rotate (B), the other vane has rotated past the inlet opening. Now the air that has just been drawn in is entrapped between the vanes. This entrapped air is then transferred into a smaller area and thus compressed. As the vane continues to rotate (C), it passess the outlet cavity in the pump housing bore and exhausts the compressed air into the remainder of the system.

carburetor or throttle body. During closed-throttle deceleration when engine vacuum is strongest, it directs air flow through a small muffler, which is usually mounted on the pump.

The valve has a vacuum chamber, diaphragm and spring arrangement that moves the stopper from one of its seats to the other and so controls the switching operation. During deceleration, the diaphragm in the valve's chamber overcomes the force of the spring and dumps air flow to the muffler. This prevents the backfiring in the exhaust system that would occur if the extra air the pump provides were available to allow the rich mixture that is present in the exhaust stream during deceleration to ignite explosively. To put it simply, the diverter valve directs the air pump's output away from the exhaust system during deceleration.

A typical diverter valve directs pump output to the exhaust ports or to the atmosphere according to a vacuum signal from the intake manifold (Chevrolet).

Gulp Valve—Another anti-backfire device is known as the *gulp valve*. It has a diaphragm chamber, a spring, and a normally closed valve inside. In a typical specimen, during deceleration when intake manifold vacuum reaches 20 to 22 in. Hg, the vacuum signal pulls the diaphragm against spring pressure, which opens the valve. This allows some of the air pump's output to flow into the intake manifold to dilute the rich mixture present in this mode and eliminate the possibility of extra gasoline vapor exploding in the exhaust system. Usually, the addition of pump air to the intake stream lasts from 1 to 3 seconds.

Some systems use a gulp valve with no air pump. They simply open a passage between the air cleaner and the intake manifold on deceleration and let the engine draw as much as it can.

VDV—There is yet another means of preventing backfiring: the Vacuum Differential Valve (VDV), which works together with an air bypass valve. The VDV shuts off the vacuum signal to the bypass valve momentarily whenever intake manifold vacuum rises or drops sharply. In normal operation, while the bypass valve is receiving vacuum, it allows air pump output to flow freely to the air injection manifold. When the VDV stops the vacuum signal, a spring inside the bypass valve opens a vent port and pump flow is directed to the

The check valve keeps exhaust from backing up into the pump.

atmosphere.

If there is excessive pump pressure or a restriction in the check valve or air injection manifold, a relief valve inside the bypass valve opens, allowing some of the pump's output to be dumped and the rest of it to take the normal route.

During all modes of engine operation except deceleration, air from the pump flows into the hose to the check valve. This is a simple one-way device, which lets air enter the air injection manifold, but keeps exhaust from backing up into the pump if a belt should break or the pump should otherwise stop working (normally the pump's pressure is high enough to overcome exhaust pressure). From the check valve, the air flows into the air injection manifold, which directs it into each exhaust port. Here, as we said above, it gives the HC and CO in the exhaust oxygen to combine with.

NEW CAT REQUIREMENTS

The main reason computer-controlled closed-loop fuel systems have been adopted is to supply the three-way catalyst with a perfect stoichiometric (14.7:1 air/fuel) ratio, which is what allows it to reduce NOx. It would be self-defeating to upset the delicate balance achieved at intake by force-feeding air into the exhaust. But air injection is still needed for catalytic oxidation of HC and CO, so it has to be carefully controlled.

In a typical late-model system, air flow goes to the exhaust manifold only while the engine is cold. Very little NOx is produced then anyway, so the efficiency of the reduction process isn't important. Once normal operating temperature is reached, however, air pump output is switched downstream of the three-way section of the converter so it doesn't affect it, yet boosts the performance of the oxidation catalyst.

Ford's Early System

Some cars employ a computerized system that switches air injection flow from the exhaust ports to a line that runs directly into the catalytic converter. An early Ford Electronic Engine Controls system (EEC-III) is an example of this. Here, the ECA (Electronic Control

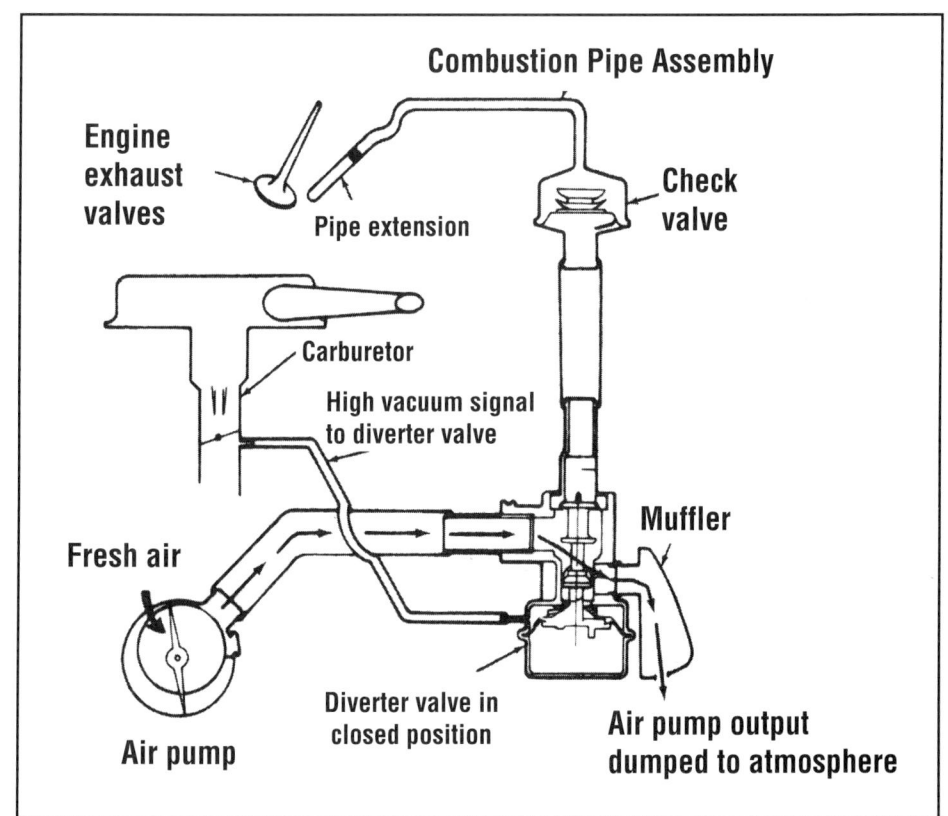

During deceleration, the strong vacuum signal to the diverter valve causes it to dump air pump output through the muffler, thus preventing the rich decel mixture in the exhaust system from exploding in the phenomenon known as backfiring (Chevrolet).

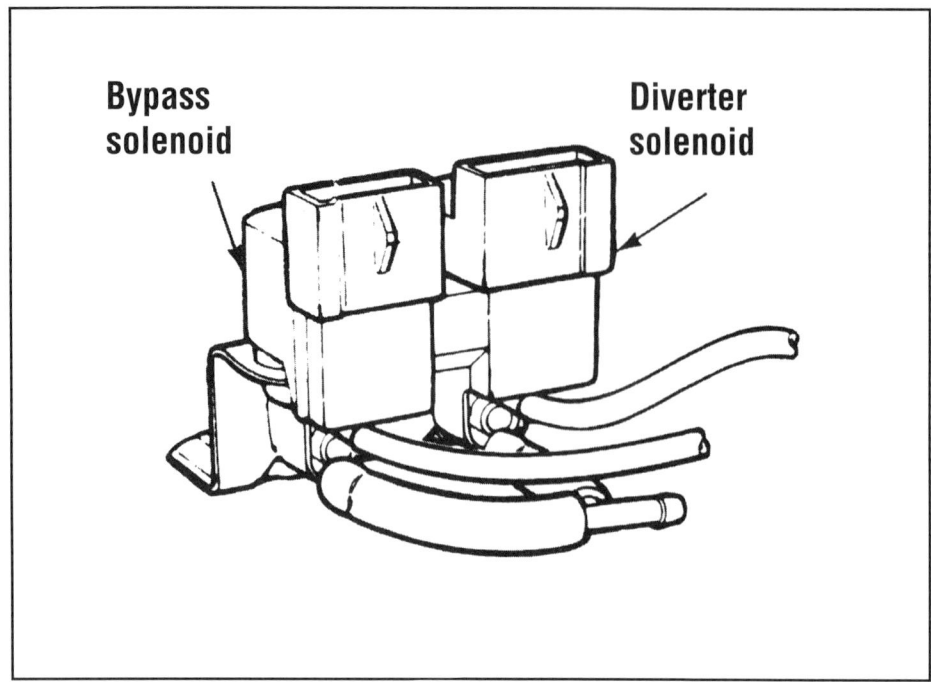

Through the use of solenoid-operated vacuum valves, the engine management computer can control air injection (Ford).

Some vehicles, such as this Mazda, use aspirator valves instead of a pump.

Assembly, Ford's name for what is more commonly known as the "Electronic Control Unit," or "ECU") makes decisions on where to send Thermactor (air pump) output on the basis of input from various engine sensors, notably that for coolant temperature, and according to certain time calibrations. It works in conjunction with a pair of solenoid valves: the bypass solenoid, which directs air pump flow to the atmosphere when energized; and the diverter solenoid, which switches air flow to either the exhaust ports or the catalytic converter.

How It Works

During normal engine operation, air pump output is routed to the catalytic converter (there's a mixing chamber inside the shell between the reduction catalyst and the conventional two-way oxidation catalyst, and this is where the air enters). The computer energizes the bypass solenoid when time at closed throttle exceeds a specified time in its memory, when the interval between a lean and a rich signal from the oxygen sensor is longer than a set time value, and during wide-open throttle. This protects the cat from overheating because of an overly rich mixture, and guards against backfiring.

The computer energizes the diverter solenoid when the coolant sensor tells it that the engine is cold. This directs air pump flow upstream to the exhaust manifold during engine warm-up, giving HC and CO more time to oxidize.

Chrysler's Version

This essential air switching is accomplished on some Chrysler cars without the help of electronics. An ordinary diverter valve is used to dump air pump output into the atmosphere during deceleration. But the directing of flow from the exhaust ports downstream to the exhaust pipe ahead of the catalytic converter is done by an air switching valve and a coolant control engine vacuum switch (CCEVS).

The purpose of this setup is different from that of the Ford system described above: When the engine is cold, fresh air directed at the exhaust ports is very helpful in reducing HC and CO emissions. But after normal operating temperature is reached, the high heat levels generated here start producing NOx. If the air is injected downstream in the exhaust system where it's cooler, HC and CO will still be oxidized, but NOx emissions won't be adversely affected.

The switching action is as follows: As long as there's a vacuum signal from the CCEVS to the air switching valve, the valve stays open, allowing air pump output to flow to the exhaust ports. When

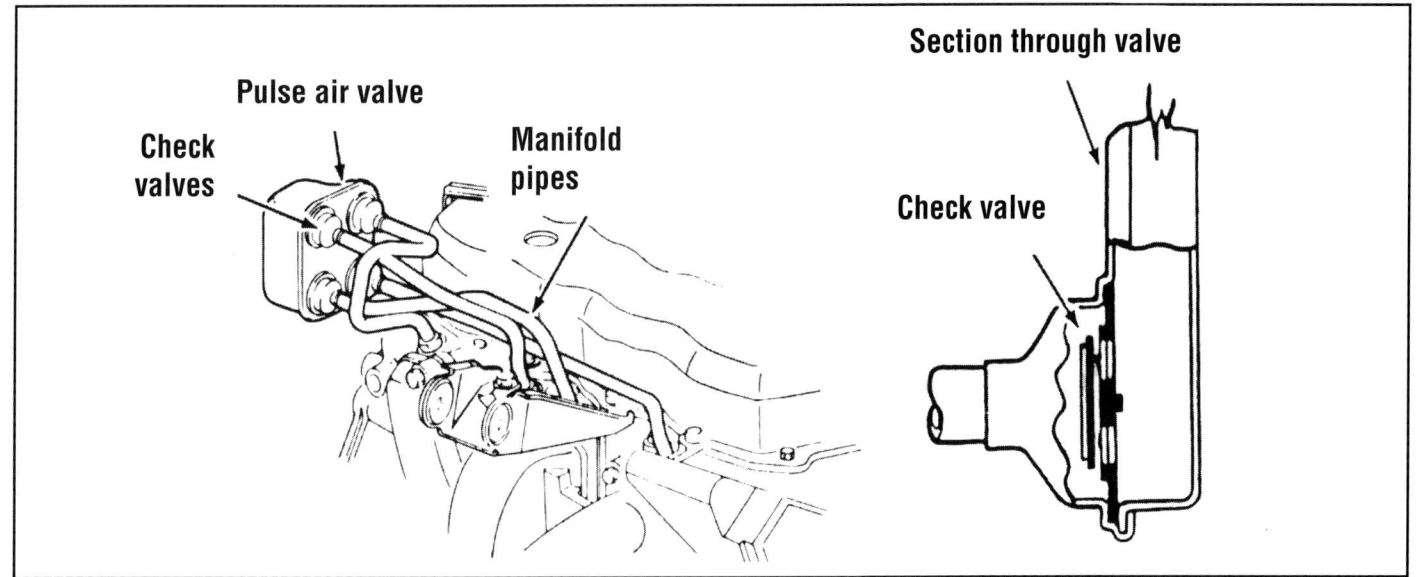

GM's Pulsair system uses one aspirator valve for each cylinder (GM).

the engine warms up, the CCEVS shuts off the vacuum signal and the switching valve directs most of the air flow to the downstream injection point. A bleed hole in the valve allows a small amount of air to be routed to the exhaust ports at all times, which assists in the reduction of HC and CO, but isn't enough to promote the formation of NOx.

ASPIRATORS

Although most people don't realize it, vacuum is present in the exhaust manifold momentarily after each cylinder's exhaust stroke. This is the result of the exhaust valve's closing and the inertia of the column of spent gases as it speeds through the exhaust system. After the valve has closed, the column continues to travel away from the exhaust port, leaving a negative pressure area behind it. This is easier to understand if you picture the exhaust not as a steady stream, but as a series of pulses.

This little-known phenomenon is the basis of pumpless air injection systems, generally known as aspirators. The main component is a one-way valve (similar to the check valve in pump-type air injection) that allows fresh air picked up from the air cleaner to enter the exhaust stream whenever vacuum opens it, but stops the hot gases from escaping by closing when it encounters pressure. There may be one aspirator valve for the whole engine, or one for each cylinder.

Aspirators have been used to eliminate the air injection pumps because they also cost less, reduce complication, and save space in that crowded engine compartment. They're quite effective at idle and low-speed operation, but much less so at high rpm.

GM's *Pulsair* is a common aspirator system with a housing containing one valve for each cylinder, a tube going into each exhaust manifold runner, and a hose that picks up fresh air from the air cleaner. ■

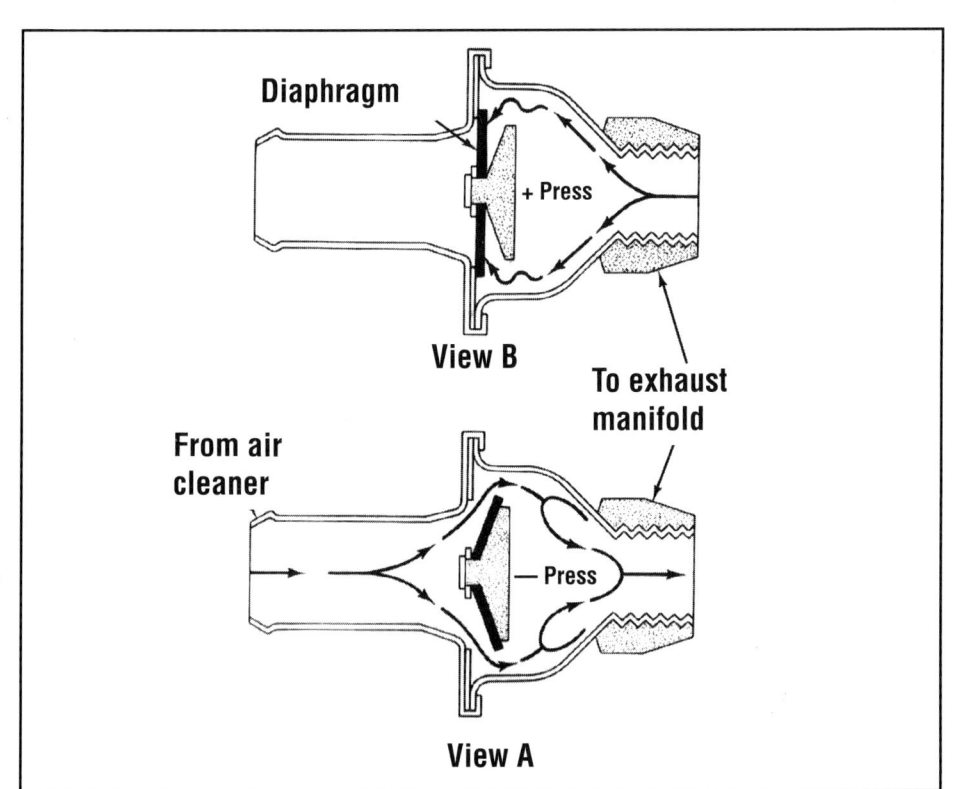

An aspirator valve is simply a one-way device that allows air to enter the exhaust stream during the vacuum moments between pressure pulses (Chrysler).

CATALYTIC CONVERTERS

The high-pressure efforts of the automakers to eliminate the air pollution their vehicles caused paid off handsomely. The redesigns and controls worked, so much so that by the early 1970s, a typical domestic car was emitting 85% less HC, 70% less CO, and 50% less NOx than a comparable vehicle from the pre-emissions-control era.

But the Clean Air Act of 1970 demanded more. The goals it stipulated seemed impossible to reach, and the auto manufacturers started to feel they were fighting a losing battle. If they couldn't meet the standards, they wouldn't be allowed to sell their products in the United States, and our economy's basic industry would be crippled. The ramifications would be too terrible to contemplate. The situation looked hopeless.

But Detroit had a card up its sleeve that it didn't play until things got critical: the amazing catalytic converter. With some cooperation from the oil companies and the government, this gave the car companies a winning hand, or at least allowed them to stay in the game.

HOW CONVERTERS WORK

Catalytic action is an odd chemical phenomenon, seemingly outside the realm of natural occurrences. The catalytic agent greatly promotes a reaction without being altered in any way itself. Only its proximity is required to cause pollutants to miraculously flare up and oxidize.

The basic automotive oxidation catalytic converter helps HC and CO in the exhaust combine with O_2, resulting in the production of harmless CO_2 and water—with soil and sunlight, just what

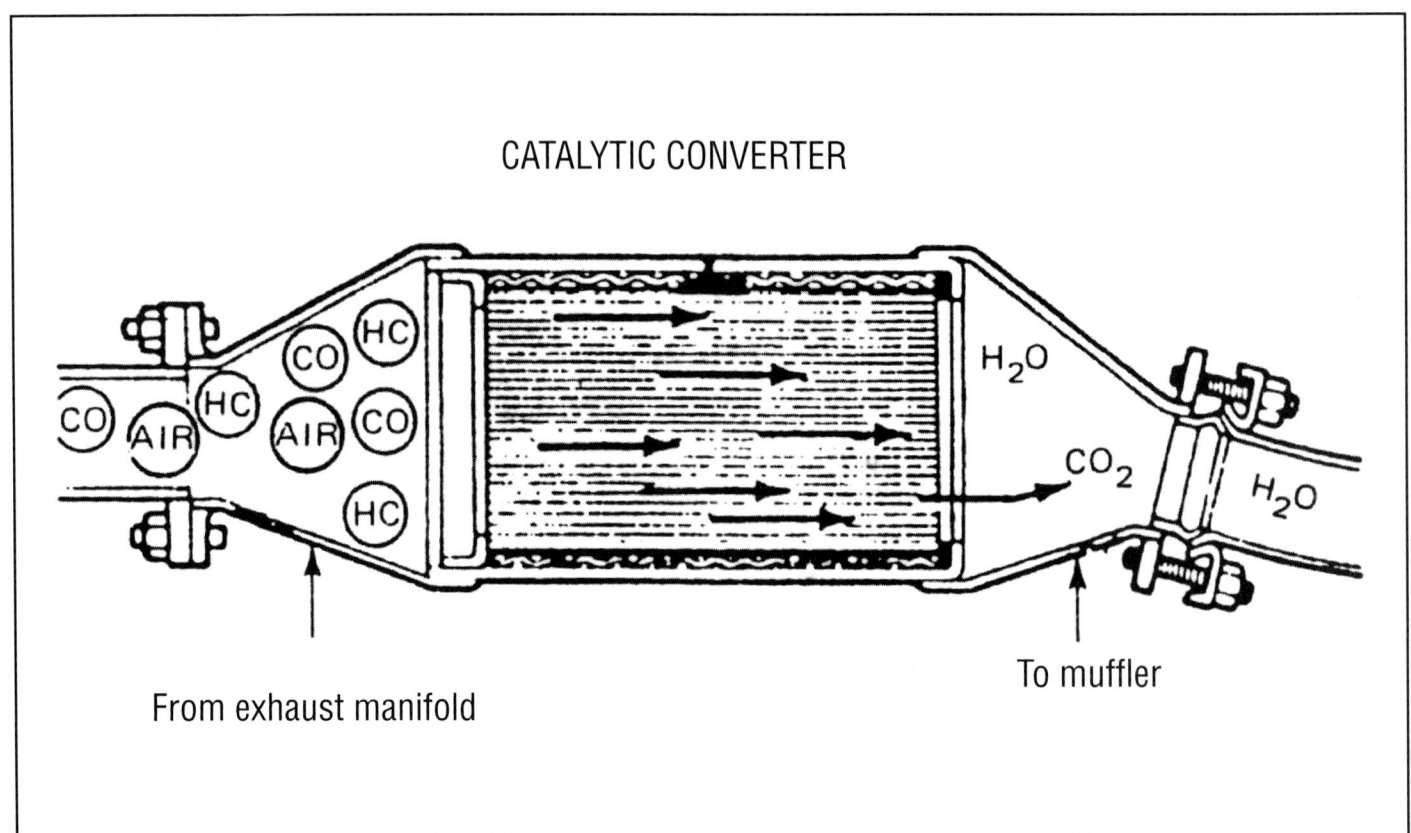

The oxidation catalytic converter came first, appearing in 1975. It is sometimes called a "two-way" cat because it burns up two pollutants—HC and CO (Ford).

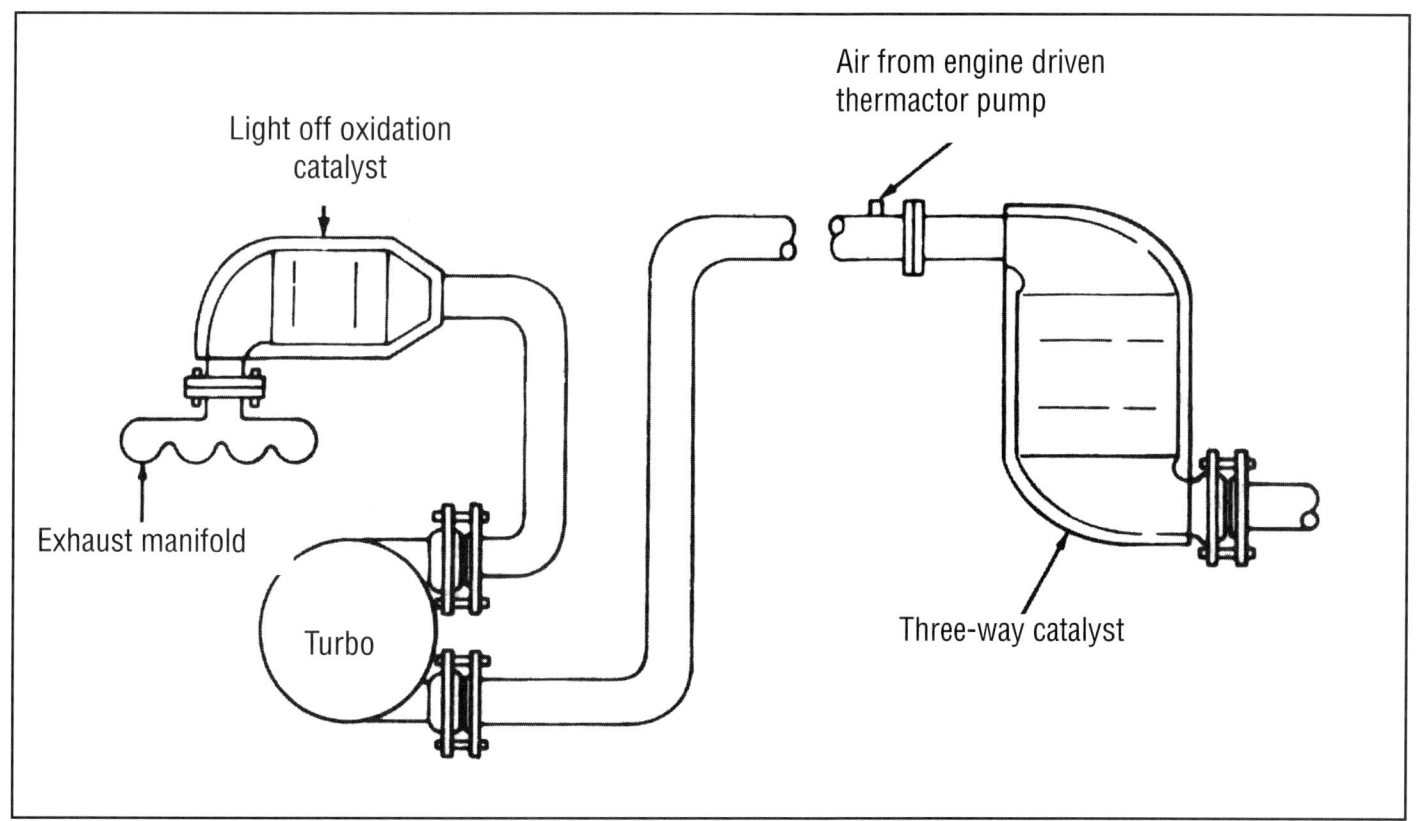

On some applications, such as this turbocharged Ford, a small light-off or mini-oxidation catalyst that gets to work quickly after the engine is started is added to cut emissions during the rich warm-up mode (Ford).

you need to grow plants. It works so well that the calibrations of other emissions controls can be loosened up, resulting in better performance, driveability and fuel economy.

As the gases flow through the pellets or honeycomb, they start to burn rapidly, and the temperature can reach 1600 degrees F. Extra oxygen in the exhaust stream is needed to support the blaze, and that may be provided by a very lean air/fuel ratio, pump-type air injection, or an aspirator valve.

Construction

The actual catalytic elements are platinum and palladium. Fortunately, only a thin film is needed because these metals are extremely expensive.

Types—There are two basic types of two-way oxidation catalytic converters: monolithic honeycomb and pellet bed. The former consists of a ceramic core that's coated with a microscopically thin layer of the catalytic agent. The configuration of the core gives it immense surface area so that the pollutants are bound to come into contact with the platinum and palladium. The latter, a GM design, is filled with ceramic pellets. They have a thin platinum/palladium coating and a large composite surface area, too. The difference is that if they become contaminated or otherwise rendered inoperative, they can be dumped and replaced with a new load, although this job (which requires special vibrator/aspirator equipment) is not done very often in the real world. If the honeycomb type becomes broken or clogged, it must be junked altogether.

Both kinds of converters, which are always mounted in the pipe between the exhaust manifold and the muffler, have a stainless steel skin so they won't rust out. Since the temperature inside gets to 1600 degrees F in normal operation, most installations include heat shields of one kind or another to avoid the possibility of making the vehicle's floor too hot.

Along with the main converter, a smaller "light-off," "mini-oxidation," or "pre-cat" catalyst may be used just behind the exhaust manifold. It gets hot very quickly after the engine is started, so it begins working in time to neutralize much of the extra pollution that's produced during cold running.

Three-Way Converters

In 1978, a different type of catalytic converter started to appear, the three-way or reduction variety. Besides oxidizing HC and CO, it also breaks NOx down into ordinary nitrogen and oxygen through the reaction of another precious metal, rhodium. The reaction that takes place is called reduction catalysis.

The only trouble with reduction catalysis is that it's a very delicate reaction that can only occur to a useful

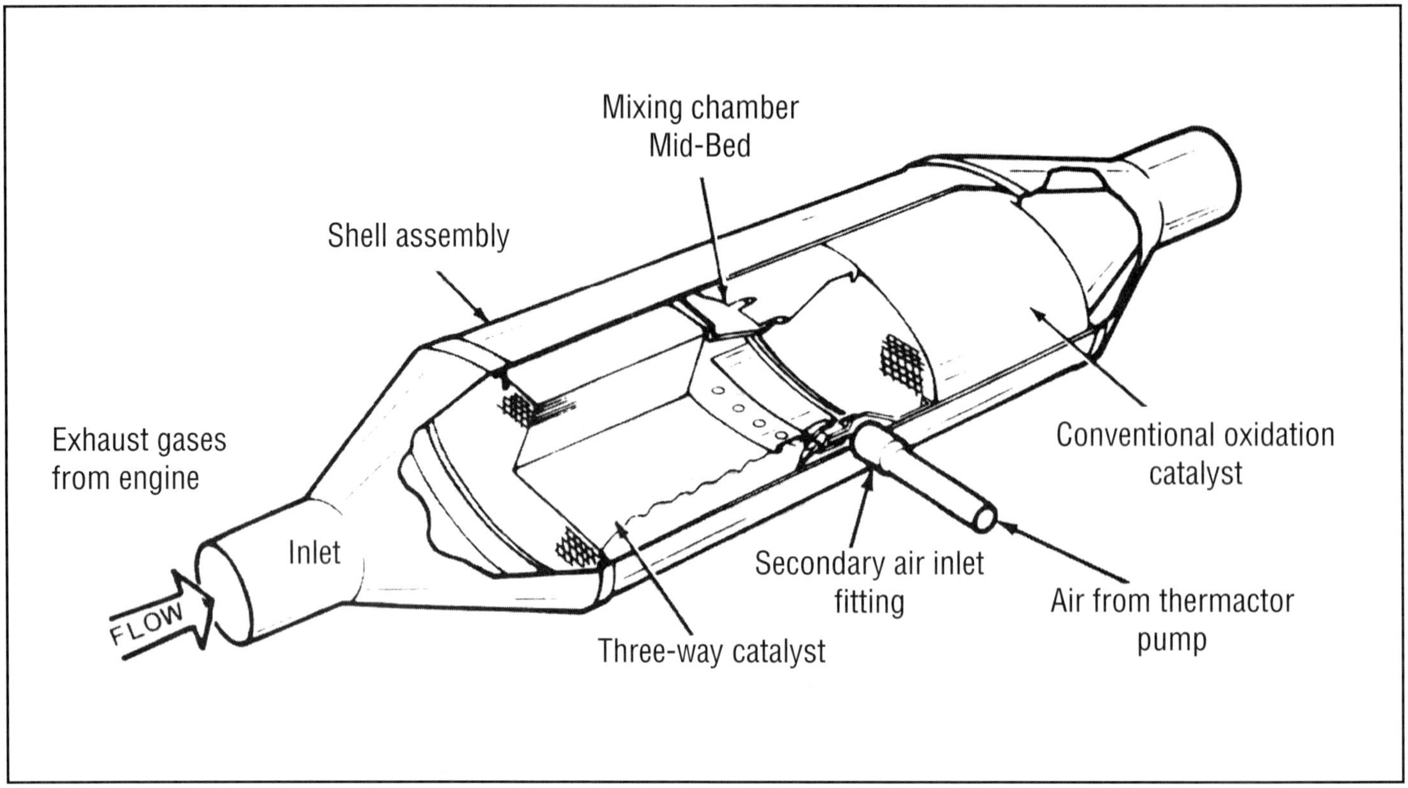

In a three-way catalytic converter, the front section handles NOx reduction, and the rear HC and CO oxidation. Air is injected into the mixing chamber downstream of the reduction cat so it doesn't upset the proper combination of gases (Ford).

degree if the air/fuel mixture fed into the engine's cylinders is kept extremely close to the ideal stoichiometric ratio of 14.7:1 (air to fuel). Ordinary carburetors or even early fuel injection systems are just too crude to hold the mixture at that level of accuracy, so electronic intelligence had to be added in the form of a computer, a network of sensors (most notably the one that reports on the oxygen content of the exhaust stream), and a carburetor or injection system that can tailor the blend according to commands from the microprocessor. The whole system is named feedback or closed-loop.

Some cars, the first FWD Mazda GLC for instance, manage to get sufficient reduction from the three-way catalyst without resorting to closed-loop, but they are exceptions. For really efficient catalysis, it's absolutely necessary.

How It Works—A typical three-way unit has the NOx-reducing section mounted upstream of the oxidation cat, and the output from the air injection pump or aspirator valve routed to a

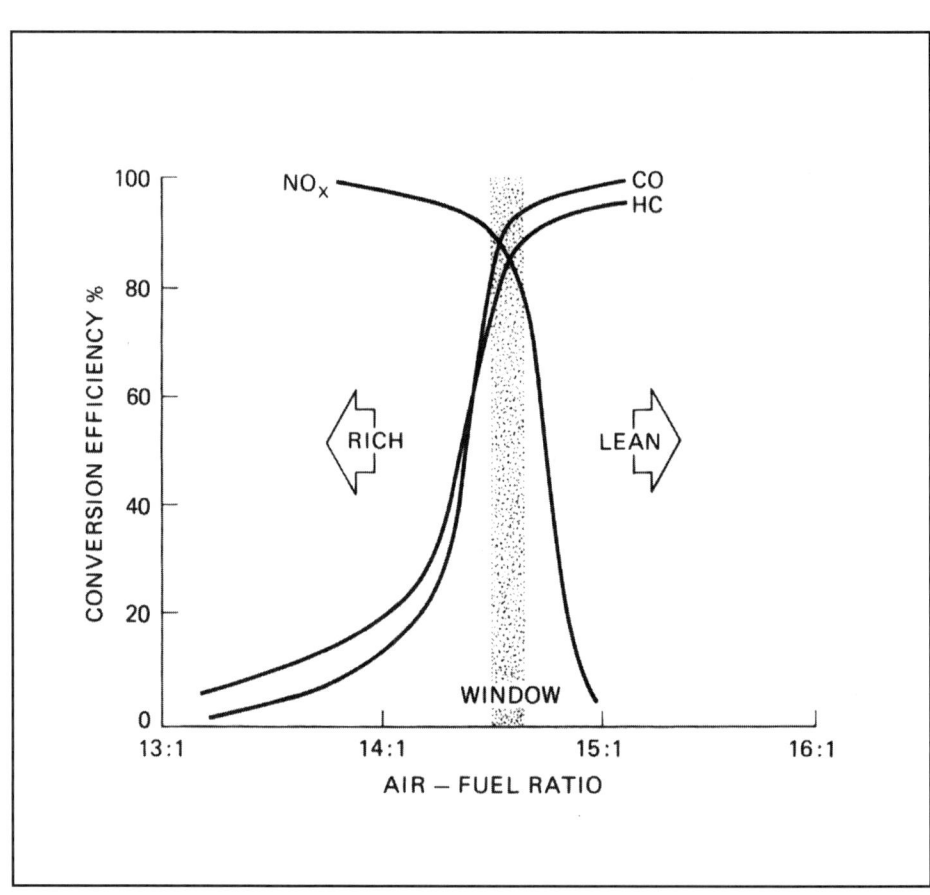

For acceptable efficiency, the three-way cat's NOx reduction reaction requires that the air/fuel mix be kept at the stoichiometric ratio (Volkswagen).

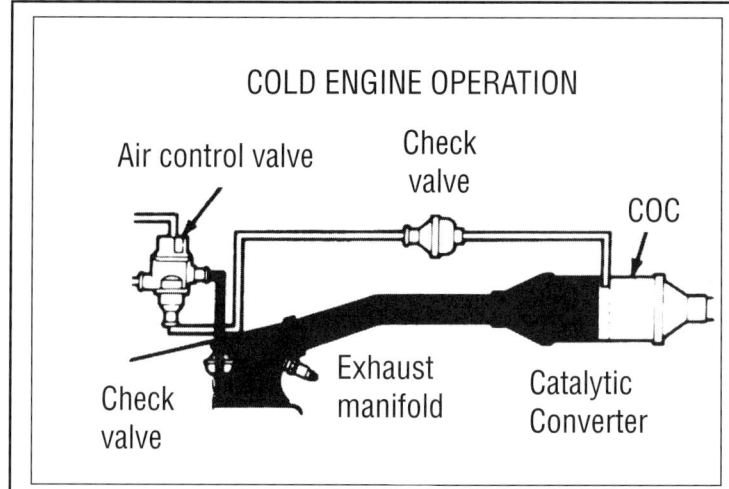
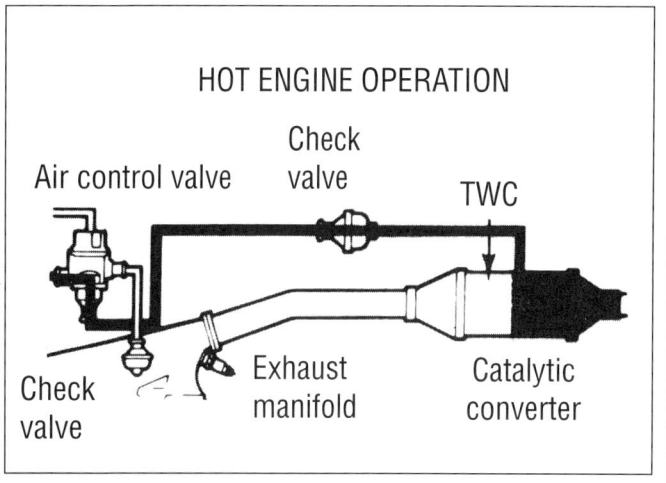

Very little NOx is produced when the engine is cold, so the air control valve switches air injection to the exhaust manifold during warm-up to help oxidize the high levels of HC and CO that are present then. When normal operating temperature is achieved, air flow is shunted to a point downstream of the cat's reduction section (Ford).

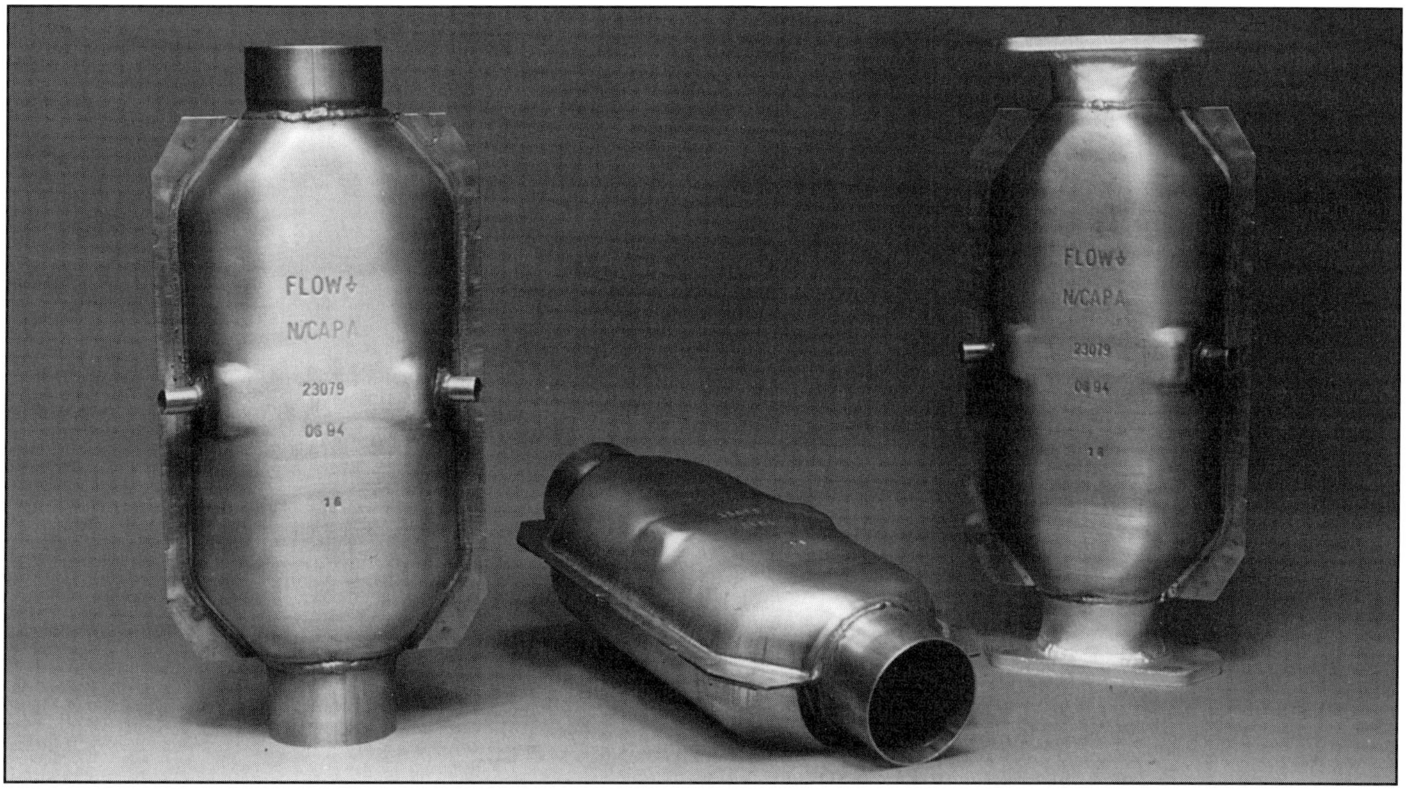

Random Technologies' Super High Flow catalytic converters increase horsepower while meeting EPA and CARB emissions requirements. They can be added to vehicles that were not originally equipped with a catalytic converter, or installed as replacements when required on late-model emissions-controlled cars and pickup trucks. (EPA regulations have strict guidelines as to when a converter can be replaced. See the sidebar on p. 30). For more information, contact Random Technologies, 2136 West Park Court, STE. L, Stone Mountain, GA 30087. 404/978-0264.

chamber between them. This arrangement is necessary to keep the extra air from upsetting that critical mixture balance created with such care at the intake. Also, a switching system is employed that directs air injection to the manifold during warm-up to help burn HC and CO in that high-emissions mode, then shifts it to the middle of the converter once normal operating temperature is reached (that is when NOx production begins) and the system enters closed loop.

As any lead in the gasoline would quickly coat the catalytic surface of any type of converter and kill the reaction, fuel without this octane enhancer (unleaded gas) must be used. At first there were problems with cross-fueling, but since 1990 the total ban on leaded gas has eliminated this possibility. ■

10
EXHAUST GAS RECIRCULATION (EGR)

The exhaust gas recirculation (EGR) system is one of your engine's basic emissions controls. Its purpose is to reduce NOx emissions. As long as the EGR system is functioning properly, you won't even know it's there. But when something goes haywire with the EGR system, you'll know it. The most common symptom is knocking or pinging when accelerating (as a result of detonation). Other symptoms can include a rough idle, stalling, hard starting, and elevated NOx and even HC emissions in the exhaust.

WHY EGR?

Exhaust gas recirculation reduces the formation of NOx by allowing a small amount of exhaust gas to "leak" into the intake manifold. The amount of gas leaked into the intake manifold is only about 6% to 10% of the total, but it's enough to dilute the air/fuel mixture just enough to have a "quenching effect" on combustion temperatures. This keeps temperatures below 2500 degrees F, which is the threshold point at which nitrogen reacts with oxygen to form NOx.

The need for some type of engine control technology to lower or reduce NOx emissions became apparent when scientific studies proved the link between NOx emissions and smog. So the Environmental Protection Agency decided to add NOx emission standards to its list of things to regulate. But in the early 1970s, the automakers were designing engines with later ignition timing, leaner carburetion and higher combustion temperatures to lower HC

The first production EGR system appeared on 1972 Buicks, such as this Gran Sport. It was then added to all passenger cars in 1973. Photo by Michael Lutfy.

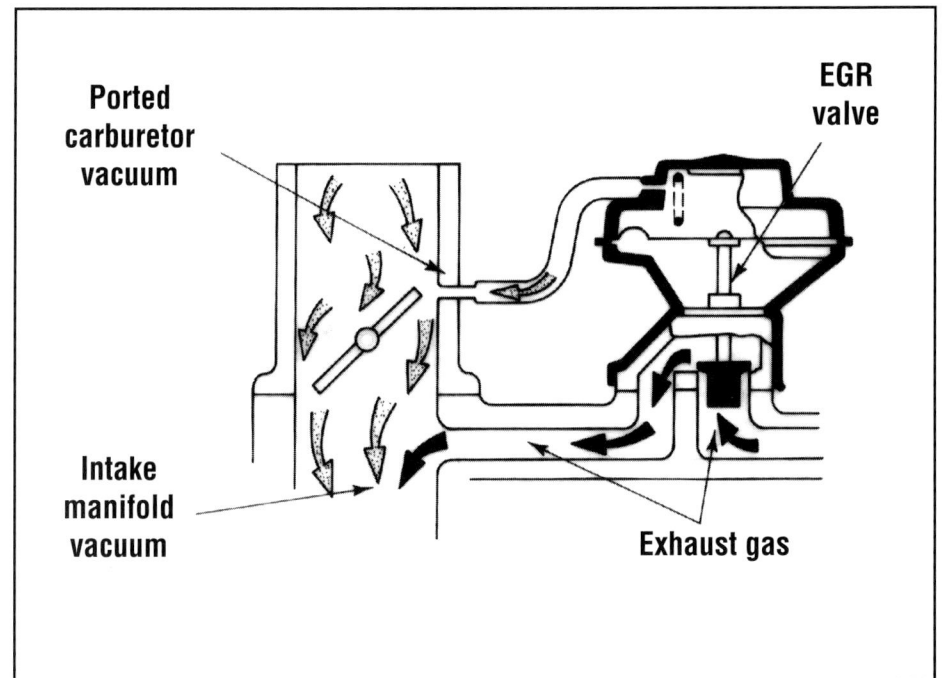

The basics of EGR. Ported vacuum pulls open the EGR valve so intake vacuum can siphon exhaust gas into the intake manifold. This dilutes the air/fuel mixture and lowers combustion temperatures to reduce the formation of NOx.

and CO emissions in compliance with EPA regulations. Unfortunately, the changes that reduced HC and CO emissions tended to increase NOx emissions. Figuring out how to lower HC and CO without affecting NOx (or even lowering it) seemed like an impossible task—until engineers discovered that slowing down the combustion process slightly and lowering combustion temperatures did the trick. They could achieve lower HC and CO emissions and lower NOx emissions too. How did they do it? By recirculating a small amount of "inert" gas into the air/fuel mixture to dilute it slightly. It worked great in a laboratory, but the next problem was how to provide the engine with an inexhaustible supply of inert gas. The answer was found in the exhaust itself. Exhaust is mostly carbon dioxide and water vapor, with nitrogen, small amounts of unburned hydrocarbons, carbon monoxide and other trace gases. There's very little free oxygen in the exhaust because most of it is burned in the engine. So for all practical purposes, exhaust is essentially inert as far as recombustion is concerned. By recirculating small amounts of exhaust back into the intake manifold, engineers could achieve the same results in everyday engines as they did in the laboratory. Thus, the new "exhaust gas recirculation" or "EGR" emissions-control system was born. The first production EGR system appeared on 1972 Buicks. It was then added to most passenger car engines in 1973 to meet federal NOx emission standards.

How It Works

To recirculate exhaust back into the intake manifold, a small calibrated "leak" or passageway is created between the intake and exhaust manifolds. Intake vacuum in the intake manifold sucks exhaust back into the engine. But the amount of recirculation has to be closely controlled, otherwise it can have the same effect on idle quality, engine performance and driveability as a huge vacuum leak. As mentioned earlier, most EGR systems restrict maximum exhaust gas recirculation to no more than 6% to 10% of the total air/fuel mixture. Higher rates of flow can dilute the air/fuel mixture excessively, causing the fuel mixture to misfire. This, in turn, can create a very rough idle, a stumble or hesitation upon acceleration, and contribute to hard starting. Engine misfire can also allow unburned fuel to pass through into the exhaust, creating elevated HC readings in the exhaust.

Some of the early V-8 EGR systems had small jets in the base of the intake manifold to regulate exhaust flow into the manifold from the exhaust crossover passage. Though simple in design, the "floor jet" EGR system had several drawbacks. One was that there was no way to "fine tune" exhaust flow. It varied in direct proportion to intake vacuum. Under some driving conditions (such as wide open throttle under load) there was not enough exhaust gas recirculation to keep NOx under control. At idle, there sometimes tended to be a bit too much exhaust gas recirculation, which made for a rough idle. Another problem with the EGR jets in the bottom of the intake manifold was that they often plugged up with carbon and varnish.

EGR Valve—A better approach proved to be that of using a vacuum regulated valve to control exhaust gas recirculation. The "EGR valve" thus became the heart of the EGR system and made possible a significant improvement in NOx reduction and overall system performance. The EGR valve meant that EGR could now be regulated more in sync with the demands of the engine: no EGR at idle, and maximum EGR when the engine was working hard under load. More on these valves in a moment.

In addition to EGR, other changes in engine design and operation also contributed to minimizing NOx. These included increasing camshaft valve overlap, redesigning combustion chambers and modifying ignition advance curves. With the addition of three-way catalytic converters that could also reduce NOx in the exhaust, it even

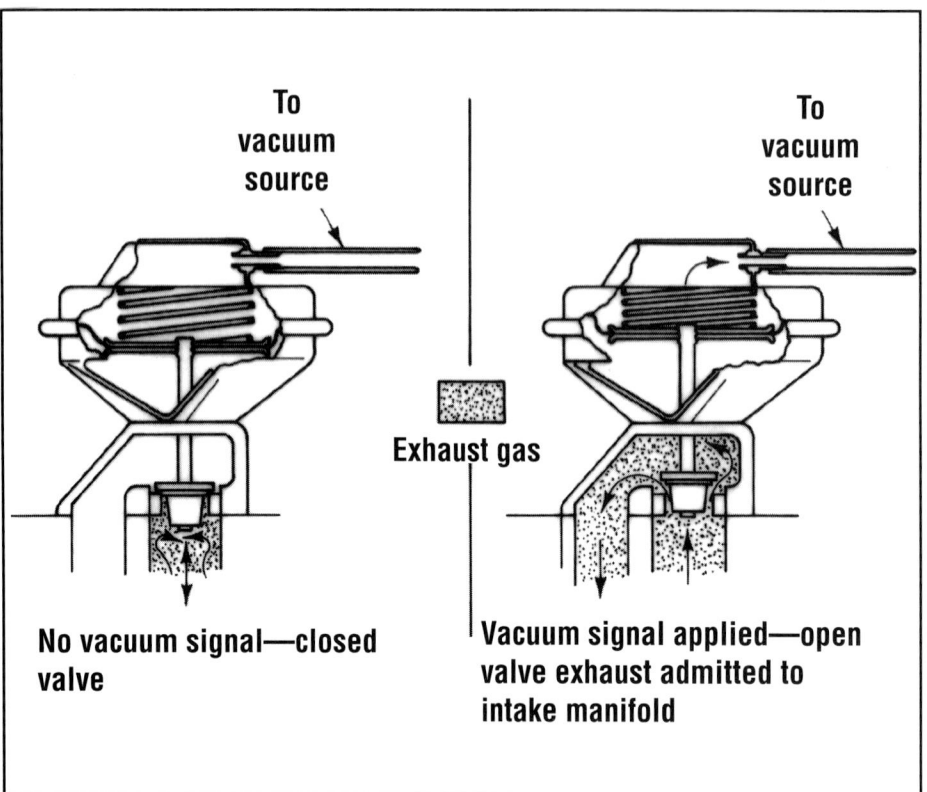

At idle (left), the EGR valve is closed and no exhaust passes through the valve. At part throttle (right), the valve opens so exhaust can enter the intake manifold.

became possible on some engines to eliminate the EGR system altogether.

NOx or Knocks?

Though diluting the air/fuel mixture with exhaust may not sound like the right prescription for making horsepower, EGR does have a beneficial effect on at least one aspect of engine performance. The quenching effect it creates in the combustion chamber helps the engine resist detonation (spark knock). That means the engine can tolerate more spark advance and compression on lower octane regular fuel. And the more spark advance an engine can handle without detonating, the more power and economy it will deliver.

Bypassing EGR—If you're thinking about bypassing the EGR system, think again. If an EGR system is rendered inoperative because someone illegally unhooks or plugs the vacuum hose to the EGR valve, removes the EGR valve or installs an aftermarket intake manifold that lacks the factory EGR connections, the quenching effect that was formerly provided by the EGR system will be lost. Without EGR, the engine will probably have too much spark advance for the stock timing setting and will likely knock and ping during hard acceleration or when the engine is heavily loaded. If the problem isn't corrected, severe detonation can lead to serious damage such as broken rings, cracked pistons, blown head gaskets and rod bearing failure!

If ignition timing is retarded relative to stock specifications to "offset" the loss of EGR, performance and fuel economy usually suffer. Switching to a higher octane gasoline or using an octane boosting fuel additive may or may not help. So the best advice is to make sure your EGR system is intact and functioning properly.

EGR VALVES

The typical EGR valve consists of a vacuum diaphragm connected to a poppet or tapered stem flow control valve. The valve opens a small passage between the exhaust and intake manifolds. The EGR valve itself is usually mounted either on a spacer under the carburetor or on the intake manifold. A small pipe from the exhaust manifold or an internal crossover passage in the cylinder head and intake manifold carries exhaust to the valve. When the EGR valve opens, intake vacuum pulls exhaust into the engine where it mixes with and dilutes the incoming air/fuel mixture. This normally occurs only when the engine has reached normal operating temperature (little NOx is formed when the engine is cold so EGR isn't needed until the engine is warm), and only when EGR is needed to reduce combustion temperatures (as when accelerating under load or driving at part-throttle to full-throttle).

The amount of exhaust that enters the intake manifold is determined by the size of the EGR valve orifice, how far the valve opens and how long it is held open.

EGR, remember, is not a full-time thing. Because it has the same effect on driveability as a vacuum leak, it is used only during part-throttle operation when the engine can tolerate a diluted fuel mixture. EGR is not used at idle because it would cause the engine to run rough and possibly stall (a classic symptom of an EGR valve that is stuck in the open position). Nor is EGR allowed when the engine is being cranked because it would act like a vacuum leak and make the engine hard to start.

How does the EGR valve know when to open and close? In most applications, it operates in response to engine vacuum either from a ported vacuum source above the throttle plates or from venturi vacuum at the carburetor throat. With ported vacuum systems, engine vacuum pulls the EGR valve open when the throttle plates are cracked open far enough to expose the vacuum port to intake vacuum. The use of ported

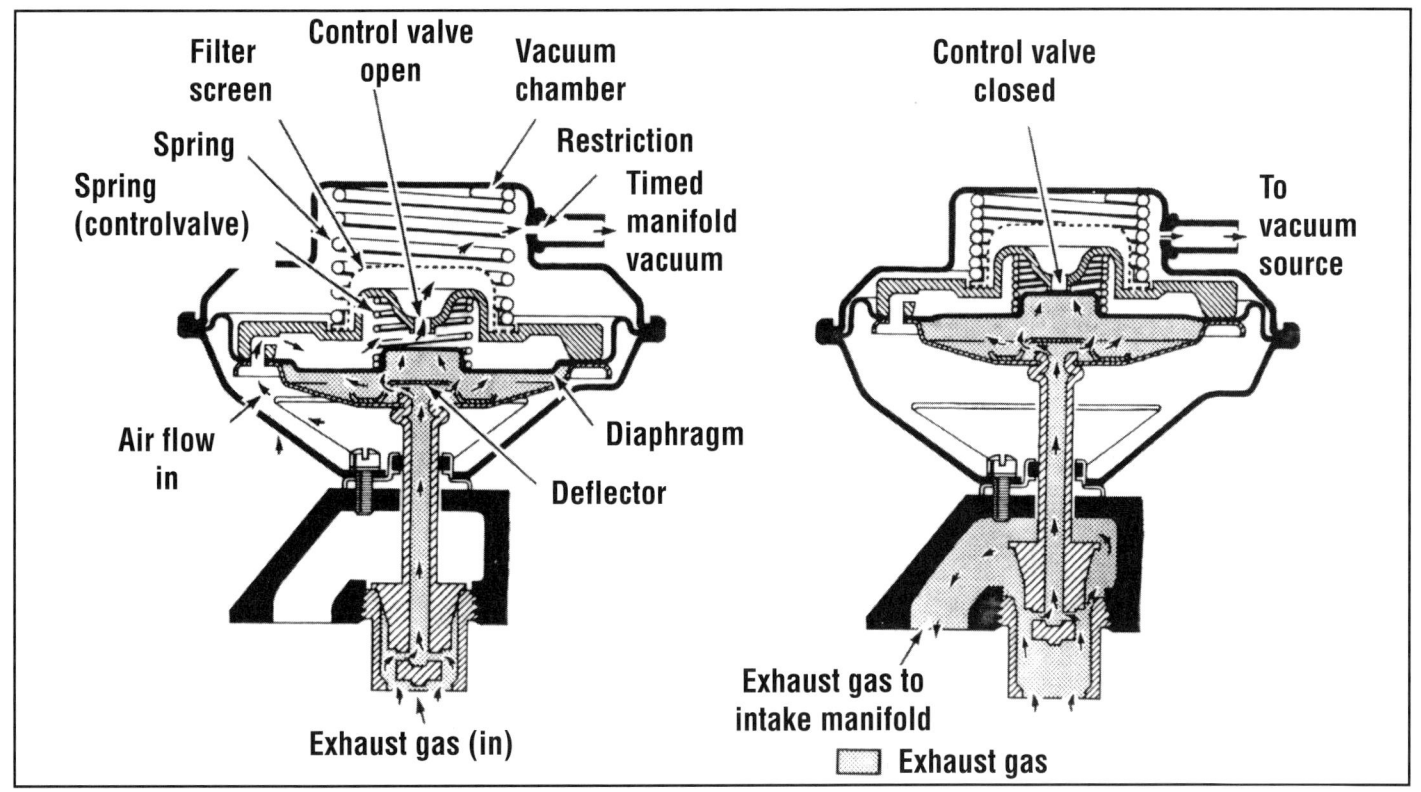

This shows how positive backpressure EGR valves operate. When there is little backpressure (left), the valve remains closed. But as backpressure increases (right), exhaust passes up the hollow valve stem, pushes the inner diaphragm up to block the vent hole so vacuum can open the valve.

vacuum prevents EGR at idle. A spring inside the EGR valve pushes the valve shut when there's no vacuum (as when the throttle plates are closed) or when engine vacuum drops below a certain level.

EGR systems using venturi vacuum require a vacuum amplifier to boost the relatively weak vacuum signal at the carburetor venturi. Reading vacuum from this point allows precise measurements between EGR action and airflow through the carburetor. The amplifier is connected to a vacuum reservoir in many such applications, and contains a check valve to maintain an adequate vacuum supply regardless of variations in engine vacuum. A relief valve may also be used to dump or cancel the output EGR signal whenever the venturi vacuum signal is equal to or greater than intake vacuum. This allows the EGR valve to close at wide open throttle on some applications when maximum power is required.

Backpressure

Another feature that has been incorporated into EGR systems is the ability to change EGR flow in response to changes in exhaust system backpressure. NOx formation increases when the engine is under load and during acceleration, so EGR flow should also increase during these times to compensate for higher combustion temperatures. Exhaust backpressure is a good measure of engine load, so a pressure-sensitive diaphragm that reacts to changes in backpressure is an effective means of regulating EGR operation. The backpressure diaphragm, which may be located inside the EGR valve itself together with the main control diaphragm, or in a separate housing, opens and closes a small vacuum bleed hole in the main EGR vacuum circuit or diaphragm chamber. Opening the bleed hole reduces the vacuum to the EGR valve and prevents it from opening fully. Closing the bleed hole allows full vacuum and maximum EGR flow. This allows the backpressure diaphragm to increase or decrease EGR vacuum in direct proportion to changes in exhaust backpressure. There are two types of backpressure EGR valves: positive and negative.

Positive—The positive type uses positive exhaust backpressure to regulate EGR flow. As pressure increases in the exhaust, the valve begins to open, allowing increased EGR into the intake manifold. This reduces backpressure somewhat, allowing the backpressure diaphragm to bleed off some control vacuum. The EGR valve begins to close and exhaust pressure rises again. The EGR valve oscillates open and closed with changing exhaust pressure to maintain a sort of balanced flow.

Negative—The negative backpressure type of EGR valve reacts in the same way, except that it reacts to negative or decreasing pressure changes in the exhaust system to regulate EGR action.

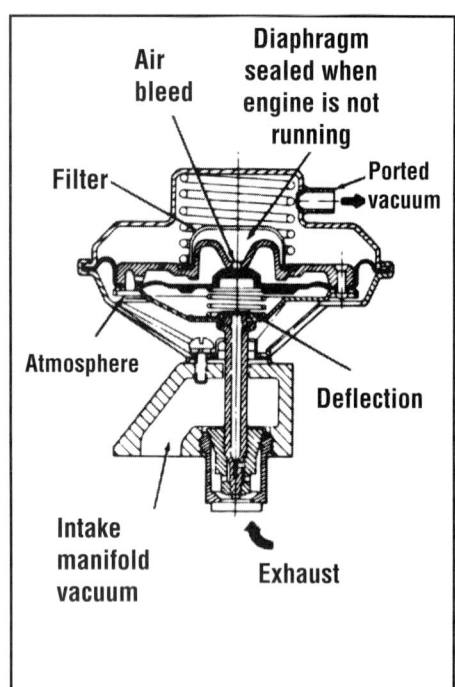

A negative backpressure EGR valve relies on negative backpressure to regulate the opening of the valve.

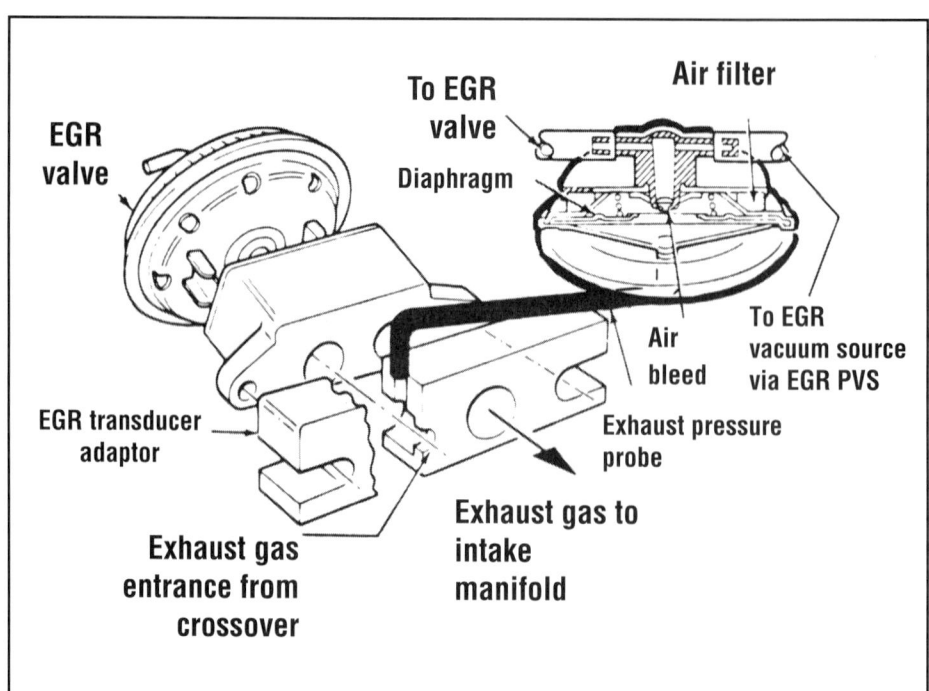

Some EGR systems have a separate backpressure diaphragm to control the operation of the EGR valve (Ford).

Decreasing backpressure signals decreased engine load, and the backpressure diaphragm opens a bleed hole to reduce EGR flow. It's the same principle as with the positive type except that the control function reacts to decreasing pressure rather than increasing pressure.

Backpressure EGR valves, both positive and negative, provide better control over EGR flow because they have the ability to react directly to changing engine loads. There are a couple of drawbacks, however. For one, the small bleed hole can become clogged. On backpressure EGR valves with the backpressure diaphragm inside the valve itself, a hollow valve stem is used to carry exhaust pressure to the diaphragm. This stem can also become plugged quite easily.

Aftermarket Exhaust Systems— One point to keep in mind with respect to engines with backpressure EGR valves is that the original equipment EGR valve is calibrated to the backpressure in the stock exhaust system. If you've replaced the stock muffler with a low-restriction aftermarket performance muffler, such as a turbo-style muffler, glass pack or similar design, or you've replaced the stock exhaust system aft of the converter with a freer-flowing aftermarket performance exhaust system, the change may reduce exhaust backpressure enough to adversely affect the operation of your engine's EGR valve. Pinging (detonation) when accelerating would be a good clue that reduced backpressure is preventing your EGR valve from opening. Reduced EGR because of a change in exhaust backpressure may cause you to fail an I/M 240 emissions test because of elevated NOx emissions. So make sure any replacement mufflers or exhaust systems you install are "emissions certified" and do not adversely effect EGR operation.

CONTROLLING EGR

Most pre-computer EGR systems have a temperature vacuum switch (TVS) or ported vacuum switch between the EGR valve and vacuum source to prevent EGR operation until the engine has had a chance to warm up. The engine must be relatively warm before it can handle EGR. If an engine runs rough or stumbles when cold, it may indicate a defective TVS that is allowing EGR too soon after startup. A TVS stuck in the closed position would block vacuum to the EGR and prevent any EGR operation. The symptom here would be excessive NOx emissions and possible pinging or detonation.

W.O.T.

Many EGR systems use some type of wide open throttle (W.O.T.) switch or valve to cut out or eliminate EGR action during those times when maximum power and acceleration are needed. On some systems, a diaphragm vents EGR vacuum to the atmosphere when intake manifold vacuum drops to zero. On later model engines with computerized engine controls, a throttle switch or throttle position sensor (TPS) signals the computer when the throttle is wide open so it can temporarily turn off the EGR system.

Other EGR Modulating Methods

Some EGR systems use an air bleed orifice or solenoid to modulate EGR action. According to engine operating conditions, the air bleed may be opened to reduce EGR vacuum, which in turn reduces how far the EGR valve opens. On some General Motors applications, for example, the amount of EGR is reduced when the automatic transmission torque converter clutch (TCC) is engaged. The same thing may be used on vehicles with manual transmissions when running in high gear. The reason for doing this is to provide smoother engine operation. Too much EGR flow can cause roughness and hesitation.

With diesel engines, there is no intake vacuum for EGR operation or control, so vacuum is usually created by an auxiliary vacuum pump. The pump provides a steady amount of vacuum for opening the EGR valve, and operation is regulated by computer-controlled vacuum solenoids and input from an electrical backpressure sensor in the exhaust system.

EGR & COMPUTER CONTROLS

On most engines with computerized engine control, a temperature vacuum switch is not used because EGR vacuum is controlled by the computer. The computer monitors engine temperature through the coolant sensor, and when the programmed operating temperature is reached, the computer opens the EGR vacuum solenoid, allowing intake manifold vacuum to pass through to the valve. On some systems, the control solenoid is normally closed. Energizing it opens the solenoid and allows vacuum to reach the EGR valve. On other systems, the EGR solenoid may be open in the normal position. It is energized (closed) only when EGR is not wanted, as when the engine is cold, during cranking, or at wide-open throttle. In either case, once

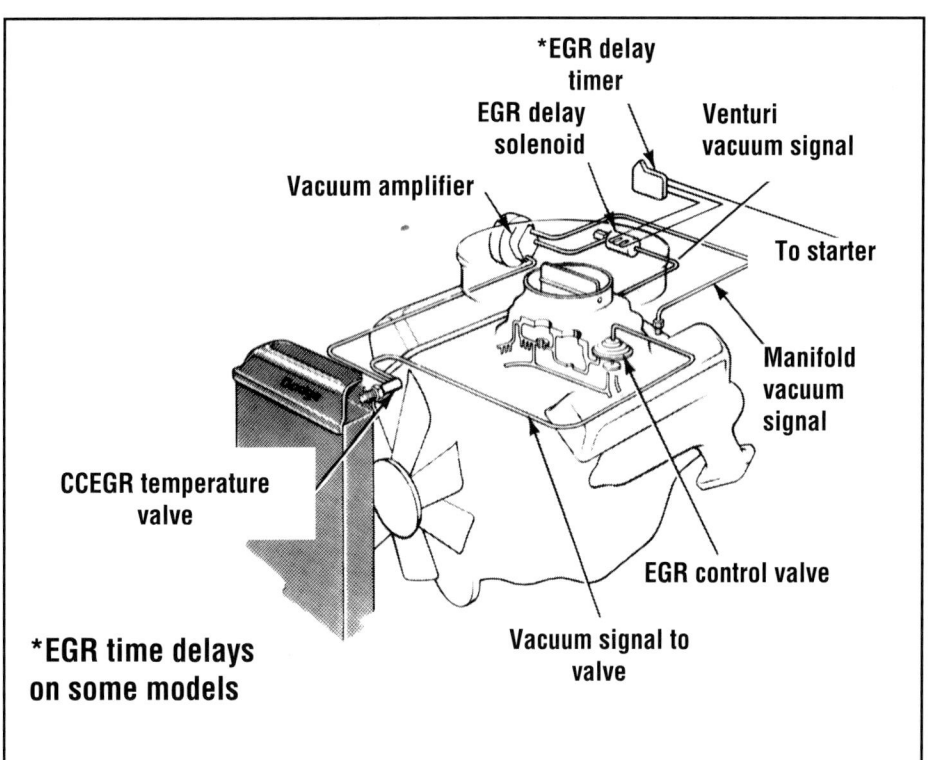

This is an example of a venturi vacuum EGR system. A vacuum amplifier is used to boost the relatively weak venturi vacuum signal so it can open the EGR valve (Chrysler).

the engine warms up, EGR flow is controlled as usual by ported vacuum and exhaust backpressure.

Regulating EGR Flow

Some computer systems use air bleeds or vents in conjunction with the EGR solenoid to regulate EGR flow. Both Ford and General Motors do this on certain systems. Early Ford Electronic Engine Control systems (EEC-II & EEC-III) have two EGR control solenoids and an EGR valve position sensor mounted on top of the EGR valve. The EEC computer monitors various engine functions as well as EGR position to determine how much EGR is needed. Opening the normally closed EGR vacuum control solenoid (EGRC) allows manifold vacuum to pass to the EGR valve. The normally open EGR vent solenoid (EGRV) vents some air into the vacuum line to reduce vacuum and limit how far the EGR valve opens. If more EGR is needed, the computer energizes (closes) the EGRV to stop the air leak so that full vacuum can reach the EGR valve. The computer can modulate EGR action by "dithering" (opening and closing) the two solenoids to achieve the amount of EGR needed. Energizing both provides maximum EGR (full vacuum), energizing only the EGRC but not the EGRV provides a sort of "midrange" EGR (part vacuum), while not energizing either solenoid prevents any EGR (no vacuum).

Pulse-Width Control—Another approach to EGR control (first introduced in 1984 by General Motors) is to use a pulse width-modulated EGR control solenoid. With this technique, the engine control module cycles the EGR vacuum control solenoid rapidly on and off. This creates a variable vacuum signal that can regulate EGR operation very closely. The amount of "on" time versus "off" time for the EGR solenoid ranges from 0 to 100%, and the average amount of "on" time versus "off" time at any given instant determines how much EGR flow occurs.

Digital Control—On some applications, a "digital" EGR valve is

73

used. This type of valve also uses vacuum to open the valve but regulates EGR flow according to computer control. The digital EGR valve has three metering orifices that are opened and closed by solenoids. By opening various combinations of these three solenoids, different flow rates can be achieved to match EGR to the engine's requirements. The solenoids are normally closed, and open only when the computer completes the ground to each.

Linear—The latest innovation in EGR systems is a "linear" EGR valve that uses a small computer-controlled stepper motor to open and close the EGR valve instead of vacuum. The advantage of this approach is that the EGR valve operates totally independent of engine vacuum. It is electrically operated and can be opened in various increments, depending on what the engine control module determines the engine needs at any given moment in time. GM started using this type of valve on many of its passenger car engines in 1992.

EGR PROBLEMS

If you think your EGR system isn't working properly either because you're experiencing engine detonation under load, you have a rough idle or hesitation problem, or because your vehicle flunked the NOx part of an I/M 240 emissions test, then it will be necessary to diagnose the EGR system to find out what's wrong.

With so many different EGR systems in use today, the first step in troubleshooting a suspected EGR problem is to find out what type of EGR system is used on your engine. For this, you'll probably have to refer to a service manual.

Does your EGR system use ported vacuum, venturi vacuum or is it computer-controlled? Does it have a temperature vacuum switch, a computer-controlled EGR vacuum or vent solenoid, or a wide open throttle (W.O.T.) switch or valve? Is the EGR valve the backpressure type, and if so which type, positive or negative? Is there an EGR valve position sensor? Are there other systems, such as canister purge, plumbed into the EGR vacuum circuit? These are some of the things you'll have to know if you're going to troubleshoot the system yourself. For step-by-step details, see Chapter 16, page 129.

EGR VALVE REPLACEMENT

If your EGR valve is defective and your vehicle is still under warranty (5 years or 50,000 miles federal emissions warranty, or three years and 50,000 miles on 1990 & newer California vehicles), you can take your vehicle back to your new car dealer for free repairs. Otherwise, you pay for any repairs out of your own pocket.

A word of caution about replacement EGR valves: With so many variations from one vehicle application to the next in emissions-control systems and calibration, it is extremely important that you obtain the correct replacement EGR valve for your engine. Two EGR valves may look identical but be calibrated differently in terms of flow and the amount of vacuum and/or backpressure it takes to open the valve. Therefore, you may have to provide your vehicle's VIN number as well as year, make, model and engine size when ordering a replacement EGR valve. It may also be necessary to refer to the OEM part number on your old EGR valve (if possible) when ordering a replacement, so don't throw the old EGR valve away until you have the new one, have installed it and made sure it's working correctly.

Many aftermarket EGR valves are "consolidated" so fewer part numbers are necessary to cover a wider range of vehicle applications. Some of these valves use interchangeable restrictors to alter their flow characteristics. Follow the supplier's instructions as to which restrictor to use for the correct calibration. ∎

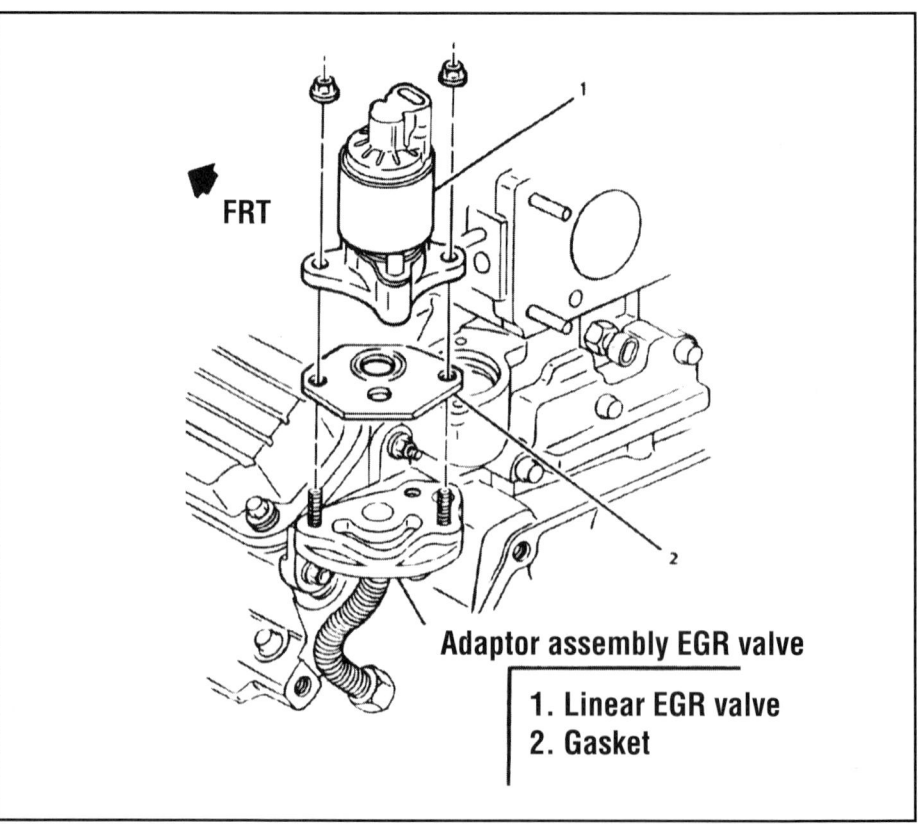

Adaptor assembly EGR valve
1. Linear EGR valve
2. Gasket

A GM "linear" EGR valve. It uses a small electric stepper motor instead of vacuum to open the valve.

11
HOW AIR/FUEL RATIO AFFECTS EMISSIONS

An engine can have any number of add-on emissions-control devices, but if the air/fuel ratio is way off, it will still be a gross polluter. Of course, performance, driveability and fuel mileage will suffer too. Except perhaps for ignition misfiring, nothing affects the composition of the gases a powerplant pumps into its exhaust manifold more than the mixture. The automakers recognized this very early in their fight for clean air, so they focused a great deal of their attention on fuel delivery. In those days, of course, that meant the carburetor in the overwhelming majority of the cases (the number of mixture problems was vastly reduced with fuel injection, but that wasn't to become popular for years).

Some of the carburetor modifications they made were simple and easy to live with, while others caused driveability problems—hesitation, stalling, rough idle and so on—but all were aimed at providing the engine with a blend of air and fuel that would be burned as completely as possible inside the cylinders. This meant that the mixture had to be leaner in all modes of operation than what was typical in the pre-emissions-control era.

COMBUSTION BASICS

Before we go on, it would be a good idea to define some basic principles and terms that are related to the science of internal combustion. To begin with, an engine always burns gasoline at a ratio of air to fuel of 14.7:1 by weight, regardless of the ratio that is drawn into the

Pre-emissions-control carburetors, such as this old AFB, gave good performance and driveability, but were calibrated on the rich side, which produced excessive amounts of HC and CO.

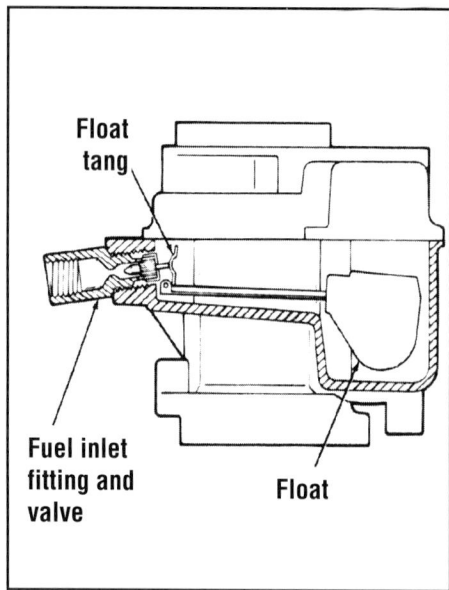

The float and needle-and-seat valve system keeps fuel in the bowl at the proper level. Leverage gives the float enough force to overcome fuel pump pressure (Chrysler).

Not only is float level critical, but if the float should become too heavy to do its job (from leaks in the brass pontoon type shown here, or absorption by the plastic foam type), emissions will soar while fuel mileage drops.

cylinders. In other words, the chemical reaction itself (the rapid oxidation of gasoline, or the fire in the cylinders) takes 14.7 pounds of air to one pound of fuel. Note that we said by weight, not volume. Since air is a whole lot lighter than gasoline, it takes a huge volume of it to oxidize the fuel—9000 gallons to one gallon of gas.

Engineers have their own name for this sacred ratio. They call it "stoichiometric." As we said earlier, this is the ideal theoretical ratio, but put anything from about 8:1 to 18:1 into an engine and it will run. With a rich mixture (a larger amount of fuel to the same amount of air), there's more gasoline than necessary, so some of it won't find any oxygen to combine with and will just be pumped raw out the exhaust pipe. With a lean blend, on the other hand, there is an overabundance of air, so all the fuel is consumed. That is, as long as it's not so lean that it misfires.

To put it another way, the reaction always takes place at the stoichiometric ratio, no matter whether the carburetor or fuel injection system is supplying 8:1 or 18:1. It should be obvious that in terms of emissions control and fuel efficiency, the leaner you can get the mixture and still have it fire dependably, the better. That's what Chrysler's Lean Burn, Honda's CVCC, and Ford's PROCO program were all about—getting a thin blend to burn.

Physics Made Painless

Now for some physics. Maybe you've never thought about it, but it's still true that only fuel vapor will burn, so gasoline has to change its state from a liquid to a gas or the engine simply won't run. To do this, it must absorb enough heat to boil.

You may very well be wondering how it can possibly boil when the engine is cold, and that's certainly a good question. The answer is simple: Just as water boils at less than 212 deg. F at high altitudes because of the lower atmospheric pressure, so the vacuum in the intake stream helps to boil gasoline. In other words, the boiling point is reduced enough to vaporize the atomized droplets that enter the intake manifold or ports even if it's way below zero outside. Until the engine warms up, however, only a small part of the gasoline that's provided will turn to vapor, so a choke, a cold start injector, or longer injection pulse is needed to provide a terrifically rich mixture, and the fuel that remains in a liquid state is just wasted. That's why it's so hard for an engine to pass the emissions certification test during the very first part of the running cycle.

CARBURETOR BASICS

No matter what company makes it or how many barrels it has, all carburetors share the same basic operating principles. So, if you understand the theory involved you'll have the key to diagnosing any specimen you happen to encounter even if it's unfamiliar.

Any time air is forced through a tube, a pressure drop occurs, and this phenomenon is what gets gasoline to move into a carburetor's throat. The speed of the passing column of air determines the strength of the vacuum, so, especially at low rpm, it's necessary to give it a boost. That's what the venturi is for. The principle is that whenever a restriction is placed in a tube, air rushing past it must speed up, and this acceleration causes an extra pressure drop that helps atmospheric pressure in the bowl force enough fuel out of the carburetor to result in a burnable blend. Also, this additional vacuum makes the gasoline more enthusiastic about vaporizing.

Then there's atomization. You would be unhappy if you pushed the button of a household aerosol can and the contents came out in a solid stream. Well, an engine doesn't like it much either when it gets fuel in a hard-to-digest, unbroken form, so one of the carburetor's functions is to help smash gasoline into tiny droplets. These expose a lot more area to the air than a stream would, which aids in vaporization. And a mist or vapor has less inertia than big chunks of liquid, so it negotiates the curves in the intake manifold more easily.

All carburetors depend on the theories just described, and if you're familiar with it, no specimen you are apt to encounter will be able to totally mystify you—provided you also understand the systems that put them into practice.

Reservoir

A carburetor needs a reservoir of constant depth to supply its circuits evenly, and that, of course, is the bowl. Most of the time, the engine can't possibly use all the gasoline the fuel pump can supply, so the float and needle-and-seat setup stops the flow. It may not seem possible that a tiny metal pontoon or a little piece of plastic foam could have enough buoyancy to shut down the 5-7 psi (pounds per square inch) the pump puts out, but leverage gives it plenty of strength to do the job. So, it's not the fuel pump that forces gasoline into the intake stream. Atmospheric pressure does it by bearing down on the surface of the liquid in the bowl. Naturally, there has to be a vent to let in the weight of all those miles of air above us. In the old days, this was right on top of the bowl, but modern carburetors have it in the form of a tube that sticks up out of the throat and takes advantage of the ram effect of the moving air column to push on the gasoline.

Idle and Up

At idle, the velocity of the air entering

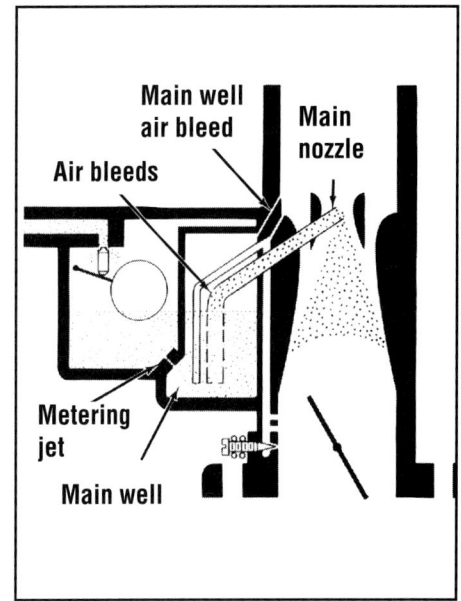

The cruising circuit gets its supply of fuel from the main well, and the amount is controlled by the size of the hole in the metering jet (Chevrolet).

the carburetor and hence the vacuum generated is too low to get fuel moving through the cruising system, so it has to come in somewhere else. That's the function of the idle circuit, which is made up of a port below the throttle plate where there's lots of vacuum, a passage from the bowl with little air bleed holes in it to aid in atomization, and an adjustment screw.

Other, similar outlets are used to provide a smooth transition from idle to moderate rpm, and these are called transfer or off-idle ports. They're positioned higher up in the throat than the idle ports and are progressively uncovered and exposed to vacuum as the throttle plate opens. In most specimens, they get their supply of fuel from the same tube as the idle port, but they aren't affected by the mixture screw.

Cruising

The cruising or main metering system is the next to come into action. Its nozzle projects into the part of the venturi where there is the highest vacuum, and the maximum amount of gas it can spray is controlled by the diameter of the main jet

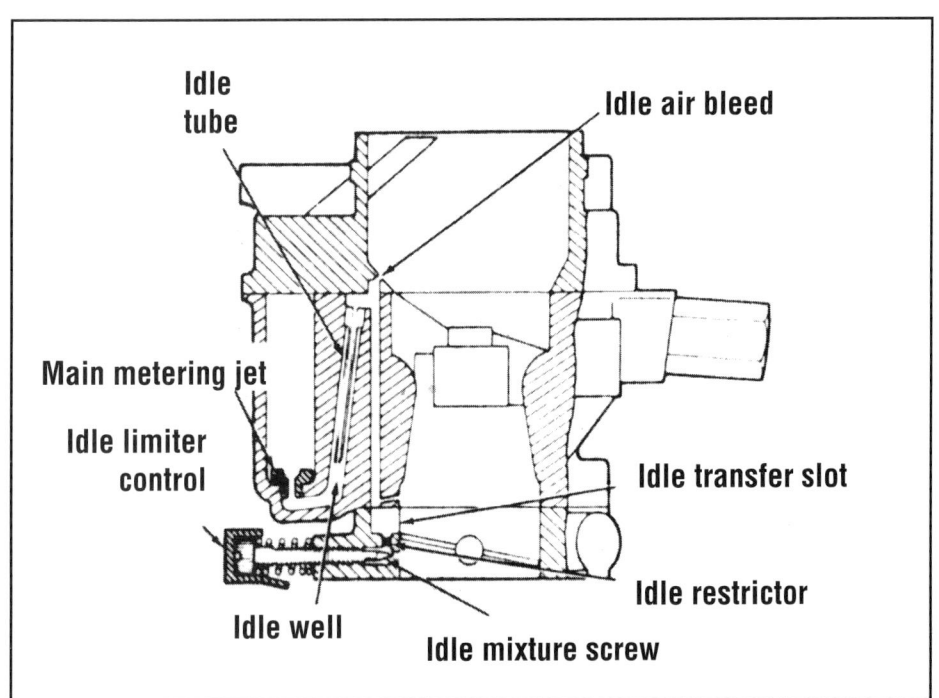

The idle circuit is controlled by the mixture screw, and the transfer ports admit extra fuel as the throttle opens and exposes them to vacuum (Chrysler).

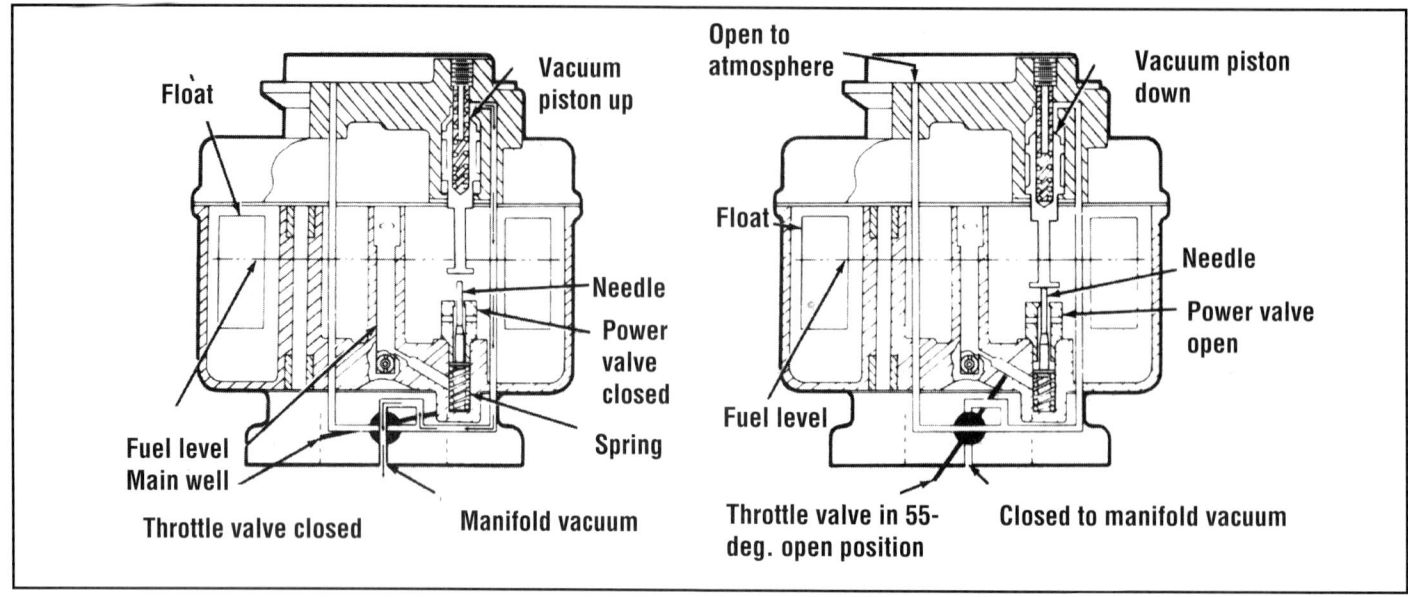

When the throttle approaches wide open, vacuum drops. This puts the power enrichment system into action. High vacuum holds the piston up, but the low vacuum of heavy acceleration lets it fall, opening the power valve and admitting extra gasoline (Chrysler).

through which it is supplied. This is the most efficient circuit in a carburetor, usually calibrated to provide the ideal 14.7:1 air/fuel ratio (although an ordinary carburetor can't actually provide this ratio under the varying conditions of normal driving). It starts working before the last off-idle port is uncovered, then takes over completely.

W.O.T.

During hard acceleration, when the throttle is open wide, far too much air enters the carburetor for the cruising system to handle. Without help, the mix would lean out and acceleration and top speed would be way below what the engine is actually capable of producing—if it could keep running at all.

Enter the power circuit, which opens up and allows extra gasoline to pass through the carburetor. It can be either a separate fuel valve, or a metering or step-up rod that normally blocks some of the main jet's flow, but is pulled up out of the way mechanically or by vacuum when the engine needs more fuel. The vacuum-operated type commonly opens at 5-7 in. Hg.

You could drive a car that was only supplied by the circuits described above, but it wouldn't be much fun. If you stepped on the gas just a little too rapidly, a blast of wind would rush in and put out the fire before enough fuel could get moving to make a flammable mixture.

So carburetors have an accelerator pump to compensate for gasoline's inertia. It squirts an extra shot of fuel into the intake stream whenever the throttle is opened and, since it works mechanically, does it before the engine gets a chance to choke on those big chunks of plain air. It's a simple piston pump with one-way inlet and outlet valves and an air bleed, weight or spring arrangement that keeps vacuum from drawing fuel through the nozzle while idling or cruising.

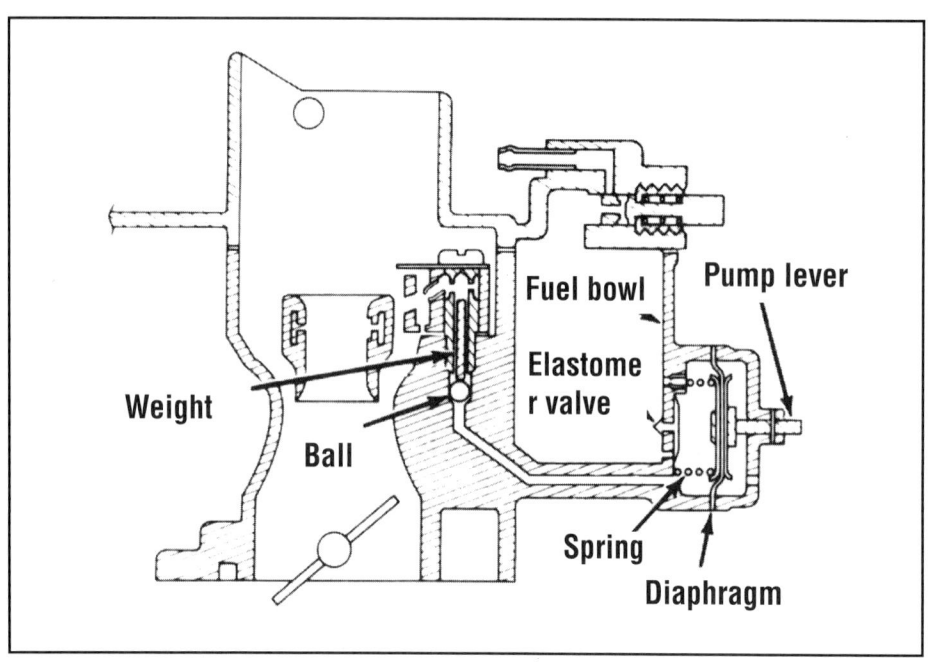

The accelerator pump squirts extra gasoline into the throat whenever the throttle is opened (American Motors).

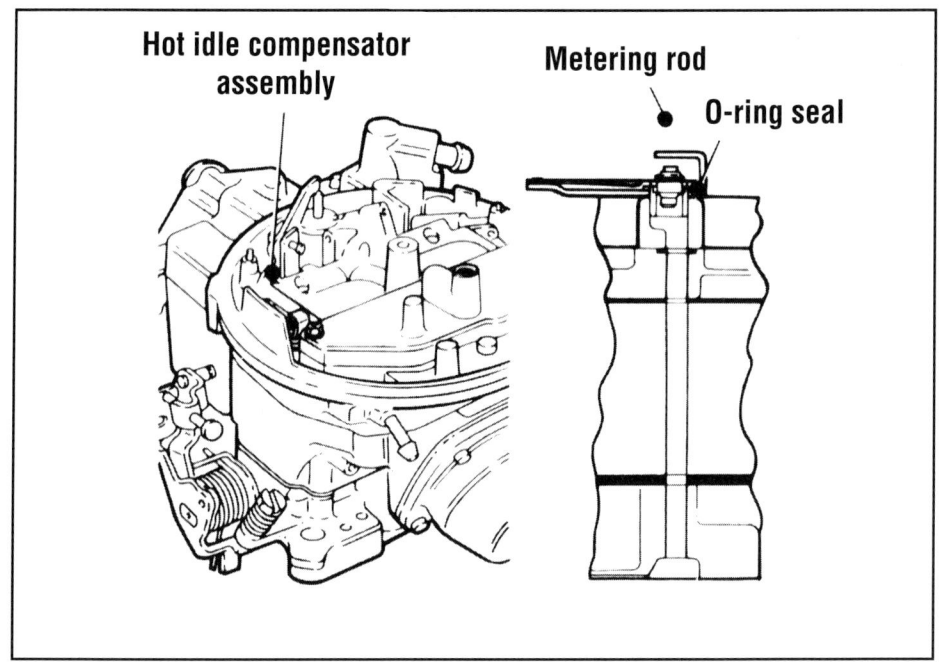

The hot idle compensator opens to admit extra air to keep the mix from getting too rich when the air is hot and thin (Ford).

Electrically heated choke coils work well to cut pollution during warm up (Ford).

Super Rich

When the engine and the weather are both cold, gasoline is very reluctant to vaporize, so an extremely rich mixture is needed to get that blaze going in the cylinders. That's where the choke comes in. It closes off the mouth of the carburetor so that fuel is pumped through every possible orifice as the starter is cranked.

Once the powerplant fires up, vacuum pulls on a diaphragm or piston (known as the vacuum break or choke pull-off), which opens the choke a crack so enough air enters to allow combustion. Then a thermostatic coil, which is heated by exhaust, electricity, or coolant, continues to move the plate until it no longer impedes air flow.

Finally, there's the hot idle compensator that's used on many carburetors to make up for the difference in density between warm and cool air. On a hot summer day, the air that gets past the throttle plate at idle is so thin that the mixture becomes excessively rich. The compensator is simply an extra air passage that opens up to bring the blend back into line. It's controlled by a tiny bimetal valve.

Leaner and Later

The "leaner and later" strategy appeared in the late 1960s. One of the basic ideas was to keep idle rpm down by retarding the spark timing, then to bring the speed up to acceptable levels by having the carburetor throttle plates open wider than in earlier models. This allowed extra air in to reduce rich running. Later spark and leaner mixtures were also used in the accelerating, decelerating and cruising modes.

This approach helped lower HC and CO levels at the expense of some performance, efficiency and driveability problems. For example, cars started to "run on" or "diesel" when switched off if idle speed was even slightly above specifications, because there was enough air available to allow this phenomenon to occur.

Idle Stop Solenoids—Dieseling is an unpleasant result of the engineers' efforts to cut pollution. Its biggest contributors are the increased air flow at idle just mentioned, and the higher temperature thermostats used to improve combustion efficiency. The idle stop solenoid was invented to eliminate this run-on condition.

It's actually a simple device, yet many people who work on cars never really learned its purpose—it's still common to

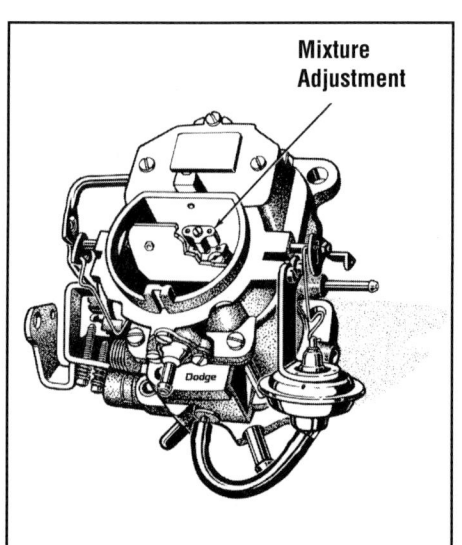

Then came carburetors that compensated for the thin air of high altitudes. This 1977 two-barrel had a manual adjustment, but most models changed the air/fuel mix automatically (Dodge).

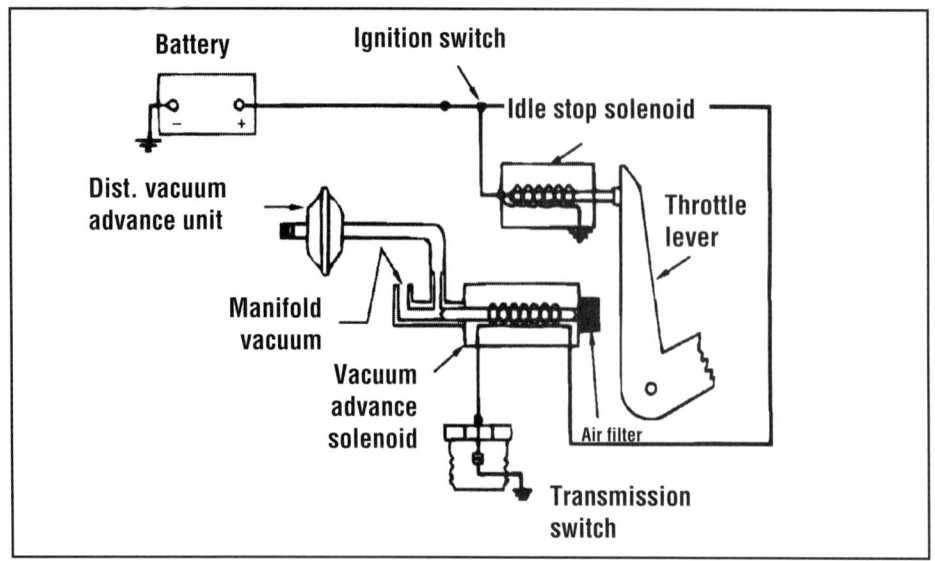

The idle stop solenoid eliminated dieseling or run-on. That is, if it was set correctly (Chevrolet).

find older cars with the idle rpm set by the regular throttle stop screw, the result being that the engine continues to run after the key is switched off.

Mounted so its plunger contacts the throttle lever, and wired so that it's energized when the ignition is on, the idle stop solenoid is supposed to hold the proper idle speed while the engine is running. When the key is switched off, the plunger retracts, and the throttle closes so much that the engine strangles instantly. Whenever you're confronted with a car that diesels or idles very slowly, see that the solenoid is operating, and that the throttle stop screw is backed out sufficiently. Two idle speed specifications are commonly given, one each for the energized and de-energized conditions.

Occasionally, you may find that someone has set idle speed with the plunger retracted, not realizing that the solenoid does not have the strength to move the throttle lever, and can only hold a position. This results in a dangerous condition—when the driver steps on the accelerator pedal, the plunger extends and keeps the throttle open too far, so the vehicle doesn't slow down when he lifts his foot from the accelerator.

An interesting variation on this theme is GM's Combined Emission Control (CEC) system. Here, a solenoid holds the throttle open a predetermined amount during high-gear deceleration, and this helps lean down the mixture in that normally rich, high-vacuum mode.

Computer-Controlled Carburetion

Feedback carburetors first appeared in the late 1970s. They provided much more precise control of the air/fuel ratio than any ordinary carb has ever been able to give, but without the expense of resorting to fuel injection.

The basic reason for the introduction of this concept is the three-way or reduction catalyst, the kind that fights NOx along with HC and CO. This extra reaction is very fragile. It takes place only if the amount of oxygen in the exhaust stream is exactly right, the result of the perfect 14.7:1 air/fuel ratio mentioned above. As we said, old-fashioned carburetors just can't guarantee that kind of accuracy.

The first feedback carb appeared on 1978 2.5L Pontiac Sunbirds built for California, and most GM cars were similarly equipped from '81 until EFI took over. A typical unit has a solenoid inside that pushes down a metering rod when it's energized by the ECM (Electronic Control Module—see Chapter 14), and this reduces the flow of fuel to the main metering and idle circuits. Also, when the rod is down, it opens an auxiliary air bleed to the idle system, further leaning the mixture.

The solenoid cycles 10 times per second and the amount of "on" time

The feedback carburetor was an interim solution to providing closed-loop fuel control before everything went EFI. Most feedback carburetors use a mixture-control solenoid, which operates on a duty cycle provided by the engine management computer. This is a Ford variable venturi unit.

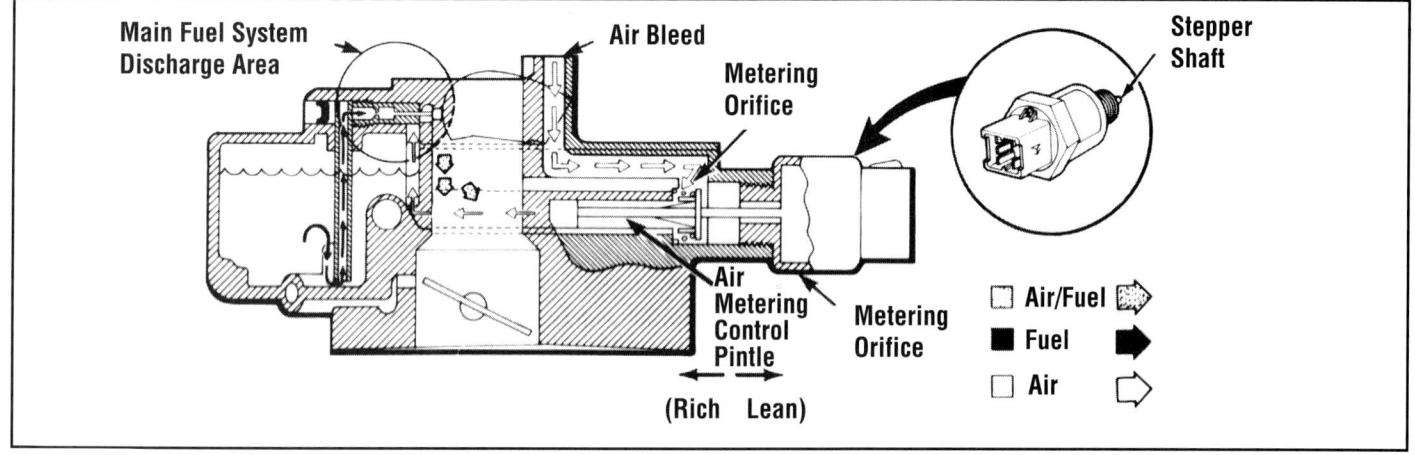

Instead of controlling the mixture by opening and closing a fuel jet, this back-suction feedback carb uses a stepper motor that varies the amount of atmospheric pressure that bears on the gasoline in the bowl (Ford).

relative to "off" time is what determines the richness of the blend. This is called the *duty cycle*, and can be measured with an ordinary dwell meter, which is available at most auto supply stores.

Ford Back-Suction—Other ways of getting the carburetor to obey computer commands were used, too. For example, the Ford back-suction type (essentially a 2700 Variable Venturi) has a stepper motor mounted on its right side that moves a vacuum metering rod according to signals from the EEC-III (Electronic Engine Control) computer. The position of the metering rod in its orifice regulates the amount of vacuum that's applied to the fuel bowl. The larger the opening, the greater the vacuum and the leaner the mix because it is harder for the gasoline to leave the bowl.

The motor has 100 steps within its 0.4-inch range and four separate coil windings. The microprocessor energizes the windings sequentially to put the metering rod where it wants it.

Adjustments—Besides setting the choke and fast- and curb-idle speeds, the only adjustment that's normally available on a carburetor is that of the idle mixture. But starting in the early 1970s, limiter caps were installed over the idle mixture screws at the factory. They limit the amount the screws can be turned, thereby keeping the blend within acceptable parameters. These caps were typically pried off during service, usually to make the setting richer (although restrictions in the passages still kept things within reason compared to pre-control carbs), but a reasonably smooth idle was actually obtainable within their limits.

On later models, especially those with feedback/closed-loop systems, the mixture screws are far more tamper-proof. For example, they may be concealed under a plug that must be drilled out. Unless you have the proper equipment and service information, it is unwise to attempt to make idle mixture adjustments in ordinary service.

FUEL INJECTION

Whether electronic or mechanical, fuel injection is an extremely accurate means of supplying an engine with gasoline. In the case of the more popular EFI (Electronic Fuel Injection), this precision is due in a large part to computer logic, which takes several factors into account before determining how much gasoline to provide. Compared to a carburetor, EFI's advantages include:

•Better fuel economy and lower levels of air pollutants

•Total fuel cut-off during deceleration in most cases

•Total fuel cut-off when the ignition is switched off, which prevents dieseling or run-on

•Faster starting

•Quicker response to load changes

•Fewer service requirements and adjustments

•It's easier to add closed-loop controls to fuel injection systems than to a carburetor

Mechanical Fuel Injection

While most cars manufactured today have electronic fuel injection as standard equipment, there's a mechanical system that's still on the road in large numbers: Robert Bosch K-Jetronic (the "K" stands for the German word for "continuous"), also known as CIS (Continuous Injection System). Since it appeared in the early 1970s, it's gained an excellent reputation for efficiency (it allowed some makes to pass exhaust emissions tests for several years after 1975 without the addition of a catalytic converter) and reliability.

There are variations found on several late-model cars—KE-Jetronic and K-Jetronic with Lambda—which work with the engine management computer to

Robert Bosch K-Jetronic/CIS was popular on European cars because it's much more precise than a carburetor, but it's been superseded by L-Jetronic/EFI systems.

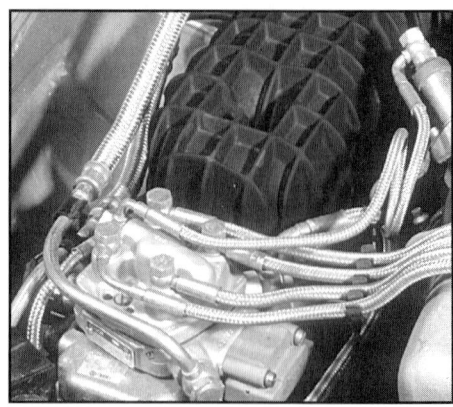

In K-Jetronic, the amount of fuel reaching the injectors is controlled by this unit.

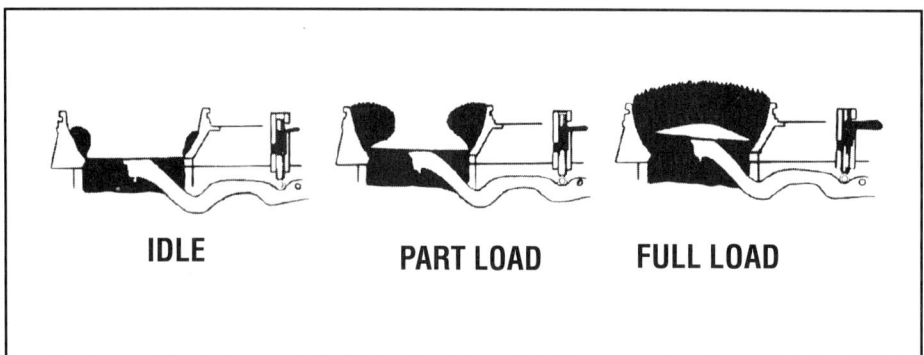

CIS senses the volume of air entering the intake by means of a floating disk, and mechanically adjusts the amount of gasoline injected (Volvo).

keep the air/fuel mixture within an even stricter range (the variation is 0.02%).

Components—The major components of K-Jetronic are an electric fuel pump, a fuel pressure accumulator, a mixture control unit, a cold start valve, and an injector aimed at the back of each intake valve. The fuel supply system includes the pump, which is mounted in or near the tank and has the capacity to supply far more gasoline than the engine can use under any condition (excess fuel is diverted back to the tank by the pressure relief valve in the fuel distributor); the spring-loaded accumulator, which helps hot starting by maintaining residual fuel pressure while the engine is off and reduces pump noise by damping pressure waves; and a large fuel filter mounted in the line between the accumulator and the fuel distributor.

The mixture control unit is what gives the system its accuracy. It contains an air flow sensor (basically, a metal disk floating in a cone) that rises as the engine demands more air. The disk is connected by means of a lever to the fuel distributor, which contains a barrel-and-slits type fuel metering valve. The valve opens according to the position of the air flow sensor, thereby providing the right air/fuel mixture for the conditions encountered. Pressure regulators (one for each cylinder) maintain a constant pressure difference between the inlet and outlet sides of the metering slits.

From there, the gasoline travels through separate lines to the injectors, which spray a mist continuously into the intake ports (a fundamental difference between K-Jetronic and other fuel injection systems such as Bosch L-Jetronic, in which the injectors spray only when pulsed by the electronic control unit). The fuel "cloud" that's always present when the engine is running is drawn in when the intake valves open. Each injector has its own fuel filter and a vibrating pin that helps atomize the gasoline droplets. Also, the pin and a spring hold about 45 psi of residual pressure to ensure quick starting.

An extra injector, the electrically operated cold start valve, is energized when engine temperature is low, which causes it to open and supply additional fuel for starting. The thermo-time switch controls the circuit to the cold start valve to keep the engine from flooding if it doesn't start immediately. While the engine is warming up, the control pressure regulator maintains a richer-than-normal mixture, and the auxiliary air regulator provides the fast idle speed needed to keep the engine running until it becomes warm enough to be efficient.

Electronic Fuel Injection

It may seem hard to believe, but electronic fuel injection has been used in considerable numbers since the late 1960s—the VW Type III was equipped with it in 1968. And 11 years before that, the very first production system appeared as a very rare option on the Chrysler 300 (only a few hundred of these Bendix systems were actually built).

In spite of this long history, a distressingly high percentage of car buffs, ordinary motorists and even some auto service technicians still don't understand how EFI works. So, we'll explain the parts and operation of a typical Robert Bosch system (such as the popular L-Jetronic EFI—the "L" stands for "Luftmengenmessung," German for "air flow management"). Many Japanese cars use Nippondenso versions, but they're manufactured under a Bosch license, so they are similar. And all other EFI types, both domestic and foreign, share the same principles.

How It Works—The most basic point to be understood is that, unlike the mechanical K-Jetronic system just described that supplies fuel through its nozzles continuously, an L-Jetronic injector is actually a little electrical solenoid valve that opens when it gets a signal from the electronic control unit (actually a computer). It's either on or off, open or closed. In other words, it either allows fuel to flow at a particular rate, or none to flow at all, according to whether the solenoid that controls it is energized or not. Fuel is supplied to the injectors at a constant pressure regardless of operating conditions, so the amount of gasoline delivered depends on how long the injectors are energized.

Just as the correct amount of light for a good exposure can be allowed to strike the film in an adjustable camera by setting the shutter speed (the "on" time), so EFI provides the right amount of gasoline by how long the injectors are held open—volume is a function of "on" time. A range of 1/1,000th to 7/1,000th of a second is common, and this interval is known as the "pulse width." In our Bosch example, there's an injector for each cylinder, and they're aimed at the intake valve. While some EFI's trigger them individually or in groups, the typical L-Jetronic fires them all at the same time (once each crankshaft revolution). It was found that the mixture was stored for such a short time in the manifold before the intake valve opened that it was unnecessary to add the complications that would be required to have each injector spray individually at the beginning of its cylinder's intake stroke.

Sensors—The method of controlling the injectors is what gives the system its accuracy. The general idea is that the computer decides how long to pulse the injectors on the basis of reports it gets from various sources. One of the most important of these is the intake air sensor, a housing containing a rotating flap, which is connected to a variable resistor. This changes the voltage signal to the control unit according to the volume of air that is actually entering the intake manifold. The greater the flow, the farther the flap rotates, and the lower the resistance. Other EFI systems may use a manifold vacuum sensor, the more recent air mass sensor or the Karmann-Vortex sensor to provide similar information.

A thermostatic resistor located in the intake air sensor gives the computer data on the temperature of the incoming air. On certain applications equipped with heated air intake systems, this is replaced by an ordinary resistor.

A temperature sensor screwed into the cylinder head gives the control unit additional information. Its internal resistance varies with engine temperature, and the computer monitors this change so it can pulse the injectors longer, thus richen the mixture, while the engine is warming up.

Another input comes from the throttle switch. This isn't variable as on some other EFI systems, but an on-off switch that signals idle or full-throttle using two sets of points connected to the throttle shaft.

Finally, the computer is informed about

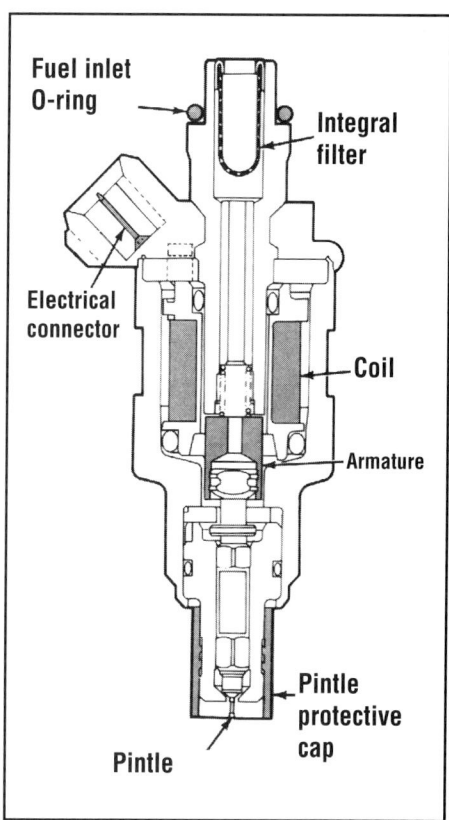

EFI injectors are really just solenoid-operated fuel valves that are either on or off (Ford).

Some EFI systems use a rotating-vane air flow meter to tell the computer how much of the atmosphere the engine is ingesting.

engine speed by the ignition system.

Fuel is supplied to the injectors under constant pressure by an electric roller cell pump and a fuel pressure regulator that responds to manifold vacuum and returns excess gasoline to the tank. A set of points in the intake air sensor assures that

83

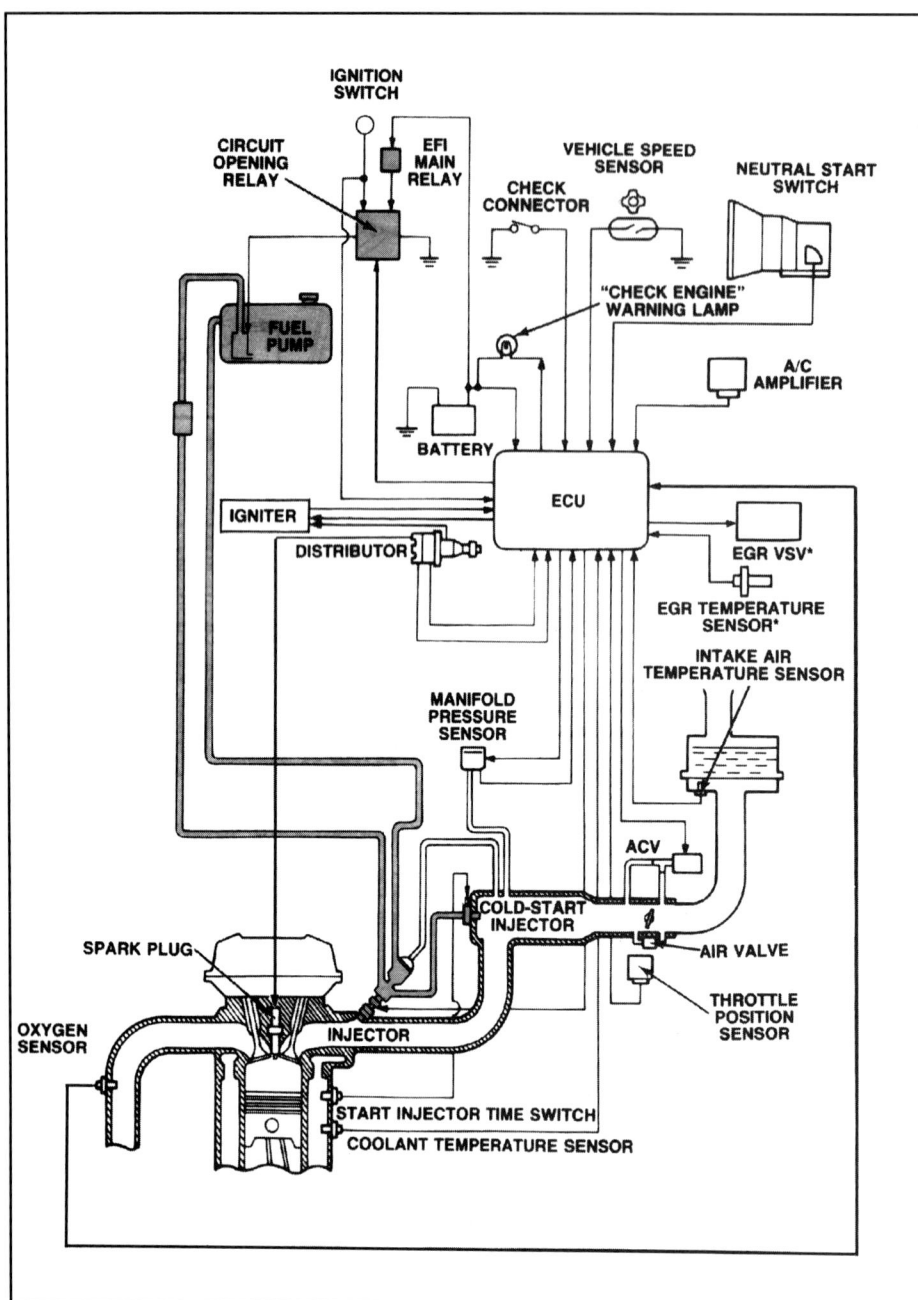

Here's a schematic of a modern EFI system (GM).

Closed-Loop EFI

Since the early 1980s, one advantage of EFI has become especially important: Since it's always used electronic computations to determine the amount of fuel that needs to be injected to maintain a proper mixture, it's not particularly difficult to incorporate a closed-loop/oxygen sensor feature into it that can keep a three-way catalytic converter (the type that both oxidizes carbon monoxide and hydrocarbons, and reduces oxides of nitrogen—see Chapter 9) operating efficiently. Where automakers that used carburetors had to engineer whole new systems with feedback solenoids and computers in order to achieve lower output of oxides of nitrogen, those who were already using L-Jetronic simply had to add an oxygen sensor (Bosch calls it a "lambda probe," lambda being the Greek letter that has come to represent the ideal, stoichiometric ratio of air to fuel) and expand and reprogram the computer so that if the sensor reports that the mixture is too rich, the injectors receive a shorter pulse, and vice versa.

THROTTLE BODY INJECTION (TBI)

Hybrid, half-breed, or compromise—whatever you may choose to call it, throttle body injection (TBI) resides somewhere on the evolutionary scale between the carburetor and individual port systems.

TBI has helped the automakers through a difficult transition because it works fairly well and is relatively inexpensive. It's simpler and does a better job than a feedback carburetor, and, while not capable of providing a perfect blend because it allows some stratification of the mixture to occur on the trip through the manifold, it's much cheaper than individual port injection.

A typical TBI system consists of many of the same components as multi-port,

the pump only gets current when the engine is running or being cranked. This eliminates flooding in the event that an injector sticks in the open position.

Three components that aren't connected to the computer assist in starting and warm-up. The cold start valve in the intake manifold injects extra fuel while the engine is being cranked. It's controlled by the thermo-time switch, a thermostatic device that energizes the cold start valve for three to ten seconds.

Its bimetal strip is affected by both engine temperature and an electrical heating coil. The auxiliary air regulator bypasses the throttle plate to provide extra air during warm-up, which prevents stalling. A heating coil inside the unit causes the air passage to close gradually.

Carbon monoxide adjustment is handled by turning a screw that controls the size of an air bypass through the intake air sensor. Backing the screw out lowers CO.

TBI (Throttle Body Injection) has been popular because it's relatively inexpensive and trouble-free. It's sort of a cross between a carburetor and MPFI (Multi-Port Fuel Injection). This is a GM version.

L-Jetronic-style systems, but instead of having an injector for each cylinder, there's just one, or, in some cases, two (as in the Corvette's Cross-Fire injection system), mounted up in the throttle body. This has the advantage of reducing the number of parts and positioning the injector away from excessive heat, which reduces internal deposit buildup.

Rochester Central Port Injection (CPI)

Originally installed on the optional L35 version of the 4.3L Chevrolet Vortec V6 in 1992, CPI is an ingenious way of getting port injection performance at a TBI price (50 more horsepower and 30 ft-lbs. more torque than the TBI version of the same engine with no increase in fuel consumption).

The breakthrough idea is to use one TBI-style injector to pulse fuel directly to the individual intake ports through some simple plumbing. That eliminates the main disadvantages of TBI: mixture stratification and condensation on the inside of the manifold runners during the long trip to the intake ports. This situation has limited the freedom of design of the manifold, so optimum performance hasn't been available with the injector placed way up in the throttle body.

Although it's not sequential, there's only a minor incremental gain to be had from that added complication because the plume of atomized fuel remains suspended in the port for a very short time before the intake valve opens—it doesn't have a chance to condense or disperse.

The single, disk-type, bottom-feed injector, which is mounted along with its fuel pressure regulator in a molded phenolic housing, receives gasoline through a 20 micron filter, and meters it through six orifices into the flexible nylon tubes that run to poppet nozzles at the ports.

Each individual nozzle is similar to that of a K-Jetronic/CIS, or even a diesel, but smaller and less expensive. It's calibrated to open at about 50 psi, and the spring that holds the valve mechanism closed acts as a final fuel filter.

Atomization is actually superior to that of a typical multi-point system, and since the poppets are much smaller than individual port injectors they can be aimed more accurately.

This is a speed-density system—no MAF (Mass Air Flow) sensor or vane-type air flow meter is needed. Theoretical mass air flow is calculated using MAP (Manifold Absolute Pressure), temperature, and rpm input. Injection is controlled by pulse width modulation. In other words, the frequency is constant and the amount of "on" time varies with the duty cycle. It's of the "peak-and-hold" variety—four amps opens the injector, one amp keeps it open.

You may hear CPI's injection process referred to as "simultaneous double fire." That means one-half of the total injection event occurs per revolution. ∎

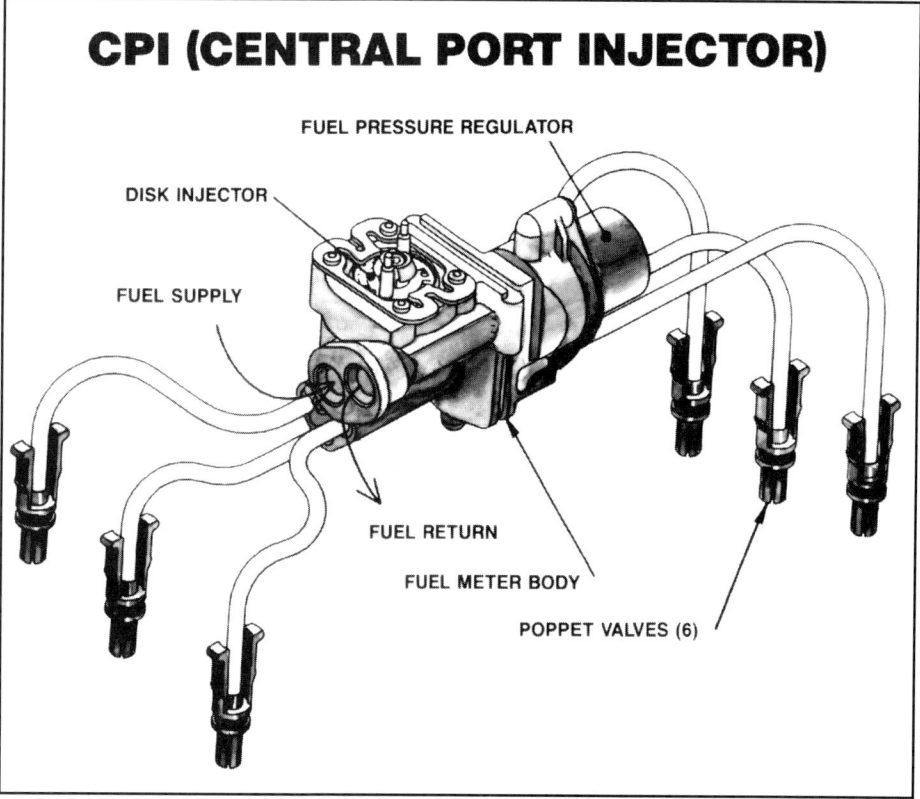

Chevy's CPI system, introduced in 1992, was so efficient that it produced an additional 50 horsepower and 30 extra ft-lbs. of torque over the TBI system, without using more fuel.

12
HOW IGNITION AFFECTS EMISSIONS

In this chapter, we're going to look at how ignition performance, timing, spark advance and retard affect emissions. To understand timing's role in emissions, we first need to review the ignition system's role in combustion.

First and foremost, when there's no ignition there's no combustion—except in a diesel engine which uses compression to ignite the fuel, or in cases of abnormal combustion, such as preignition, where hot spots cause the fuel to spontaneously ignite. And when there is no combustion in a cylinder, you're going to have raw unburned fuel in the exhaust and hydrocarbon emissions in the thousands of parts per million.

For the fuel mixture to burn, therefore, the ignition system must be capable of producing a reliable hot spark. This means the ignition coil (or coils in the case of distributorless ignition systems) must be capable of delivering sufficient firing voltage to the plugs when the spark is needed. The "secondary" ignition components (distributor cap, rotor and wires) must also be capable of transferring the voltage to each plug. Cracks or carbon tracks on the coil, distributor cap or rotor can all short-circuit the voltage before it reaches its destination. So too can an excessive air gap between the rotor and cap and excessive resistance in the plug wires. And if the spark plugs are fouled, worn out, incorrectly gapped or shorted, no spark will occur to ignite the fuel mixture. That's why the ignition system must be in perfect condition for the engine to fire reliably.

One of the most common causes of

Ignition plays a significant role in what comes out the tailpipe. Misfiring spark plugs are a common cause of excessive HC emissions.

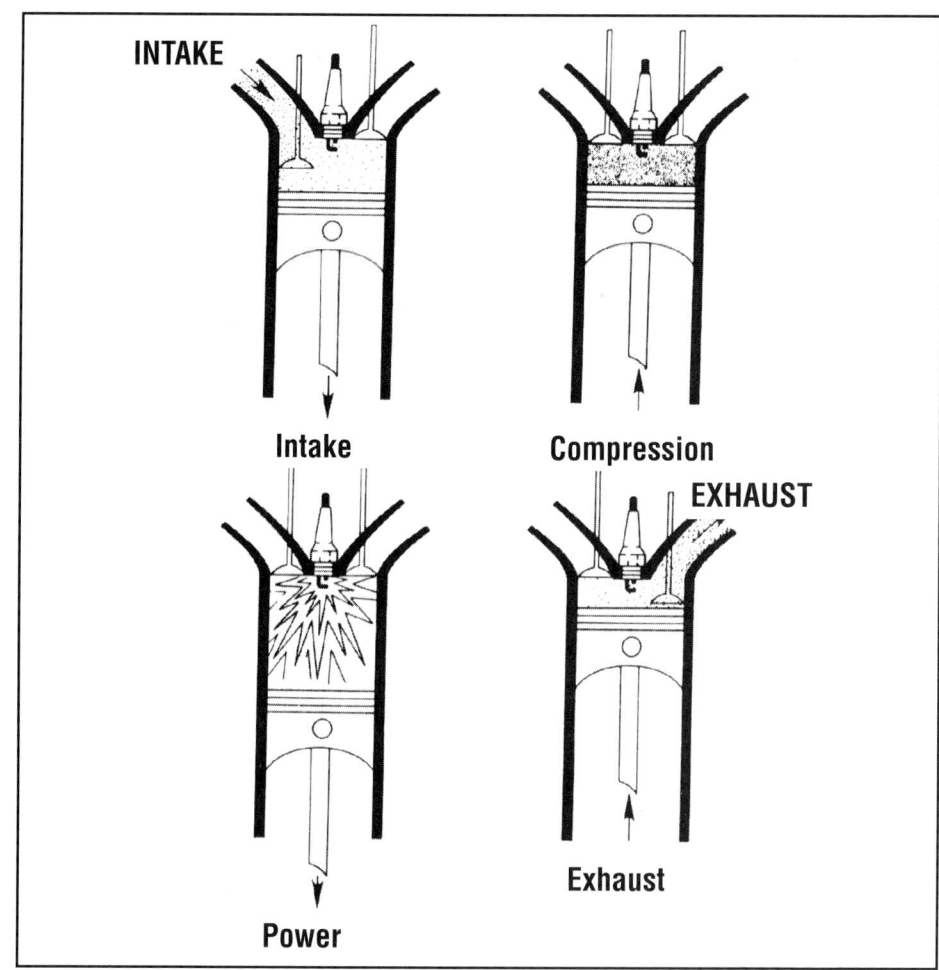

Basic four-stroke combustion process.

abnormally high hydrocarbon emissions, therefore, is ignition misfiring. So unless your engine has a bad valve that's leaking compression (which a simple compression test will reveal), or a fuel problem such as a vacuum leak or bad injector that's leaning out the fuel mixture to the point where it won't fire reliably (a condition known as "lean misfire"), check the condition of the various components in your ignition system. You'll most likely find the problem here.

ADVANCE TIMING

Another way ignition affects emissions is through spark or timing advance. The instant at which a spark plug fires, with respect to the relative position of the piston during the piston's compression stroke, determines the amount of timing advance. All engines require a certain amount of timing advance so the fuel can finish burning before the exhaust valve opens. In a four-stroke internal combustion engine, the engine goes through four basic strokes to produce power: intake, compression, power and exhaust.

The Four Strokes

On the intake stroke, the intake valve opens as the piston travels downward in the cylinder. This creates a vacuum and pulls a fresh charge of air and fuel into the cylinder. When the piston reaches the bottom of its travel, it reverses direction and starts back up. This is the beginning of the compression stroke. The intake valve closes so that the air and fuel will not be pushed back out of the cylinder. As the piston continues its upward travel, it compresses the air and fuel. This helps to mix the air and fuel more thoroughly and raises the temperature of the mixture so that it will burn more readily. As the piston reaches the top of its travel, the fuel mixture is ignited by a spark. This starts the mixture burning, which produces heat and pressure. The hot expanding gases push the piston down while they continue to burn. This is the power stroke. As the piston reaches the bottom of its travel and starts back up again, the exhaust valve opens so that the piston can shove the exhaust gases out of the cylinder. This is the exhaust stroke. The exhaust valve closes as the piston reaches the top of the cylinder (called top dead center, or TDC). The intake valve then opens as the piston starts back down again to begin the process all over.

At Idle—At an idle speed of 600 rpm, the crankshaft is spinning around 600 times a minute or 10 times a second. Each piston is also reciprocating up and down at the same 10 times a second rate. Since combustion occurs only on the power stroke every other revolution, ignition takes place five times a second at 600 rpm.

The air/fuel mixture has plenty of time to burn at idle because the duration of the power stroke at 600 rpm is on the order of 1/20 (0.05) second. The actual combustion of the air/fuel mixture takes place in about 1/200 (0.005) second, so you can see that there is plenty of time for the fuel to burn and expand before it is pushed out of the cylinder on the exhaust stroke.

At highway speeds, a typical engine might be turning 3000 rpm. At this speed, the pistons will be traveling up and down at the rate of 50 times a second. With ignition taking place every other revolution, combustion occurs 25 times a second. The time available during the power stroke at 3000 rpm is now on the order of 1/100 (0.01) second, still enough time for the fuel to complete burning on the power stroke but not enough time for cylinder pressures to reach a maximum at the right point in the power stroke.

For an engine to deliver maximum fuel economy and power, cylinder pressure should reach a maximum early in the power stroke while the piston is accelerating downward. If the instant of maximum pressure occurs too late in the power stroke, it will not produce as much horsepower. And if maximum pressures are not achieved until very late in the power stroke, some of the oomph will be lost out the exhaust valve when the exhaust stroke begins. Timing, therefore, plays a very important role in both power output and economy.

To give the air/fuel mixture sufficient time to burn so that maximum cylinder pressures can be achieved at the best point in the power stroke, the ignition timing is advanced before top dead center. Instead of happening exactly at TDC as one might assume, ignition takes place farther and farther before TDC as engine speed increases. In other words, ignition takes place toward the end of the compression stroke

For example, a typical engine at idle may be timed at anywhere from several degrees ATDC to 10 or 12 degrees BTDC. Remember that at idle there is plenty of time for combustion, so little or no initial advance is needed. At 3000 rpm, however, considerable advance is needed to achieve maximum cylinder pressure early in the power stroke. Such an engine might have 28 to 34 degrees of ignition advance, the amount of advance increasing in proportion to engine speed.

Now that you have a little better picture of what's going on inside the combustion chamber, let's review some basic terminology regarding the ignition system:

- If ignition occurs exactly at top dead center, there is zero degrees of timing advance.

- If ignition occurs after TDC, the timing is

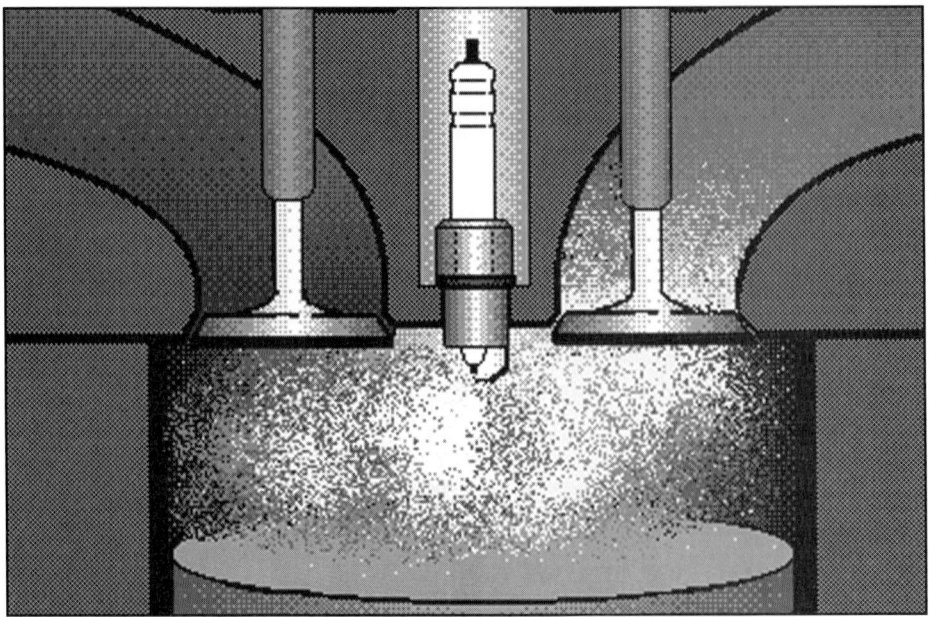
During normal combustion, the flame-front propagates outward from the spark plug like an expanding balloon.

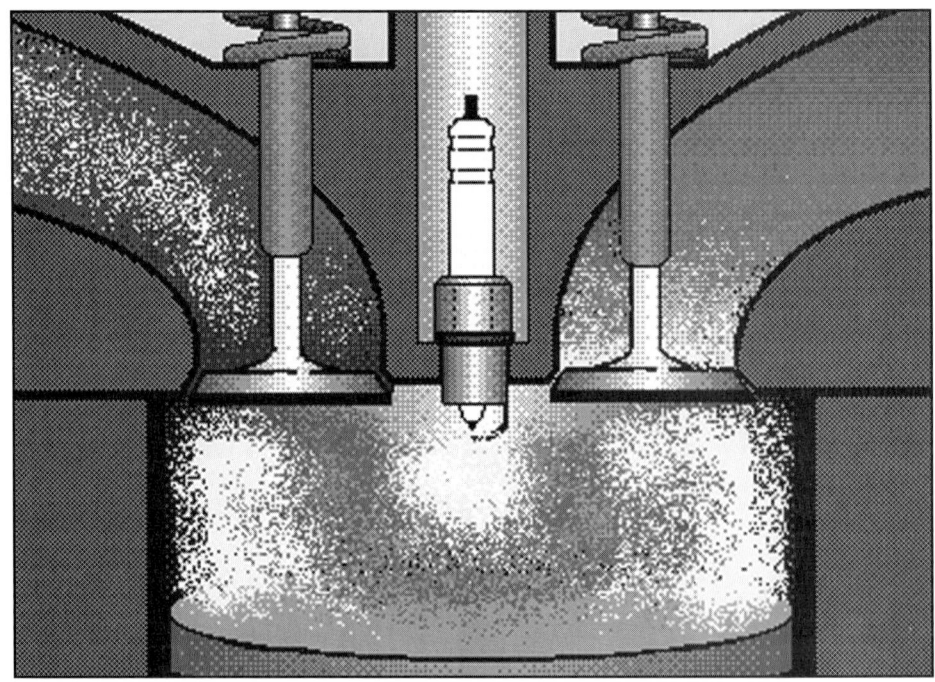
When detonation occurs, multiple flame-fronts are generated spontaneously from heat and pressure in the combustion chamber. When they collide, it produces a sharp metallic pinging or knocking noise that can be quite damaging.

said to be "retarded."

- If ignition occurs before TDC, the timing is said to be "advanced."

- "Initial" timing (also called "basic" timing) is the amount of advance or retard the ignition has at idle as set in accordance with the manufacturer's instructions. When setting timing on vehicles with vacuum advance distributors, this usually requires disconnecting the vacuum advance hose from the distributor.

- Increasing the amount of ignition advance is called advancing the timing. Decreasing the amount of advance is called retarding the timing.

Detonation

The higher the rpm, the more timing advance an engine needs to maximize power output and fuel economy. The timing must advance in proportion to the rpm rate. If the fuel is ignited too far in advance, however, the pressure of the expanding gases rises too quickly and peaks before the piston can respond. This causes detonation and increases HC emissions.

Think of combustion as an expanding balloon. Under normal circumstances, the flame kernel expands and fills the combustion chamber outward from the spark plug. When there is too much advance, though, the rapidly increasing pressure inside the combustion chamber causes fuel to ignite spontaneously in other areas of the chamber. This is akin to several balloons expanding all at once. And when the flames collide, they do so with great force. This produces the sharp knock or ping noise that is characteristic of detonation. Under severe circumstances, detonation can crack or punch holes in pistons, crack heads and piston rings, flatten connecting rod bearings, and blow out head gaskets. Hydrocarbon emissions are also increased because erratic combustion leaves little pockets of unburned fuel.

Achieving a Balance—For optimum performance and power, an engine should be timed so that it is just on the verge of detonation. The only drawback to this approach is that an engine's resistance to detonation changes. It can vary with the quality of gasoline used, the additives in the fuel or be affected by variations in humidity and temperature. High relative humidity increases the effective octane rating of gasoline while dry weather does just the opposite. What this means to real-life driving is that an engine that's tuned for optimum performance on a rainy day may tend to knock and ping during dry weather. To play it safe, therefore, the manufacturer's recommended timing settings have a built-in safety margin to account for some variation in fuel quality and weather. Late-model cars with computerized engine controls can push these limits somewhat if they are equipped with a knock sensor. This device picks up engine vibrations that are produced when detonation occurs, signaling the computer to momentarily retard timing until the detonation ceases.

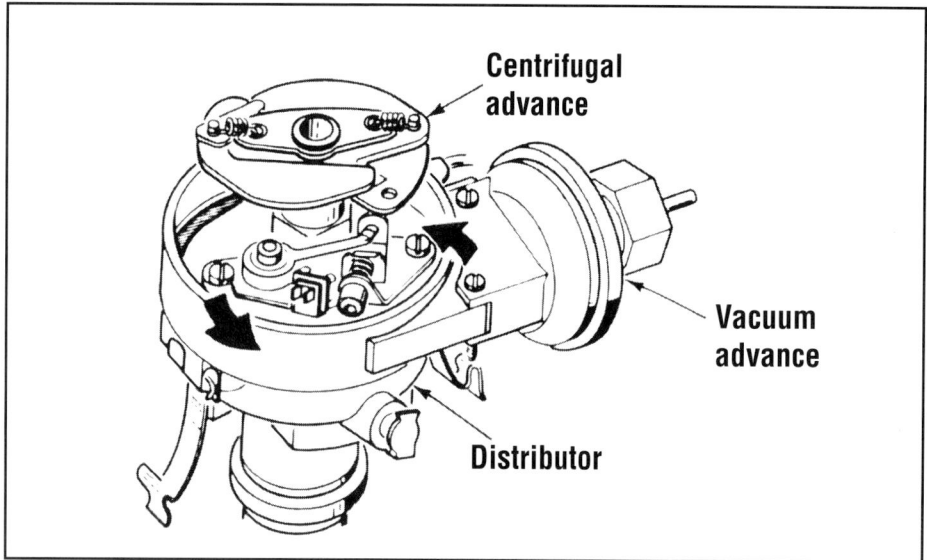

Older distributors had relatively simple mechanical and vacuum spark controls (Chevy).

Centrifugal Advance

As described earlier, the amount of timing advance must increase in proportion to engine speed. This is called centrifugal advance and it is accomplished in two ways: mechanically or electronically.

In distributors with centrifugal advance mechanisms, two small spring-loaded flyweights control the rate of advance. As engine rpms increase, the weights are thrown outward against spring tension. The movement of the weights rotates the rotor and trigger wheel (electronic ignitions) or breaker cam (point ignitions) into an advanced position, which fires the coil and spark plugs sooner to advance timing. The size of the weights and the strength of the springs that resist the weights determine the rate of advance. Changing the weights and/or springs will change the advance curve.

Computer Control—On cars equipped with computerized engine controls and electronic spark timing, no flyweights or advance mechanism are used in the distributor to advance timing. It's done electronically by the computer or ignition module. The computer or module calculates the equivalent amount of "centrifugal" advance the engine needs based on rpm, then times the firing of the coil to create the required amount of advance. The amount of advance may be further modified by inputs from other engine sensors (such as the coolant sensor, MAP sensor, barometric pressure sensor or throttle position sensor), depending on the vehicle application and operating conditions.

Vacuum Advance

To improve fuel economy, most engines also employ "vacuum" advance, which differs from centrifugal advance in that it is engine load sensitive. Vacuum advance is added when there is a light load on the engine (like when cruising or decelerating), and subtracted when the engine is under load or accelerating at wide open throttle. By comparison, centrifugal advance is speed sensitive and changes with engine rpm, not load. So the

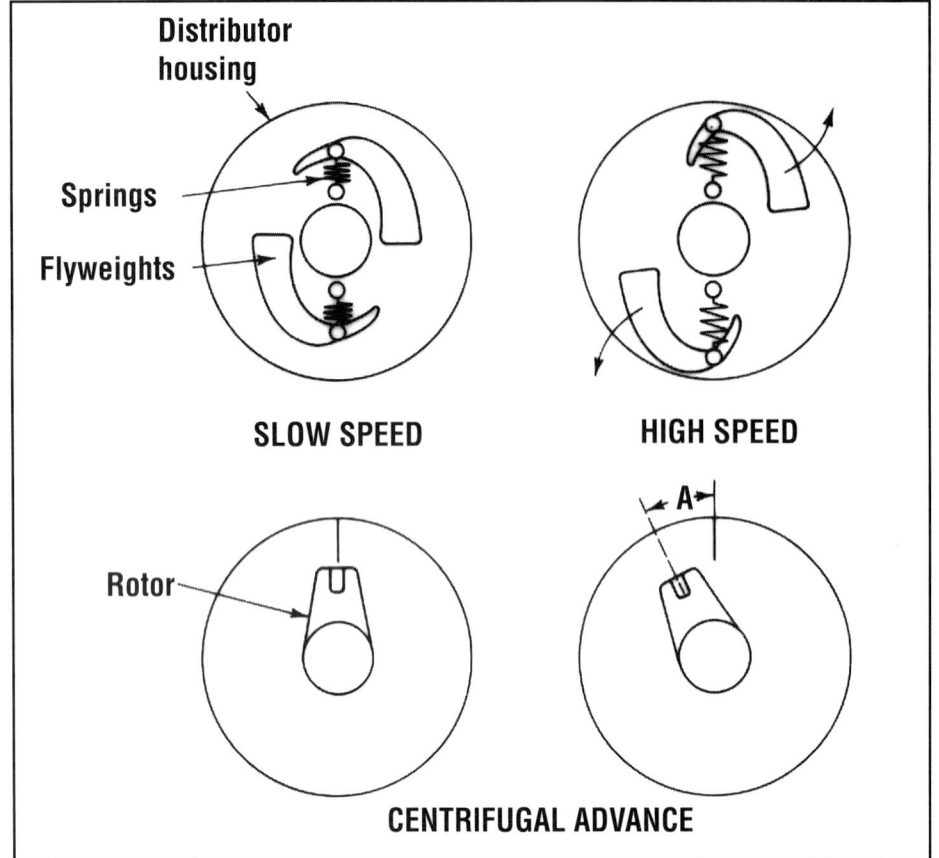

The outward movement of the weights in a centrifugal advance mechanism advances the relative position of the rotor, which opens the points or triggers the magnetic pickup sooner to advance timing.

two together allow ignition timing to be varied according to both changes in speed and load. Centrifugal advance plus vacuum advance equals the "total" amount of timing advance the engine has at any given instant in time.

Since engine vacuum drops in proportion to the load applied, using or monitoring intake vacuum to control timing advance allows the timing to respond to the conditions. At idle, light load, and during deceleration, intake manifold vacuum is very high. As the throttle is opened wider, intake manifold vacuum drops. At full throttle there is very little manifold vacuum.

Signal—The vacuum signal for vacuum advance comes from one of three sources: intake manifold, ported vacuum, or venturi vacuum. With intake manifold vacuum, the vacuum hose is simply connected to a fitting on the manifold or the base of the carburetor or throttle body. With ported vacuum, the hose is connected to a fitting just above the throttle plates. At idle, there is no vacuum signal because the port is above the throttle plates (which are closed). As the throttle is opened, the port is exposed to intake vacuum. The vacuum signal passes through the hose and timing is advanced. On some vehicles, the vacuum hose is connected to a fitting on the carburetor that vents into the venturi. Reading engine vacuum at this point produces a faster response, but the venturi vacuum signal is typically too weak to move the distributor diaphragm. The vacuum hose from the carburetor, therefore, is usually connected to a vacuum amplifier to boost the strength of the signal.

The vacuum advance mechanism itself is fairly simple. A vacuum hose from the carburetor or intake manifold is connected to a vacuum diaphragm on the distributor. The diaphragm moves the breaker point or pickup plate to change the relative position of the plate and advance timing. Many older Fords have a dual vacuum diaphragm that both advances and retards timing according to changing throttle position.

On Chryslers with Lean Burn computers and later models with a vacuum transducer on the computer, no vacuum advance mechanism is used on the distributor. The same is true on other vehicles with computerized engine controls and electronic spark timing. The computer monitors engine vacuum through a transducer or MAP sensor to calculate the equivalent amount of vacuum advance that's needed. The computer or control module then alters the firing of the coil as needed to add or subtract advance according to engine load. The amount of advance that's added or subtracted may be further modified by other sensor inputs, such as the throttle position sensor and so on.

TIMING AND EMISSIONS

Believe it or not, ignition timing has a significant impact on tailpipe emissions. Generally speaking, retarded timing at

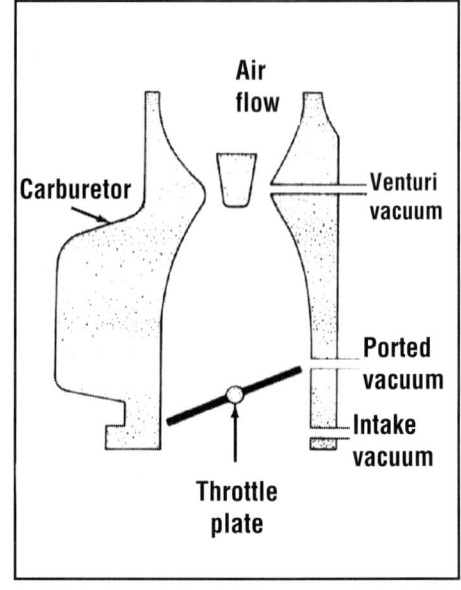

Vacuum for a distributor vacuum advance diaphragm can be taken from intake, ported or venturi vacuum.

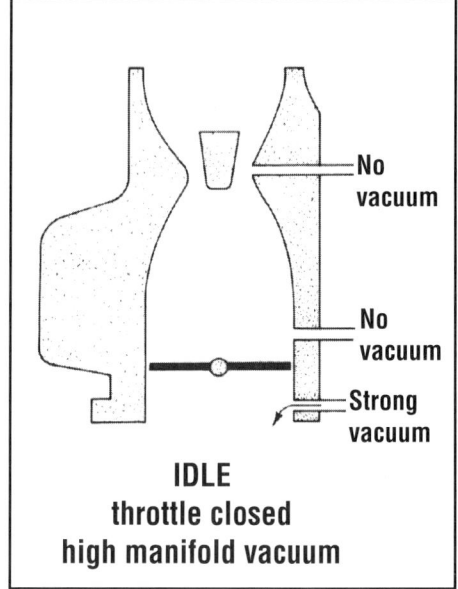

At idle, there is strong vacuum at the intake manifold connection, but no vacuum at the ported or venturi connections.

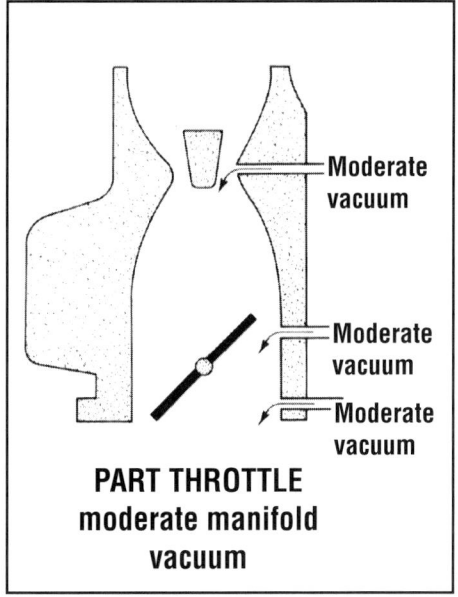

At part throttle, there is moderate vacuum at all three ports.

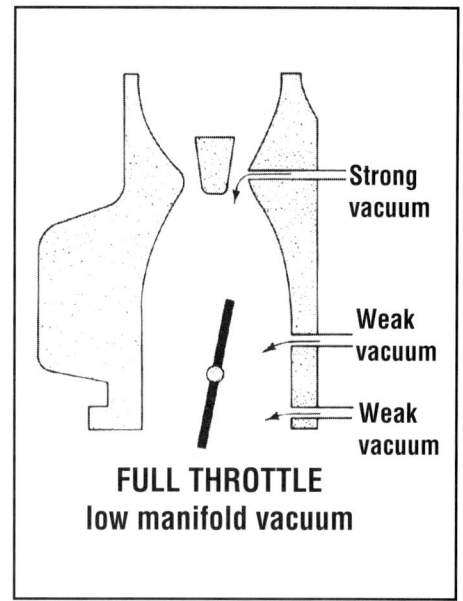

At wide open throttle, intake and ported vacuum drop but venturi vacuum is strong.

idle and during deceleration reduces hydrocarbon emissions. When the timing is retarded, combustion occurs later in the power stroke. This increases exhaust gas temperatures and promotes more complete burning of hydrocarbons in the exhaust. When the hot exhaust gases enter the exhaust manifold and meet fresh oxygen supplied by an air pump or aspirator valve, the unburned HC continues to burn. Add a catalytic converter to accelerate the process, and the result is almost complete combustion of any hydrocarbons that were not already burned inside the engine. But retarded ignition timing also requires a slightly wider throttle opening. This is necessary to increase the flow of air and fuel so that the idle speed can be maintained. The wider opening and increased flow promote better mixing of air and fuel. This aids combustion and reduces HC emissions.

Controlling Emissions

The easiest way to retard ignition timing during idle and deceleration to reduce emissions is to use ported vacuum advance. At idle there is no vacuum advance because the port is located above the throttle plates. During deceleration there is no vacuum advance either because again the throttle is closed.

Another way to retard ignition timing during idle and closed throttle deceleration is to use a combination of ported and intake manifold vacuum. The two vacuum sources are balanced against one another through a spring-loaded valve. Manifold vacuum reaches the distributor only when ported vacuum is

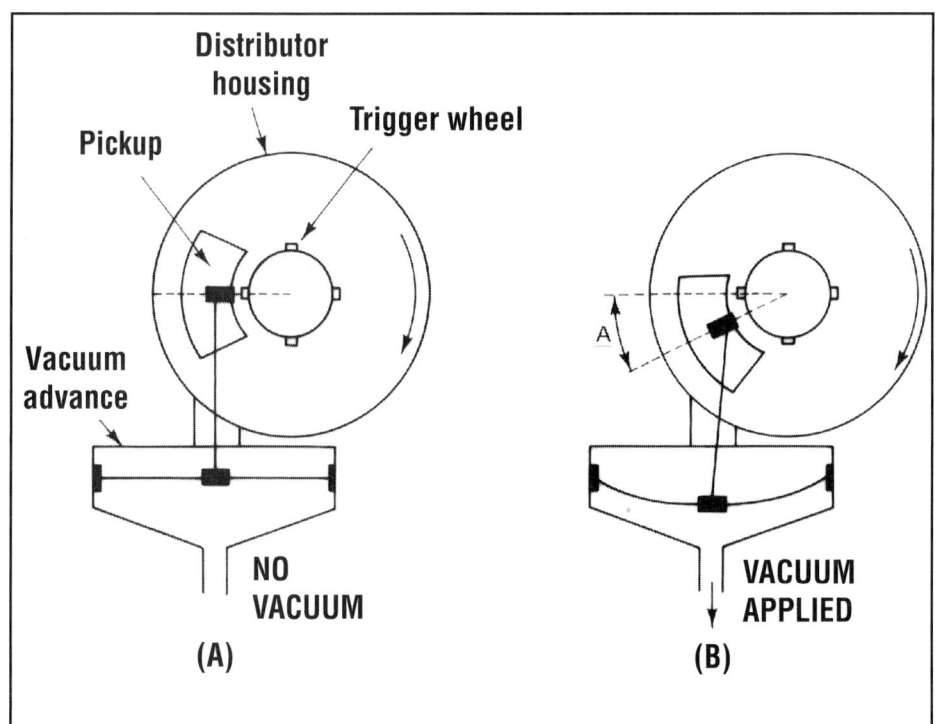

Vacuum pulls against the diaphragm and moves the breaker point or pickup plate to advance timing.

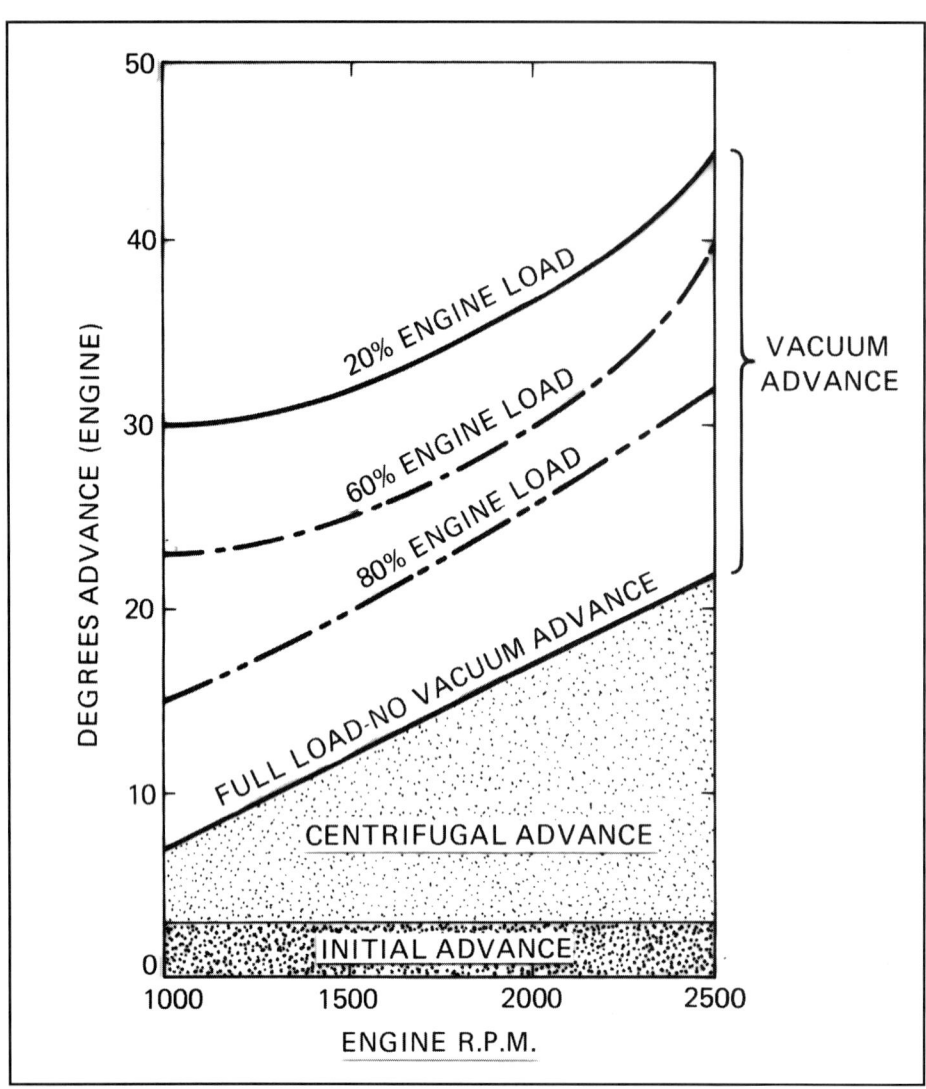

Ignition advance under various operating conditions. The total amount of advance will be the initial advance (basic timing setting), plus centrifugal advance plus vacuum advance.

sufficient to open the valve. On the older Ford dual vacuum diaphragm distributors, carburetor ported vacuum is connected to one side of the diaphragm and intake manifold vacuum to the other. At idle and during closed-throttle deceleration, ported vacuum is zero and intake manifold vacuum is high. The ported vacuum side of the diaphragm produces timing advance while the intake manifold vacuum side produces retard. Thus timing is retarded at idle and during deceleration when manifold vacuum is strongest, and advanced during other modes of operation when ported vacuum is strongest.

Ported Vacuum Switches—Many systems, including the Ford dual vacuum diaphragm unit just described, use ported vacuum switches (PVS) or temperature vacuum switches (TVS) to reduce emissions. A PVS or TVS installed in the vacuum line between the distributor and vacuum source prevents vacuum from reaching the distributor during certain modes of operation. For example, tailpipe emissions are lower on some cars if vacuum advance is restricted to cruising speeds only. A PVS is used to prevent the vacuum signal from reaching the distributor until the transmission is shifted into high gear. This approach is known as "transmission-controlled spark." A small electrical switch located on the shift linkage controls a ported vacuum solenoid. When the transmission is put into high gear, the switch grounds and energizes the solenoid, allowing vacuum to pass to the distributor.

Blocking Vacuum Advance—On many engines, emissions are reduced by blocking vacuum advance to the distributor until the engine reaches operating temperature. This is because a relatively cold engine causes droplets of fuel to condense on the cylinder wall surfaces. This increases unburned hydrocarbon emissions so that higher exhaust temperatures are needed to burn the HC. A "thermal vacuum switch" (TVS) is installed in the vacuum advance line to block vacuum until the engine reaches operating temperature. The TVS is screwed into the engine block or intake manifold so that the tip of the TVS is in contact with the engine's coolant. When the coolant reaches normal operating temperature, the wax plug inside the TVS expands and opens the line between the vacuum source and the distributor.

Some thermal vacuum switches work in just the opposite way. Some engines require a certain amount of vacuum advance when cold to improve idle quality and performance. On these applications, the TVS is designed to allow full vacuum advance until the engine is warmed up. The TVS then closes off the intake manifold vacuum line and opens a ported vacuum line.

On other applications, a TVS is used to perform yet another function. If the engine starts to overheat, the TVS opens up a line between intake manifold vacuum and the distributor temporarily to advance the timing. This increases idle speed, which in turn circulates coolant through the engine more quickly. As the engine cools back down to normal temperature, the TVS closes the intake manifold vacuum line and idle speed returns to normal. Such a switch is often called a coolant temperature override (CTO) switch.

Spark Delay Valve—Another device used to modify vacuum advance is the "spark delay valve." This device works like a restriction in the vacuum line. Depending on the application, the valve can prevent full vacuum from reaching the distributor for a few seconds up to a half a minute. The delay valve is used to prevent sudden ignition advance that can cause combustion temperatures to soar, and thus increase NOx formation. By delaying the advance for a short period of time and allowing it to build up gradually, you can avoid peak combustion temperatures and reduce NOx.

Spark delay valves go by a variety of names but all do basically the same thing. Older Chryslers use an "orifice spark advance control" (OSAC) valve. Ported vacuum from the carburetor is routed to the OSAC valve, then to the distributor. When a vacuum signal reaches the OSAC valve, it takes about 20 seconds for it to pass through the valve and reach the distributor. General Motors uses a "spark delay valve" (SDV) or a "vacuum delay valve" (VDV) for the same purpose. The SDV or VDV is located in the vacuum line between the ported vacuum connection on the carburetor and a TVS. The valves have a 0.005 in. orifice restriction, which delays the vacuum signal from reaching the distributor by about 40 seconds.

Many spark delay valves use a porous metal filter to restrict the flow of vacuum. Such delay valves are also fitted with a one-way rubber valve. The valve allows vacuum to escape from the distributor side of the line when the ported vacuum signal drops to zero (as during deceleration or idle). In other words, the valves restrict vacuum one-way only. Therefore, it is very important that these valves be installed facing the right direction. Most are marked "DIST" on the distributor side or "CARB" on the carburetor side. They are also color-coded according to the amount of delay they provide.

Spark delay valves can become clogged with dirt. If this happens, the delay period may be longer than usual, or there may be complete blockage of the vacuum signal. Delay valves can be tested with a hand-held vacuum pump. Apply vacuum to the carb side and see how long it takes for the reading to drop. Compare this to the manufacturer's specs to see whether or not the valve is doing the job it is supposed to do. When vacuum is applied to the distributor side, there should be no restriction and the reading should be zero.

ELECTRONIC SPARK TIMING

Since the 1980s, most vehicles have gone to electronic spark timing (EST) to keep emissions under control. Electronic spark timing takes electronic ignition one step further by totally eliminating the mechanical centrifugal and vacuum advance mechanisms. In many late-model cars with "direct ignition systems"

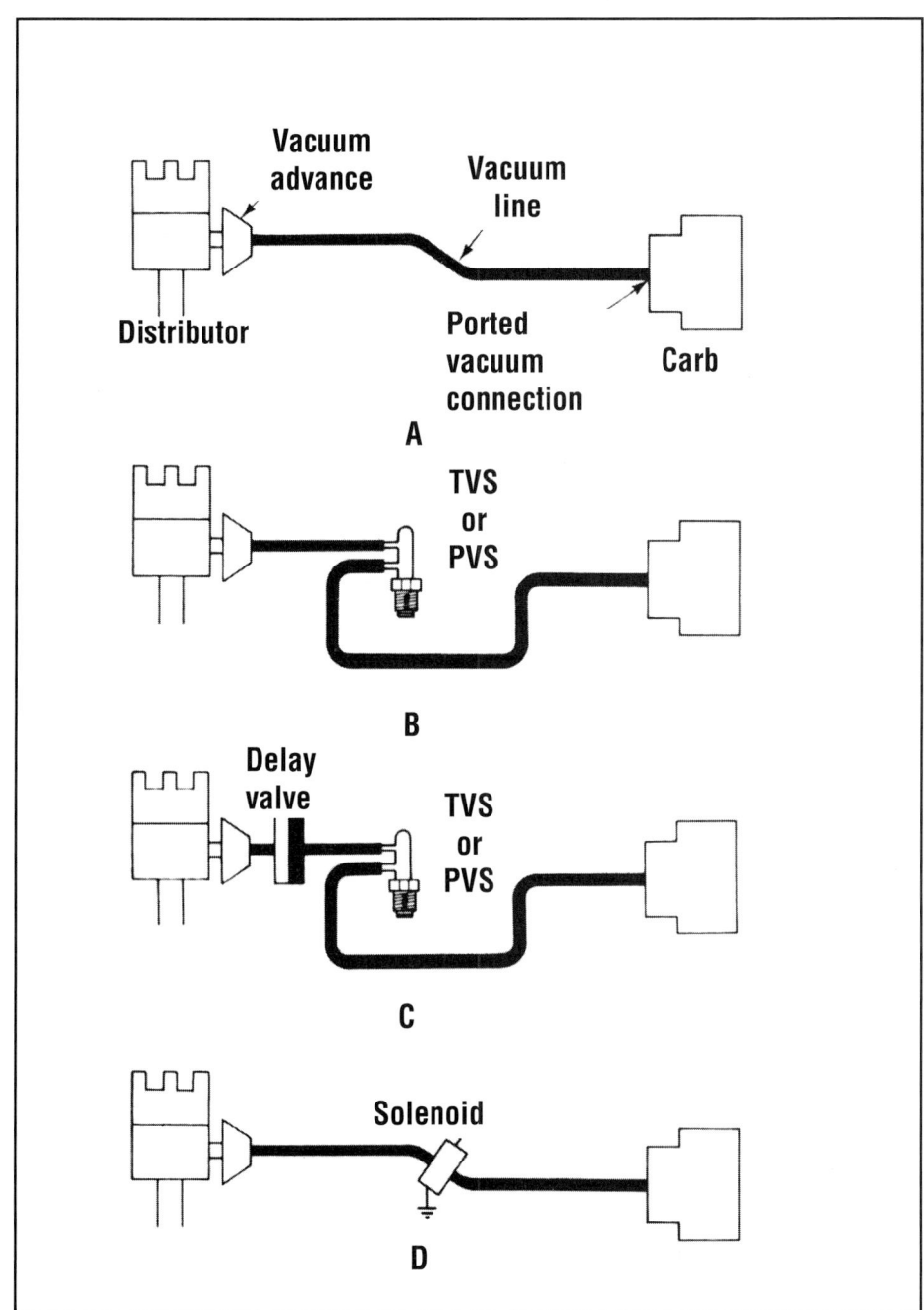

These are the various methods that may be used to control or delay vacuum advance to reduce emissions.

or "distributorless ignition systems" (DIS), it has also eliminated the need for the distributor itself.

How It Works

There are many different electronic spark control systems on the road today, but all share essentially the same basic operating principle. Various sensors keep the computer informed about engine speed, coolant temperature, manifold vacuum, throttle position, ambient air temperature, transmission gear position and anything else it needs to know to calculate the optimum spark timing based on the engine's current needs. This allows spark advance to be changed almost instantly and varied infinitely according to the pre-programmed spark curves in the computer.

Electronic spark timing also increases the reliability of the ignition system by eliminating mechanical components that can wear out. And it eliminates much of the complex plumbing associated with older mechanical vacuum spark control distributors.

Like electronic ignition, electronic spark timing is a "set it and forget it" kind of system. Once set, electronic spark timing will not change unless someone tampers with it or there's a malfunction in the control module or sensors. On some vehicles with distributorless ignition systems, base timing isn't even adjustable so all you can do is check it to see if it's within specs.

Modifying—Electronic spark timing curves cannot be altered unless the computer itself is replaced, or the computer's "Program Read Only Memory" (PROM) chip is replaced (or reprogrammed in the case of "Electronically Erasable Program Read Only Memory" or EEPROM chips), or a special "interface" control module is installed that alters the engine computer's inputs and outputs to alter ignition timing. Aftermarket performance PROMs have been a popular means of upgrading engine performance, but such products must be emissions certified for street use.

Troubleshooting—Troubleshooting electronic spark control systems is beyond the scope of this book. For that, you'll need a factory shop manual that details the specific diagnostic procedures which may in some cases also require special test equipment. Keep in mind, though, the basic principles of spark control. If the spark timing does not advance with increasing rpm (considering any built-in delays), something is wrong with the system. If an engine knocks and pings on acceleration, there is too much advance. Check basic timing, the operation of the knock sensor, the EGR valve and heated air intake system for malfunctions. If everything seems to be okay and the engine does not have too much compression due to carbon buildup, there is too much advance. A faulty throttle position sensor, coolant sensor, intake manifold vacuum sensor, or knock sensor may be at fault. ■

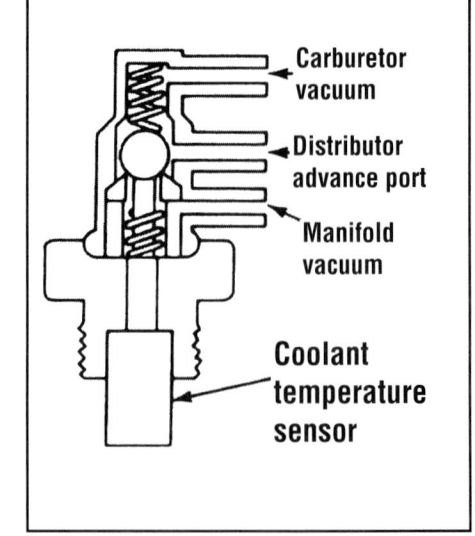

Here's what the inside of a ported vacuum switch looks like. Heat causes the wax inside the sensing element to expand and move the ball or plunger up and down to uncover or block various ports.

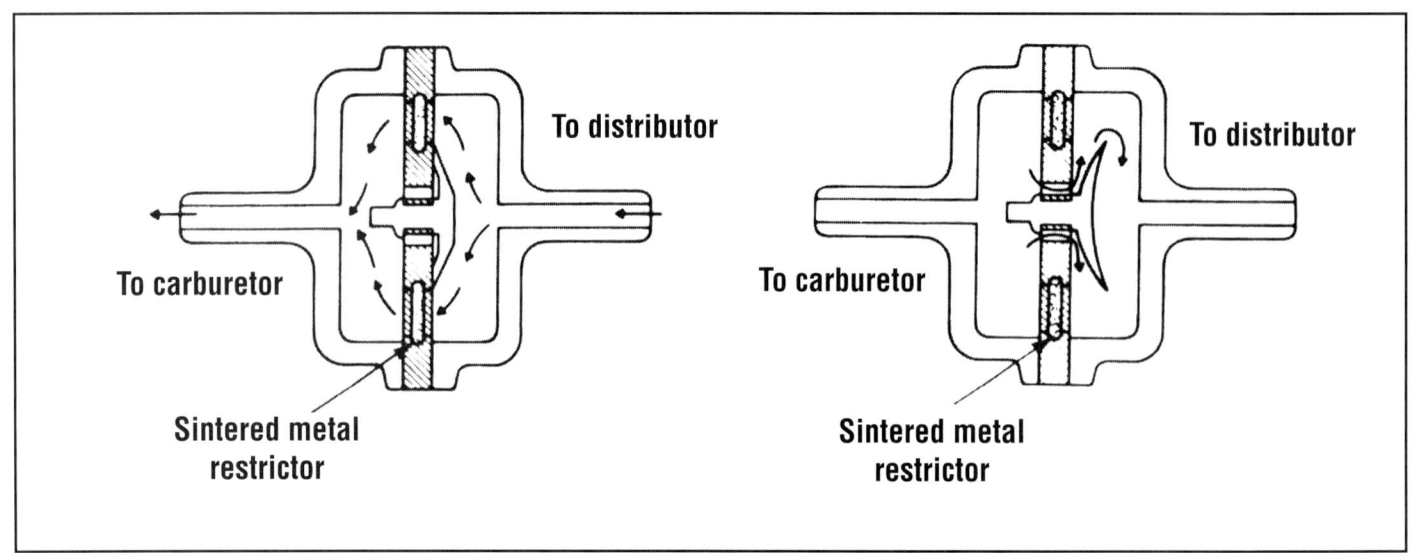

Spark or vacuum delay valves create a restriction in a vacuum line. This can delay vacuum from 1/2 up to 60 seconds or more. It's important to note that spark delay valves are directional, so make sure the arrow on the outside is pointing the right way.

13
ENGINE WEAR & OTHER FACTORS

In addition to the air/fuel ratio and ignition, there are a number of other factors that also influence the emissions produced by an engine. These include:

- Wear

- Valve and ring seating (compression & blowby)

- The type of valve guide seals used and their condition

- Valve stem-to-guide clearance

- The shape of the combustion chambers
- Camshaft duration and overlap

- Engine operating temperature

- The type of fuel used

ENGINE WEAR

As an engine accumulates mileage, internal clearances loosen up as a result of normal wear. The rings and valves don't seal as tightly as they once did. The rod and main bearings tend to throw more oil around inside the engine. Timing gears, chains and distributor drives get sloppy. Consequently, there can be a gradual rise in hydrocarbon emissions over time due to increased oil consumption and less accurate valve and ignition timing.

Regular oil and filter changes can slow the rate at which wear occurs as far as the rings, bearings and valve guides are concerned. But all engines, regardless of how much care they've had, will show some deterioration over time due to wear in the valve seats and valvetrain components.

Most engines that are in good running condition and are not burning oil should be able to pass an emissions test—

A loss of compression, either from poor ring sealing or improper valve seating, will create excessive HC and CO emissions. Photo by Jim Richardson.

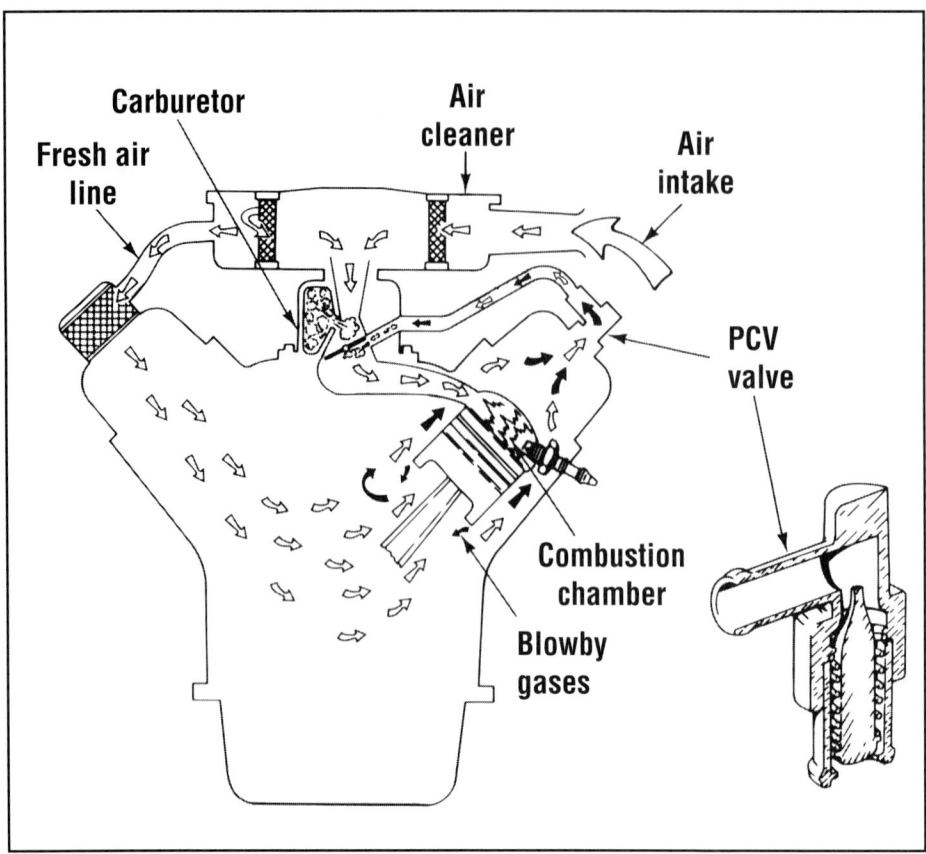

Combustion blowby in a high-mileage engine with worn rings can create serious emissions problems. The vapors will be siphoned back through the PCV system into the engine, creating a rich fuel condition and high CO emissions. HC emissions may also be elevated due to oil leaking into the exhaust through bad exhaust valve stem seals.

assuming, of course, that the fuel system, ignition system and emissions controls are all working correctly, too. But if your engine has anything major wrong with it, you're going to have problems passing an emissions test. If it's burning oil (more than a quart of oil in 500 to 1000 miles), if it has low or uneven compression, or if it's missing or running rough because of a bad valve or leaky head gasket, it will probably flunk an emissions test because of excessive hydrocarbon emissions.

Valve jobs and overhauls are expensive repairs that can cost hundreds of dollars, so if your engine has a major problem you may be able to qualify for a waiver depending on what the waiver rules are in your area. In California, for example, a state referee decides whether or not the estimated repairs that would be necessary to correct an emissions problem would exceed the waiver dollar limit. If they do, the vehicle is given a smog certificate anyway even though it cannot pass the emissions test. Many states have waiver limits that range from $50 to $300 depending on the year of the vehicle, so it doesn't take much in the way of major engine repairs to exceed the waiver limit. The new $450 waiver limit in I/M 240 inspection programs, however, makes it much more difficult to take advantage of this loophole in the law.

Compression & Blowby

Engine compression depends on three things: proper valve seating, proper ring sealing, and a good head gasket.

Valve Seating—Worn valves or seats, seats that are not concentric, valves that are not properly aligned with their seats, burned or damaged valves, sticky valves (caused by insufficient valve stem-to-guide clearance or engine overheating) and insufficient valve lash can all allow compression to leak past the valves. When this happens, unburned fuel enters the exhaust and HC emissions soar.

It's important to note that not all of these problems are the result of normal wear. Insufficient valve lash, for example, may be the result of maintenance neglect. Many older import engines do not have hydraulic valve lifters or lash adjusters. They require periodic valve lash adjustments to keep the lash from closing up—which it will over time unless normal wear is compensated for by the adjustment. If there's insufficient lash between the tip of a valve stem and its rocker arm, cam follower or camshaft (the latter two being found in overhead cam engines), the valve may not be able to close completely. This will prevent the valve from sealing the combustion chamber and also cause it to run dangerously hot because it can't conduct heat away from itself through the seat. Exhaust valves are especially vulnerable to overheating as a result of poor seating, and often "burn" or fail, causing a loss of compression and power in the affected cylinder, or in some cases a catastrophic engine failure if the valve head separates from the stem and smacks the piston.

Sometimes a freshly rebuilt cylinder head will not have as much compression as it should have because someone did a poor job refacing and matching the valves and seats. The valves and seats should be concentric (round to within about .001 inch for each inch of valve head or seat diameter), otherwise they will not mate properly when the valves close. The valve guides must also be aligned so the valves seat squarely on their seats (not off-center). Besides leaking compression, misalignment causes the valve head to flex, which can lead to cracking and valve failure. So make sure that whomever does your head repairs does quality work.

Ring Sealing—Proper ring sealing is also important. Piston rings do two

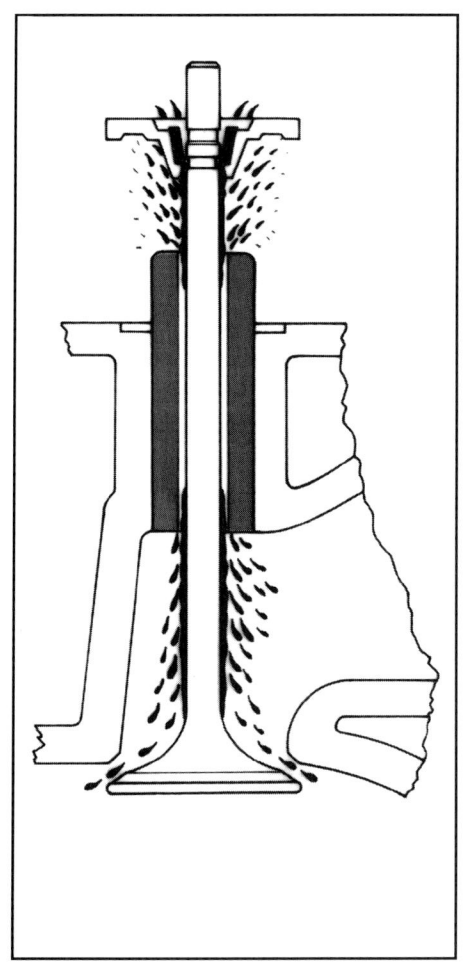

Worn intake valve guides can allow vacuum to siphon oil into the engine.

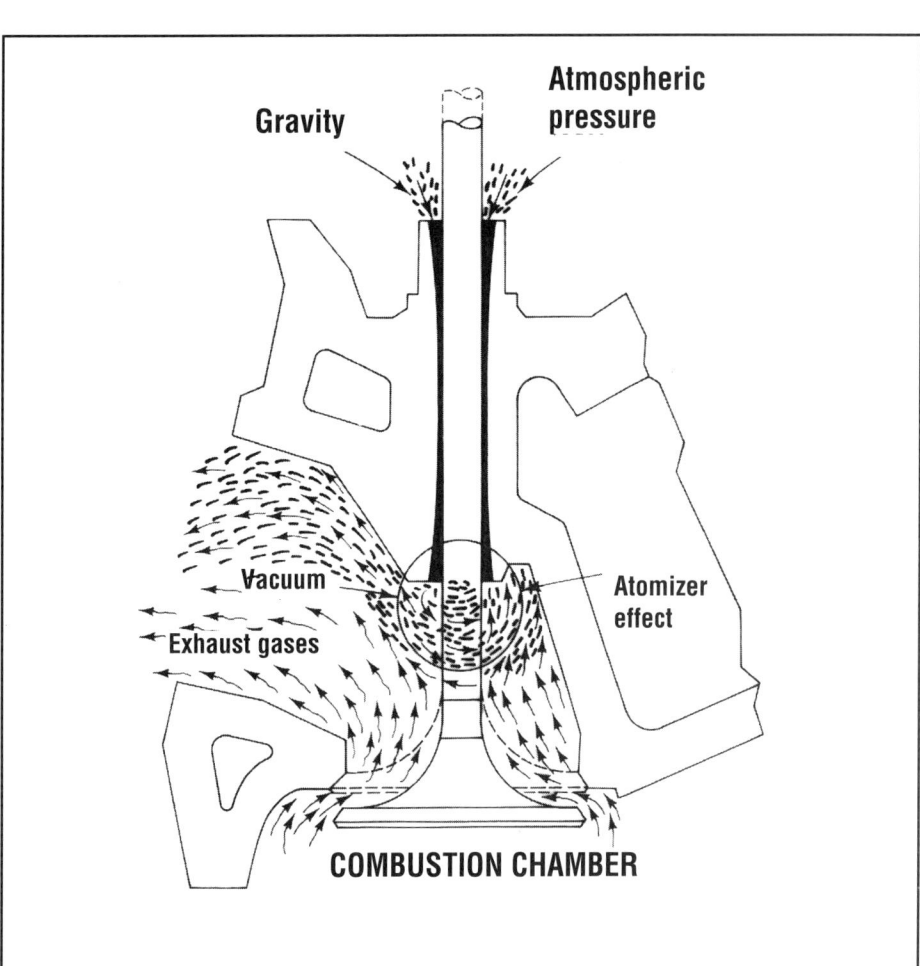

Contrary to popular belief, exhaust valves can suck oil too if the guides are worn or the valve stem seals have disintegrated.

things: the top two rings seal the combustion chamber and the bottom oil ring controls oil. The rings also prevent the hot combustion gases from blowing past the pistons and entering the crankcase. Blowby is something you don't want because it dumps moisture, carbon and unburned fuel into the oil. The blowby gases that do get past the ring are scavenged out of the crankcase by the positive crankcase ventilation (PCV) system and siphoned back into the engine where they are reburned. Even so, excessive blowby can create emissions problems by raising carbon monoxide readings. It can also turn the oil into sludge, which will result in serious engine damage unless the oil and filter are changed frequently.

As for oil control, the rings scrape the cylinder walls with every stroke of the piston to keep oil out of the combustion chamber. If the rings are worn, improperly installed (wrong side up), have too much end gap clearance, or are damaged or broken (often due to a detonation problem), they won't seal properly and will allow oil to enter the combustion chamber and excessive blowby into the crankcase. Oil in the combustion chamber will increase HC emissions, and excessive blowby will elevate CO emissions. If the rings are in really bad shape, there may be blue smoke in the exhaust and an odor that smells like burnt toast.

Ring problems can be diagnosed with a compression check. If the compression readings are low or vary by more than 10% from one cylinder to another, squirt a few drops of 30-weight oil through the spark plug hole and crank the engine over a couple of times. The oil will seal the rings temporarily. Repeat the compression test. If the readings are now noticeably higher, the rings are worn and should be replaced. If there is no change in the compression readings after squirting oil in the cylinders, the valves (or head gasket) are leaking compression. A cylinder leakdown test can be used to achieve the same results.

The only cure for ring sealing problems is a ring job. The engine must be disassembled and the rings replaced. If the cylinders are worn, out-of-round or have excessive taper, the block will have to be rebored and honed so oversized pistons and rings can be installed.

Valve Guides & Seals

Most oil consumption problems are due to worn valve guides and seals rather than worn or damaged piston rings. The guides support and position the valve stems in the cylinder head as the valves open and close. The guides help cool the valves by drawing heat away from the stems, and they keep the valve stems lubricated by allowing a small amount of oil into the gap between the stem and guide. The guides are subject to wear over time because of the constant opening and closing of the valves. As clearances begin to loosen up, the guides allow more and more oil to leak down past the stems and be drawn past the valves into the engine. The result is increased oil consumption and higher HC emissions.

If the guides are badly worn and passing a lot of oil, it can cause a buildup of heavy carbon deposits on the backside of the intake valves that interfere with proper breathing. On some fuel-injected engines, the accumulation of carbon deposits can cause rough idle and hesitation problems. Worn guides can also accelerate valve wear and sometimes lead to valve failure because of excessive valve wobble and flexing.

If the valve guides are worn, they will have to be replaced, relined or knurled.

Seals—Oil flow into the guides is controlled by the seals at the top of the guides. There are two basic types of seals: umbrella or deflector-type seals, which are little circular seals mounted on the valve stem that keep oil from splashing or running directly down and into the guides; and positive seals, which fit snugly around the valve guide boss and scrape excess oil off the valve stems. The latter are used on most late-model engines to reduce emissions.

One of the myths about controlling oil leakage past the valve guides is that it is a problem only on the intakes. The vacuum in the intake port area during the intake stroke will suck oil through the guides like a straw if the seals are bad. But the same thing can also happen with the exhaust valves. Even though the exhaust gases are pushed out of the cylinders under pressure, the flow of gases past the exhaust valve guide creates a partial vacuum, which soars right after each exhaust pulse. This can suck oil down the guides and into the exhaust system just as effectively as on the intake side. To make matters worse, exhaust valves usually have larger valve stem-to-guide clearances because they run hotter and need more room for thermal expansion than do intake valves. This can accelerate an oil consumption problem if the exhaust valve seals go bad.

Repair Options—When an engine has worn valve guides and/or seals, the problem can be cured one of two ways. The cheapest fix is to install new valve guide seals. This can sometimes be done without removing the cylinder head. By connecting an air hose to the spark plug hole and filling the cylinder with about 100 psi, the valve will be held shut so that the spring and retainer can be removed. The seal can then be easily replaced. The other alternative is to pull the head and do a complete valve job. Chances are that if the seals are bad, the guides, valves and seats will also need attention. When the head is disassembled, the valve

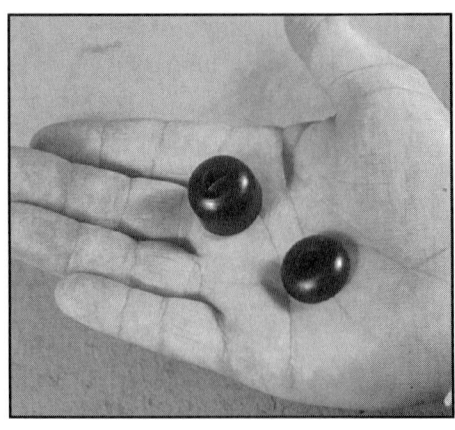

Valve stem seals are important on both intakes and exhausts.

stem-to-guide clearances should be carefully checked. If the guides are worn, they will have to be replaced, relined or knurled, or reamed out so valves with oversize stems can be installed.

HEAD CONFIGURATION

Some factors that influence emissions are an integral part of the engine design itself. One such factor is the configuration of the combustion chambers in the cylinder head.

When the air/fuel mixture is drawn into the cylinder during the intake stroke, the tiny droplets of gasoline tend to condense on the surface of the combustion chamber. This interferes with complete combustion and leaves traces of unburned HC in the exhaust. Minimizing the "quench area," therefore, is one way engineers have made today's engines cleaner.

Chamber Designs

The cleanest cylinder heads are usually those with "open" type hemispherical or pentroof-shaped combustion chambers with centrally located spark plugs. Heads with wedge-shaped chambers, on the other hand, have a large quench area that also creates a dead space with very little clearance between the top of the piston and the head. Such a design can interfere with the propagation of the flame-front and leave small pockets of unburned fuel.

Another way in which engineers have redesigned heads to lower emissions is to promote swirling of the air/fuel charge. This allows the flame kernel to grow in a more controlled manner for more complete combustion. Such heads are often dubbed "high swirl combustion" (HSC) heads.

Related efforts are Honda's CVCC (Controlled Vortex Combustion Chamber) and Mitsubishi's Jet Valve. Using a small auxiliary combustion chamber that receives a rich mixture (the spark plug is located here), the Honda system manages to get an overall lean mixture in the cylinder to fire dependably.

Mitsubishi added a tiny third valve that admits air only to churn up the air/fuel charge and promote lean running and a complete burn.

There's nothing you can do about the design of your engine's cylinder heads. But if you're rebuilding an older engine and have a choice of cylinder heads, using ones with more open combustion chambers will result in lower overall emissions.

CAMSHAFTS

Another factor that influences the emissions produced by an engine is valve timing. The camshaft that controls the opening and closing of the valves influences not only the rpm range where the engine develops the most horsepower and torque, but also such things as idle vacuum (which affects the operation of other emissions-control devices, such as PCV and EGR as well as MAP sensors on late-model cars with computerized engine controls), idle quality, idle emissions and emissions throughout the rpm range.

Because cars and light trucks have to meet strict emissions and fuel economy requirements, most stock cams are a three-way compromise that attempt to achieve as broad a usable power curve as possible while providing good fuel economy and low emissions. The requirements often conflict, so guess which ones receive priority when a car maker designs a factory cam? You guessed it, emissions and fuel economy.

Aftermarket Cams

That's where aftermarket cams come in. Everybody knows that one of the best ways to add horsepower is to install a hotter cam. And to maximize the cam's potential, it should be accompanied by the other traditional aftermarket bolt-on goodies such as an intake manifold, bigger carburetor and free-flowing exhaust. All this requires retuning the engine by recurving the spark advance and rejetting the carburetor. But if done properly with correctly matched parts, the results can be fantastic—and with little or no significant change in what comes out the tailpipe. But to be emissions-legal, a camshaft must be certified that it does not cause an increase in emissions.

Some cam grinds are inherently cleaner than others. Stock cams and torque/economy cams with relatively short duration specs and minimal valve overlap are the cleanest. High performance cams for high rpm applications are the dirtiest.

Short duration cams (less than about 214 degrees of duration) are good for high volumetric efficiency at low rpm because the intake valves close before the piston bottoms out and starts back up on the compression stroke. A short duration cam also has minimal valve overlap. Overlap between the closing of the exhaust valves and opening of the intake valves between the exhaust and intake strokes is bad for emissions because unburned fuel can pass through into the exhaust.

By comparison, high rpm performance

WEDGE-SHAPED COMBUSTION CHAMBER — Quench area

HEMISPHERICAL COMBUSTION CHAMBER

The shape of the combustion chamber can influence how "cleanly" the fuel burns in the combustion chamber. Minimizing the quench area reduces emissions.

It is possible to have your cake and eat it too. Today, you can install a "hotter" performance camshaft and still maintain emissions standards. Companies such as Crane Cams offer a variety of EO certified cams for better performance.

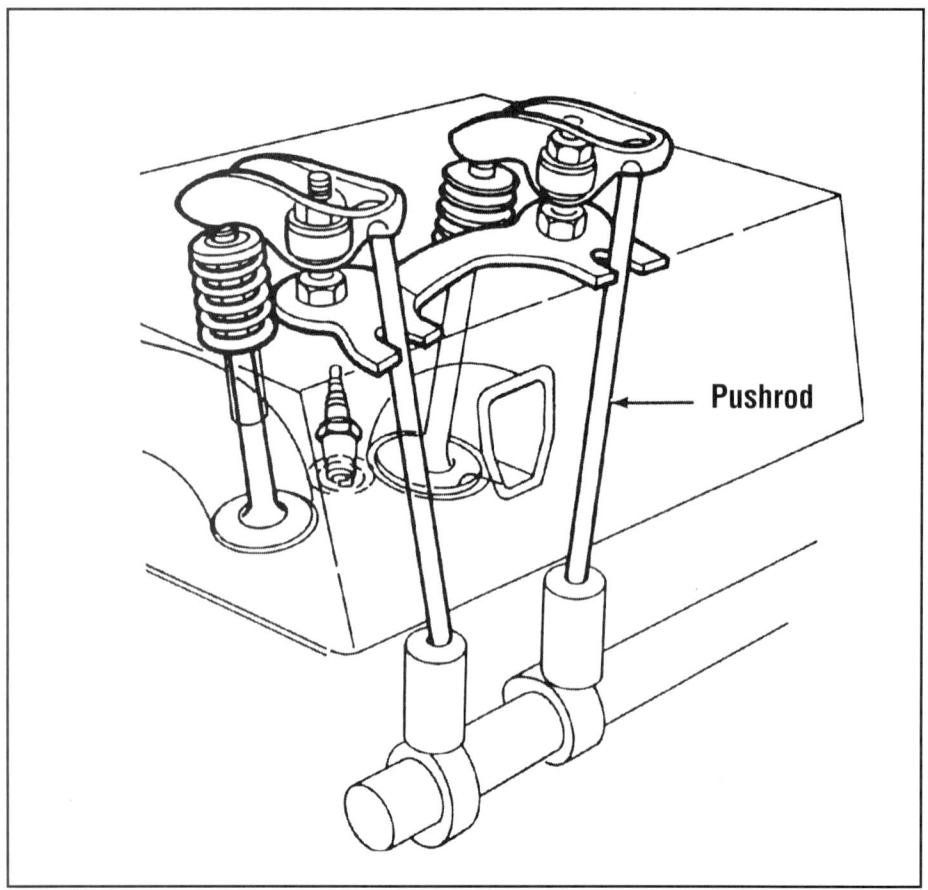

Valve timing and duration can have a big impact on emissions. That's why many performance camshaft grinds are not emissions-legal.

engines need lots of valve lift, duration and overlap to produce the most horsepower. Low-end torque isn't a concern because gearing keeps the engine working in the upper rpm range. But a wild cam can create serious idle problems and idle emissions in a street-driven vehicle because of excessive valve overlap. If it lopes at idle, you can bet it's polluting at idle. Low or erratic intake vacuum affects the operation of the MAP sensor, which in turn affects fuel calibration, spark timing, torque converter lockup and a number of other functions on computer-controlled engines. It also makes it difficult for the computer to stabilize idle speed. So for these reasons, most high performance cams don't make the cut when it comes to emissions certification.

What will happen if you run a non-certified "off-road" performance cam in a street-driven vehicle? Nothing, until you get caught in an emissions test. If the cam has more than about 214 degrees of duration or a lot of overlap, there's a very good chance it will create enough of an emissions problem to prevent your vehicle from passing a tailpipe test. Then you'll have to replace the cam with a milder one or a stock cam at your own expense—which also won't count towards any waiver limit on repairs because the cam was not emissions-certified.

Stock Cam Problems

The only emissions problems a stock camshaft can create is if it develops a flat lobe. A rounded off lobe can be caused by improper cam break-in (new engines only), lack of adequate lubrication (an oil starvation or breakdown problem), or reusing worn lifters with a new cam. Sometimes a lobe will go flat because the cam was not properly heat treated (hardened) when it was manufactured. If a cam lobe goes flat, it won't open its valve, resulting in a loss of power in the affected cylinder. If the lobe that's gone

Engine operating temperature must be close to factory specifications. If it runs too cool, the computer may produce an overly rich condition, or the rings won't seal properly. Temperature control is a function of the thermostat. Shown here is a badly corroded one that obviously doesn't work. Check yours frequently, and replace with a unit of the same temperature rating. Photo by Don Taylor.

flat prevents an intake valve from opening, it can cause the cylinder to suck oil past the rings and valve guides, increasing HC emissions. If the bad lobe affects an exhaust valve, the combustion gases will be forced past the rings into the crankcase. This will increase crankcase emissions and possibly overload the PCV system, causing an increase in both HC and CO emissions.

Worn timing gears, a stretched timing chain or a loose timing belt can have an adverse effect on valve timing as well as ignition timing. Worn gears or a loose chain typically cause retarded valve and ignition timing, which will affect HC emissions. The cure is to replace the gears and/or timing chain.

OPERATING TEMPERATURE

The engine's operating temperature depends on the temperature calibration of the thermostat (usually located in the outlet housing where the upper radiator hose connects to the engine). The operating temperature must be within the vehicle manufacturer's specifications for several reasons: For one, most computerized engine control systems won't enter the "closed loop" mode of operation for feedback fuel control until the coolant sensor signals the computer that the engine has reached a certain temperature. If the system remains in "open loop," the fuel mixture will remain fixed, which means it will probably run too rich. The result will be higher CO and HC emissions as well as increased fuel consumption. For this reason, the stock thermostat on a computer-controlled engine should never be replaced with one that has a lower temperature range.

Another reason for maintaining a minimum operating temperature is to control thermal expansion. If a "cold" thermostat (160 degrees) is run in an engine—or worse yet, no thermostat at all—the engine won't expand as much as usual, causing increased blowby past the rings, accelerated ring wear and increased oil consumption. Most engines run best at an operating temperature of 180 degrees or higher. In fact, most late-model, emissions-controlled engines require a thermostat of 195 degrees.

If the thermostat has to be replaced, install one that has the correct temperature rating for your engine application. And never run an engine without a thermostat!

FUEL CONSIDERATIONS

The type of fuel that's burned and the additives in it can also influence what comes out the tailpipe. Some fuels, such as propane and methane, are cleaner burning than gasoline because of the lighter hydrocarbons they contain. Alcohols, such as ethanol and methanol, are also somewhat lower on the emissions scale. But unless you're driving an alternate fuel, dual-fuel or "flex fuel" vehicle that's capable of burning something other than gasoline, you don't have much of a choice as to what goes in your fuel tank except what's available at the pump.

Gasoline is a complex blend of various hydrocarbons. The lighter hydrocarbons increase volatility and make winter starting easier, while heavier hydrocarbons provide added heat content for added power. Refiners adjust the various concentrations of the hydrocarbons in gasoline to "fine tune" the characteristics of the fuel for winter and summer driving as well as local climates. So the gasoline you buy in Pittsburgh may not be the same as the gasoline you buy in Phoenix.

Reformulated Gasolines

In recent years, refiners have also introduced "reformulated" gasolines that burn somewhat cleaner. Many such fuels are "oxygenated" which means they contain additional ingredients (typically alcohols) that add oxygen to lower CO emissions. Reformulated gasolines are now available in most large cities where urban air pollution is a problem.

On some cars (typically older ones without computer engine controls), filling up with a tankful of reformulated gasoline or an oxygenated gasoline that contains ethanol alcohol may lower CO emissions enough to help it pass an emissions test, if the engine is borderline to begin with from an emissions standpoint. Switching fuels won't make a big change in the readings, but a few points one way or the other can often make the difference between passing and failing an emissions test.

Octane

Octane is another factor that can influence emissions as well as engine performance. Octane is a measure of a fuel's detonation (spark knock) resistance. The higher the octane number, the better able the fuel is to resist detonation when the engine is under load. Higher octane fuels can also withstand higher compression ratios—but do not necessarily give better fuel mileage or lower emissions. When you buy premium gasoline, you're paying extra for refining a higher grade of fuel that has improved detonation resistance. Some premium fuels may also contain slightly higher concentrations of detergents to keep fuel injectors and carburetors clean, and a higher concentration of the heavier hydrocarbons that may make a slight difference in fuel economy—but there's no guarantee.

Most engines are engineered to run satisfactorily on regular 87 grade gasoline. Those with higher compression ratios (more than 9:1), turbochargers or superchargers may require a higher octane fuel to prevent detonation, which can be damaging. The erratic combustion that occurs during detonation can also increase HC emissions. Using a higher octane fuel or using additives that raise the fuel's octane rating can help minimize the danger of detonation.

The octane requirements of an engine increase with age as carbon deposits accumulate on the pistons and inside the combustion chambers. This raises the effective compression ratio of the engine, and as the compression increases, so does the danger of detonation. If your engine pings and rattles when accelerating or when it's under load (and there's nothing wrong with the EGR system, ignition timing or the air/fuel mixture), then you may have an excessive buildup of carbon in the combustion chambers. Such deposits can often be successfully removed by using a chemical "top cleaner" that is poured directly into the carburetor or throttle body. Additives that are poured in the tank are usually less effective because they can't be used in as great a concentration. Follow the directions on the product. If a chemical treatment fails to remove enough of the carbon to make a noticeable improvement, it may be necessary to pull the cylinder head(s) and physically remove the deposits by scraping with a wire brush. ∎

Engines equipped with superchargers or turbochargers generally need a higher octane fuel to prevent detonation. Photo by Michael Lutfy.

14
COMPUTERIZED ENGINE CONTROLS

You can go out today and buy quite a few new cars that will stay with, or even beat, the hottest muscle machines of the 1960s. Whether normally aspirated V8's or turbocharged V6's and fours, you can expect 0-60 times in the five-second range, and quarter-mile E.T.'s in the low 14's. And you get smooth driveability and great fuel mileage to boot, not the rough, ragged idle of a lumpy cam, jerky throttle response full of flat spots, or the need for tanker loads of super premium gasoline. Oh, yes—and the exhaust will be cleaner than the air in some cities.

What makes this seemingly impossible combination a reality? The computer, of course. So don't rant and rave about how automobiles are supposed to be mechanical, not electronic, and stop pining away for the simple old days. Computerized engine management is here to stay, and it's made cars better mechanically and driving more fun without compromising performance.

COMPUTER EVOLUTION

For most of the century-long history of the automobile, the engine's basic settings were either fixed at one value, or adjusted by mechanical means while running—manual levers, intake vacuum, centrifugal weights, etc. These arrangements, including carburetor metering rods or power valves, thermostatic bleed valves, and distributor diaphragms and advance mechanisms, served us dependably, but their accuracy

The combination of EFI and electronic engine control systems changed the way intakes look—and operate—profoundly and forever.

103

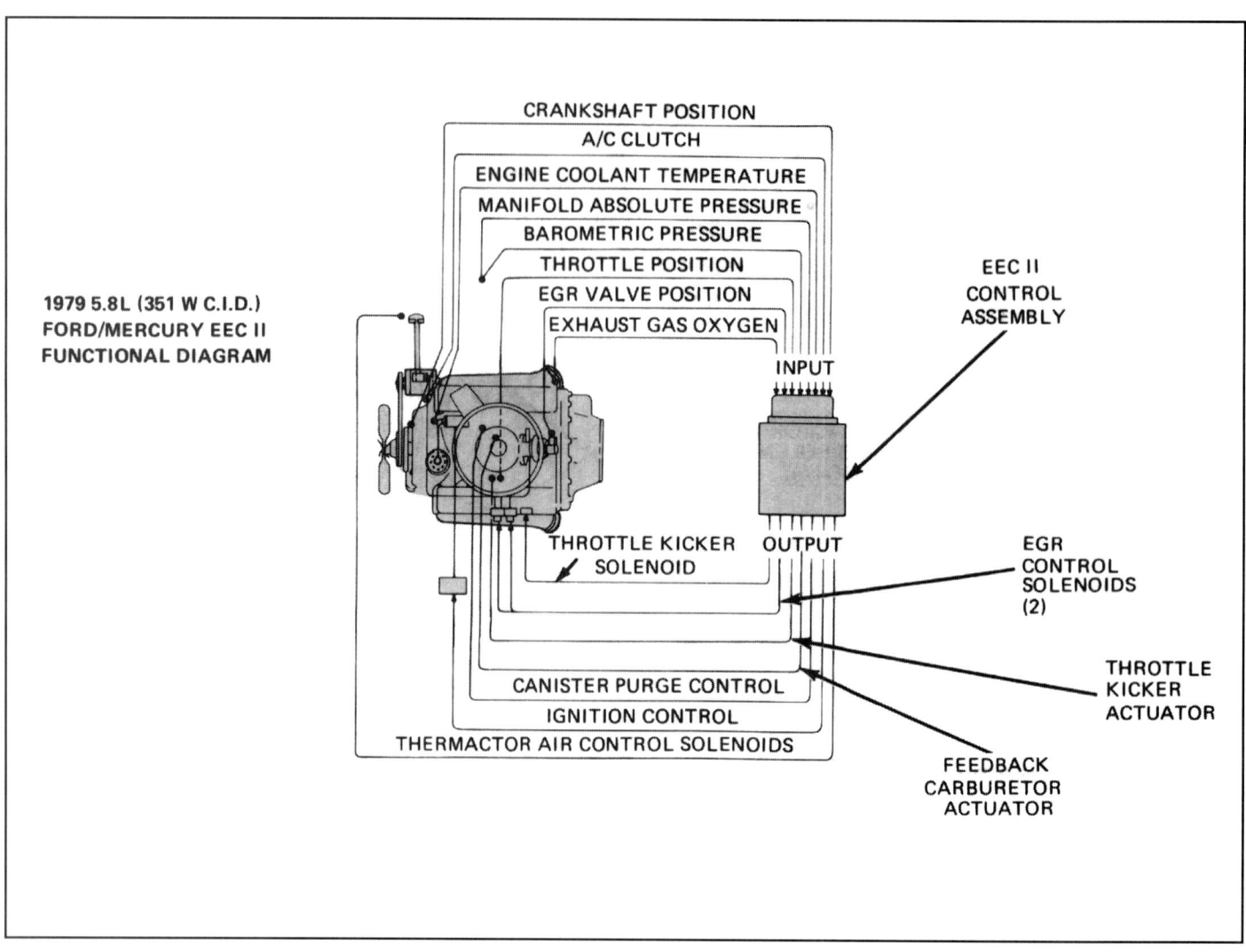

Suddenly, it was all different. Here's a schematic of one of the earliest (1979) electronic engine management systems, the EEC-II from Ford.

wasn't very good. They couldn't respond perfectly to all the different operating conditions and demands that make up the dynamics of vehicle operation. Until air pollution and fuel efficiency became critical issues, however, they were considered adequate.

Those two monumental problems made auto engineers start thinking about smarter ways to control air/fuel mixture and ignition timing. The closest thing to intelligence that could be incorporated into a vehicle was the computer, a device that can make decisions on the basis of input. So, systems were designed to apply this capability to how cleanly and efficiently a powerplant could be made to run. Automobiles would never be the same again.

System Overview

A system operation overview will aid understanding. Sensors let the ECU (Electronic Control Unit, actually a digital computer) know what is going on in the real world beyond its artificial digital environment. There are basically two types, voltage generating and voltage modifying. The former, including oxygen and detonation sensors, actually produces a small voltage signal, which is read by the computer. Those in the latter category (most other sensors) are basically variable resistors or potentiometers. The computer sends them power (called "reference voltage," commonly 5 volts), and measures what it gets back.

On the basis of this input and according to its programming, the computer makes decisions and issues commands to various actuators, such as duty solenoids and injectors. These make adjustments to tailor mixture, ignition timing, idle speed, and other settings to suit conditions. On/off functions such as charcoal canister purge and torque converter clutch lockup are also controlled. If the information the computer receives is faulty or interrupted, driveability and performance will be adversely affected.

Why Bother?

As we explained in Chapter 11, electronic fuel injection was a big advance in controlling emissions, but it still wasn't able to maintain the ideal stoichiometric air/fuel ratio (that is, 14.7:1 air to fuel by weight). This ratio

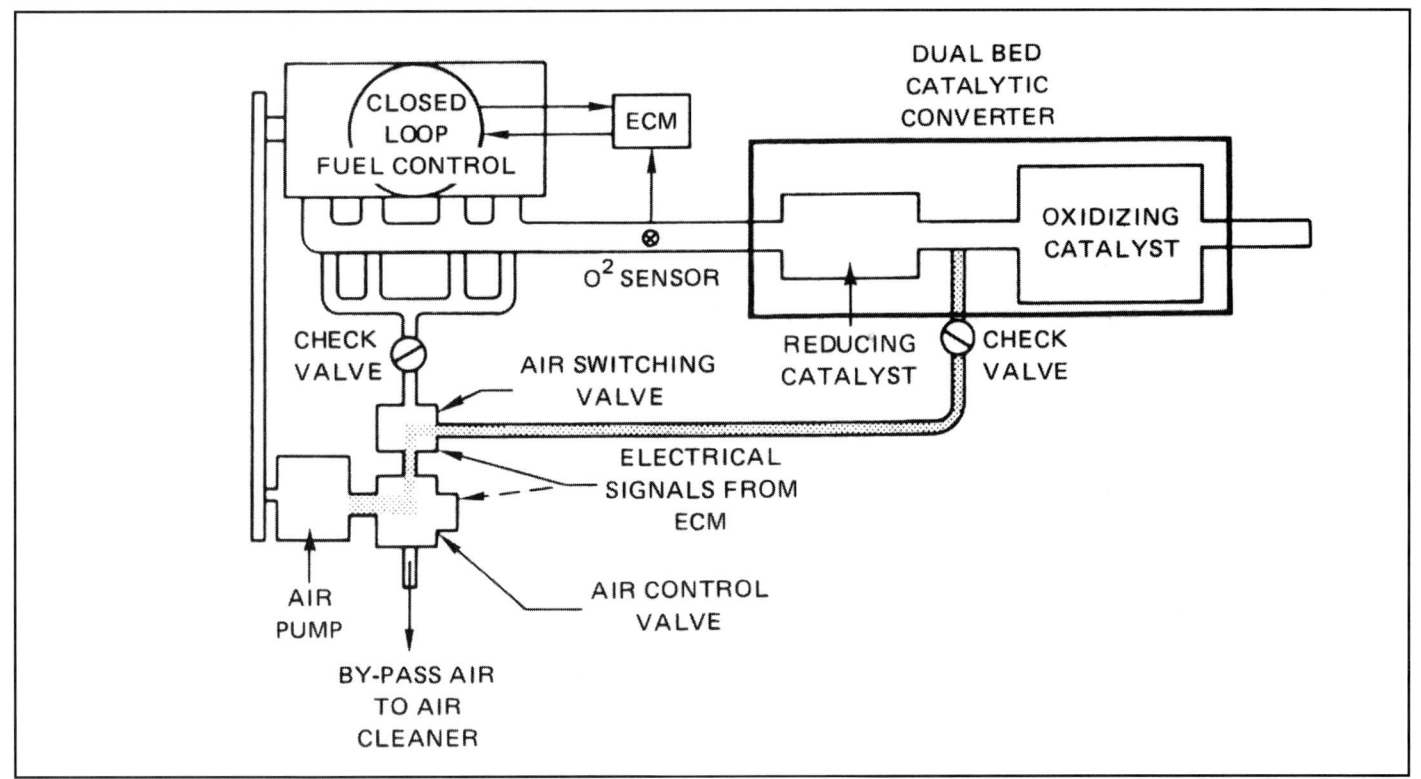

Without feedback/closed loop fuel control, there was no way to maintain the perfect 14.7:1 air/fuel ratio the NOx reduction cat needs (Chevrolet).

assures that all the fuel that enters the combustion chamber will have sufficient oxygen to combine with, thus reducing HC (hydrocarbon) and CO (carbon monoxide) emissions.

But this was not the most important reason for going to the trouble of designing systems that provided a nearly perfect mixture—after all, the two-way oxidation catalytic converter can clean up a great deal of HC and CO after it leaves the cylinders. A more difficult pollutant to eliminate is NOx (oxides of nitrogen, an ingredient of photochemical smog). While EGR (Exhaust Gas Recirculation) does a reasonably good job of keeping NOx formation down by lowering peak combustion temperatures, it can't do enough to satisfy legal requirements. So, the three-way catalytic converter was developed, but its NOx reduction reaction can only be maintained if a stoichiometric air/fuel ratio is present, and that was the prime mover for the development of feedback systems.

PROGRAMMING

The computer responds to sensor input in such a way that it sends the carburetor or injectors the signal necessary to maintain the right mixture for current operating conditions. When the engine is cold or at full throttle, its commands result in a relatively rich blend, and during deceleration, in a lean mixture. Also, if normal operating temperature is never achieved, or if a sensor stops providing input, the computer falls back on a "limp-in" or "limited operating strategy" program. This uses average values contained in memory to provide a mixture and, in some cases, a spark advance setting that will allow the car to be started and driven, but won't give very good performance, driveability, fuel economy, or emissions control. For any combination of inputs, the ECU looks into its memory to find the best timing, dwell, and pulse width. This ROM (Read Only Memory, as opposed to volatile memory) content, called a "map," is laid down through careful research on dynamometers, test tracks and all kinds of roads. Optimum settings for power, efficiency, low emissions, etc. are observed, then imprinted electronically. This entails many thousands of data bits. There's no way even the most cleverly engineered mechanical/vacuum system could even approach this kind of speed and accuracy.

Analog and Digital

Sensor input is usually analog (as opposed to digital, although there are sometimes on/off signals from switches that qualify as input). For example, an oxygen sensor produces anywhere from one tenth to about one volt. A throttle position sensor, which is actually a variable resistor, typically receives a five-volt reference signal (GM calls this "VRef") from the ECU, and its position determines the amount of resistance that signal encounters before returning to the computer.

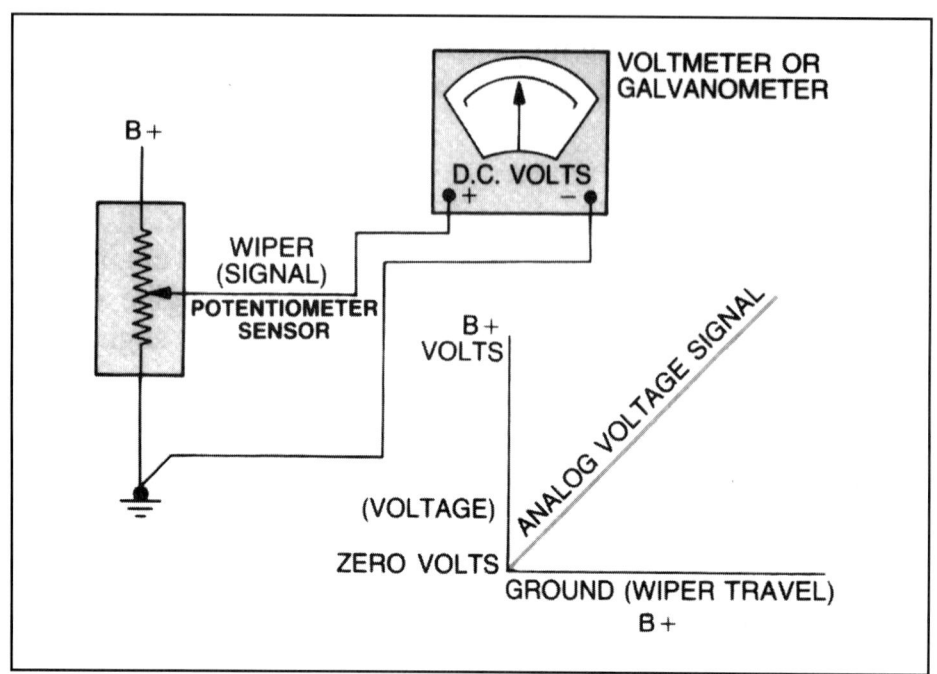

This diagram of an analog sensor circuit should help you see the difference between analog... (Ford)

But inside the ECU, things happen digitally. As analog signals are received, an input conditioner or A/D (Analog/Digital) converter assigns them the numeric values the computer can deal with (weak signals, such as those from the O_2 sensor, are also amplified, and pulses are "shaped"). This is all binary—on or off, 1 or 0, voltage present or not present. Each 1 or 0 is called a "bit," and there are eight bits in a "byte," also known as a "word." The pattern of bits in a byte represents a particular number or character, called a "BCD" (Binary Coded Decimal). The most common coding standard is ASCII (American Standard Code for Information Interchange), in which zero is represented by eight zeros (or offs), the number nine by "00001001," 15 by "00010101," etc.

Some say a computer is just a high-speed ground. Its job is to complete circuits, thus energizing various components at the right time and for the proper duration.

Memory

There are several kinds of computer memory. ROM (Read Only Memory) and PROM (Programmed Read Only Memory) hold the basic set of instructions on how information will be handled, and also vehicle-specific data—weight, axle ratio, engine type, etc. This is "burned in" as permanent programming (by the way, when you see an "E" in front of PROM, it stands for "Erasable," meaning the data can be wiped out by exposure to UV light so that it can be reprogrammed; a double "E" in front of PROM—EEPROM—means it is "Electrically Erasable") RAM (Random Access Memory) is active and volatile. Information from sensors can be put in, taken out, updated and rearranged at random. Ordinary RAM dies when the key is switched off, but there is another type called NVRAM (Non-Volatile RAM, GM's name) or KAM (Keep Alive Memory, Ford's term) that survives as long as the battery is hooked up. This allows adaptive functions. In Bosch Motronic, for instance, open-loop settings are fine-tuned according to the way the car is driven, and compensations can also be made for such variables as a small vacuum leak or high altitudes. GM has integrator and block learn, two temporary (compared to ROM) memory arrangements that make minor closed-loop adjustments in fuel delivery on the basis of how the vehicle has been

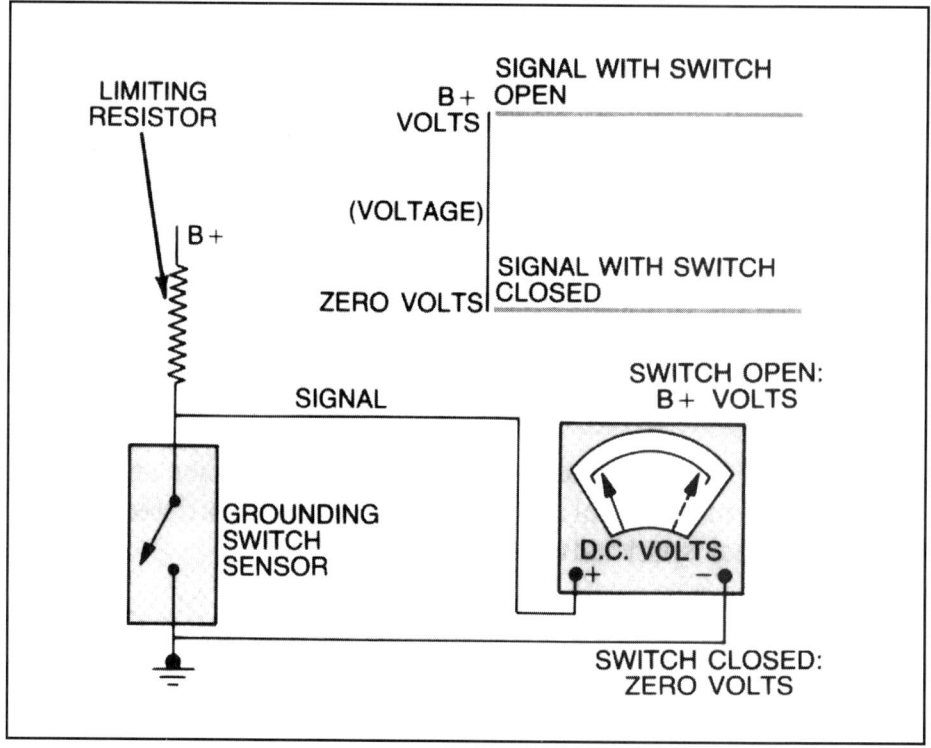

...and digital. Everything happens digitally inside a microprocessor, so analog input must be converted (Ford).

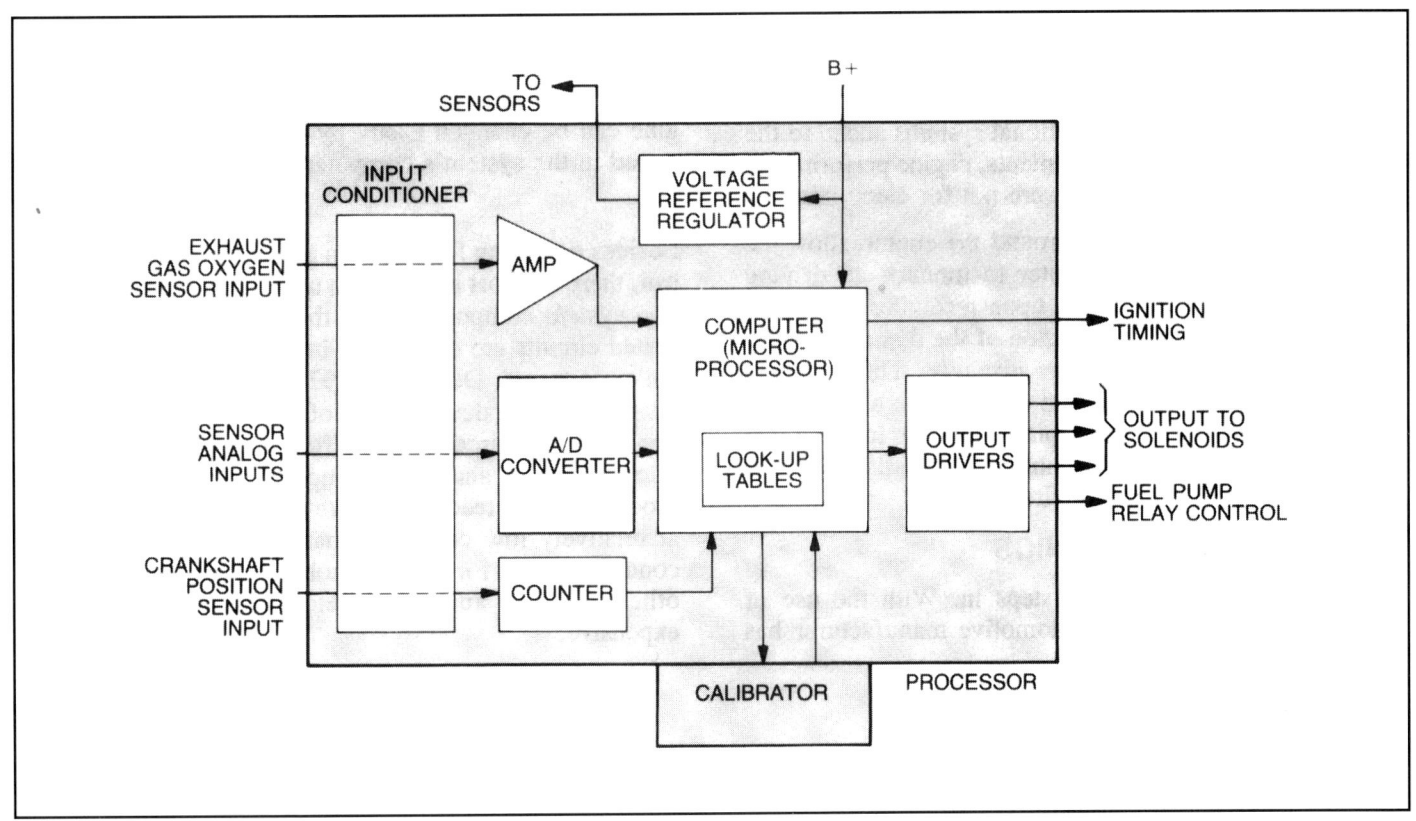

Inside the computer (usually called the ECU for "Electronic Control Unit," or ECM for "Electronic Control Module") input is amplified, converted and conditioned as necessary before it goes to the microprocessor. The look-up tables or "map" are ROM (Read Only Memory), the calibrator is PROM (Programmed Read Only Memory), and the output drivers are power transistors (Ford).

operated in the recent past.

How It Thinks

Basically, a digital computer's logic functions are carried out by a complex combination of on/off binary switches, called "gates." Ordinary mechanical switches in a simple DC electrical circuit will illustrate the principle.

Suppose you have two switches in series leading to a bulb. If neither is closed, or just one is closed, the bulb won't light. You have to close both to provide a current path. In electronics jargon, this would be called an "AND" gate (one switch AND the other switch). Now, hook up this circuit with the switches in parallel. If both are open, no light. But if either is closed, the bulb will glow, and this is known as an "OR" gate (one switch OR the other switch). There are other types of gates too, but they don't lend themselves so easily to our DC switch illustration.

In a microprocessor, huge numbers of various gates are combined in logic circuits, which can make decisions on the basis of programming and input. It's similar to the game Twenty Questions.

To oversimplify again, suppose the ECU is designed to allow vacuum to reach the EGR valve only after the coolant temperature sensor indicates the engine is warm AND when the throttle position sensor says rpm is above idle. This is roughly similar to the switches-in-series circuit mentioned above. If both conditions are met, the computer grounds a vacuum switch, opening it and permitting engine vacuum to operate the EGR valve.

Integrated Circuit—Although ordinary semiconductor components are small, a great many of them are required to do any complex decision making, which would result in a large package. That's where the IC (Integrated Circuit) comes in. Through clever and precise photographic chemical etching processes, hundreds of thousands of transistors, diodes, and resistors can be created on one tiny chip of silicon. Impossible as that may seem, just accept it and you will have altered your consciousness to fit the electronic era. IC's with that much capacity make computers of a practical size and reasonable current draw possible. They're mounted on a circuit board along with individual capacitors, resistors, transformers, etc. as necessary.

Conductors—The tiny conductors in a microprocessor couldn't begin to carry enough current to power a real-world component. So, the minuscule signals control output drivers, which are actually power transistors. The application of a small amount of voltage to a driver's base switches the relatively large volume of current needed to actuate anything. You can think of it as an electronic relay or switch. You may've heard the term "quad driver," or "QDR," which is simply a

Here's the oxygen sensor, the component that makes closed-loop efficiency possible. The O_2 sensor is usually screwed into the exhaust manifold.

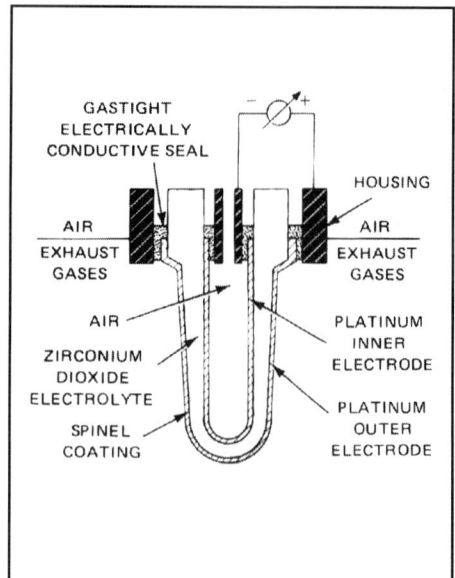

The operation of an O_2 sensor is similar to that of a battery cell. (Volkswagen)

power transistor that can handle four loads at once.

SENSORS

The incredible ability of an ECM to process, calculate and adjust for the best possible air/fuel mixture under all driving conditions is wasted if it does not receive the proper information. For that, it relies on a network of sensors to relay data necessary for it to calibrate for optimum driving performance under a wide variety of conditions, with limited emissions.

Oxygen Sensor

The device that gives the system the ability to maintain stoichiometry is the oxygen sensor, also called the "lambda" sensor or probe. It provides the computer with information on the oxygen content of the waste gases escaping from the cylinders.

Construction—The oxygen sensor itself is an interesting device. Resembling a spark plug and screwed into the exhaust manifold, it comprises a steel housing with a hex and threads, a louvered shield over the tip, a hollow internal element made of zirconium dioxide (ZrO_2, a ceramic material) and coated inside and out with a thin layer of micro-porous platinum. A wire connects the element's inside layer of platinum to the computer.

How It Works—The action involved is similar to that of a battery cell. The zirconium dioxide performs the function of an electrolyte, and the platinum layers operate as electrodes (the outer one is exposed to the exhaust stream, and the one inside the element is vented to the atmosphere).

Once the ZrO_2 reaches about 600 deg. F, it becomes electrically conductive and attracts negatively charged ions of oxygen. These ions collect on the inner and outer platinum surfaces. Since there's more oxygen in plain air than in exhaust, the inner electrode will always collect more ions than the outer electrode, and this causes a voltage potential—electrons

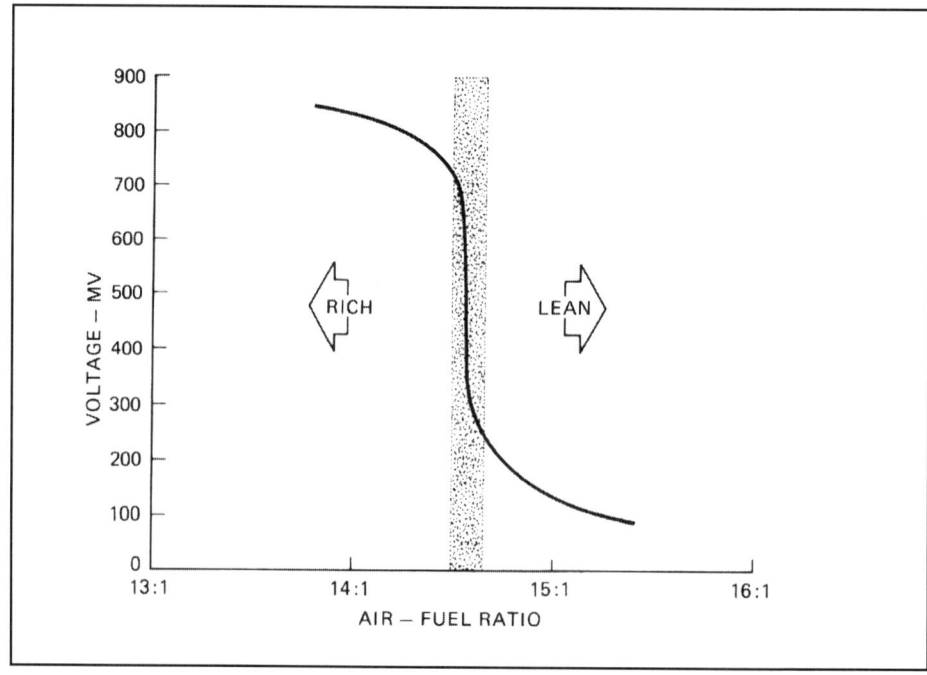

The typical zirconium dioxide-type oxygen sensor produces a small voltage that rises as the O_2 content of the exhaust falls. Just remember "lean equals low (voltage)." (Volkswagen).

closed-loop operation (i.e., while the computer is adjusting the mixture on the basis of the oxygen sensor signal), it's desirable to get the sensor up to operating temperature as soon as possible after the engine is started, and to keep it there while idling, so some oxygen sensors are equipped with an electrical heating element (these have three wires, one to the computer, and a pair that supplies the heating element with 12 volts and ground).

Many perfectly good O_2 sensors are replaced because some people test them when the engine is first started, or after it has idled for a long period, which allows the sensor to cool off. Also, never blame this component for driveability problems that occur before normal operating temperature is reached.

TPS

The TPS (Throttle Position Sensor) informs the ECU of what the driver's doing with his right foot. A signal that indicates a quick transition from cruise to wide open, for instance, tells the ECM

The TPS (Throttle Position Sensor) is important, too. The one on this Corvette is adjustable.

will flow. The amount of voltage produced depends on the number of ions on the outer electrode, which is determined by the oxygen content of the exhaust. If the engine is running rich, little oxygen will be present in the exhaust, few ions will attach to the outer electrode, and voltage output will be relatively high. If a lean condition prevails, more oxygen will be present, and that translates into more ions on the outer electrode, a smaller electrical potential, and less voltage. Remember "L=L" for "lean equals low."

The voltage is always small, never exceeding about one volt (or, 1,000 mV), with a typical operating range being between 100 and 900 mV. It is, however, sufficient for the computer to read. If it receives a sensor signal of less than about 450 mV, it recognizes a lean condition, and if it receives more than that amount of voltage, it registers a rich condition. Either way, it instantly corrects accordingly by altering the air/fuel mixture.

It's important to remember that the oxygen sensor won't produce voltage until it has reached the 600 deg. F mentioned above. So, when the engine is first started, or if it's idled for a long period, the computer will see that it isn't getting a signal from the sensor and will operate in the "open-loop" mode. That is, it'll hold the mixture at a fixed setting. Since good fuel efficiency and minimum exhaust emissions can only be had during

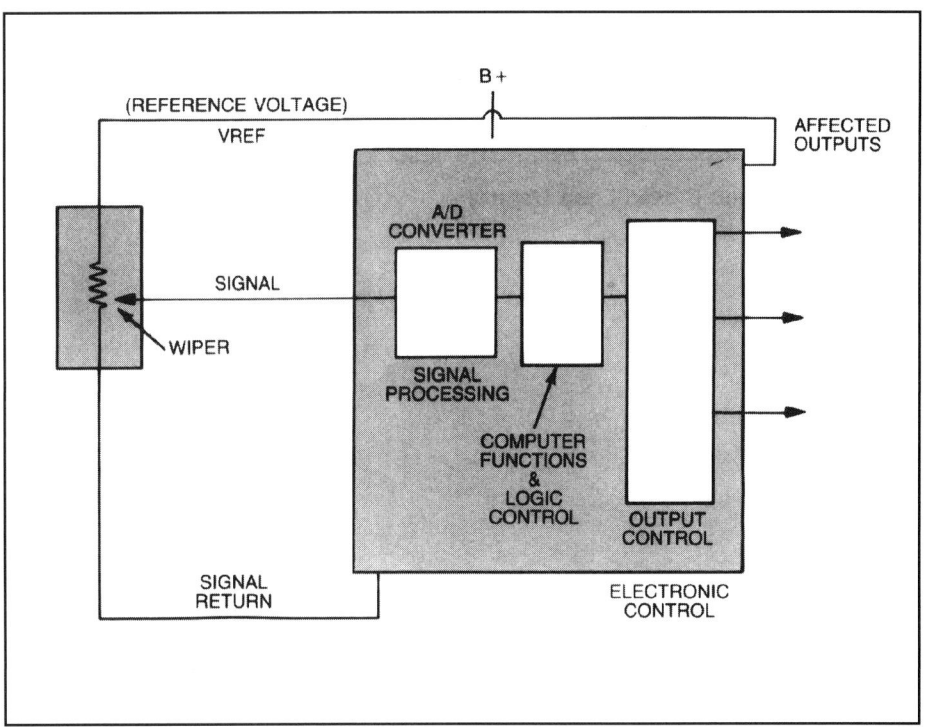

Most other sensors, such as those for throttle position and coolant temperature, change resistance to modify the reference voltage the ECU sends them (Ford).

109

The rotating vane air volume meter was very popular. Inside, the rotation varied the position of a potentiometer.

that a richer mixture is needed.

A TPS is a variable resistor or potentiometer—the same thing that adjusts the volume of a stereo. A strip of carbon provides resistance, and a brush or wiper moves along its face as the throttle is opened. The farther the brush is from the source of current, the more carbon the electrons have to travel through, and the higher the resistance. It's a three-wire device with terminals for supply voltage (connected to one end of the carbon strip), output back to the ECU (attached to the brush), and ground (the other end of the resistance element).

Coolant Temperature Sensor

The computer can't do a good job of orchestrating efficiency and performance unless it knows if the engine is warm or cold, so all systems, regardless of make or vintage, include a CTS (Coolant Temperature Sensor). This is a thermistor—its resistance changes with temperature—and there are two varieties. In the PTC (Positive Temperature Coeffient) version, ohms go up with temperature. The NTC (Negative Temperature Coefficient) type works just the opposite—resistance goes down as heat goes up (for example, a common GM range is 25,000 ohms at 0 deg. F to 185 ohms at 210 deg.).

On the basis of the amount of resistance the sensor is presenting to the reference voltage signal, the ECU learns the state of the engine, and makes decisions accordingly. Temperature affects air density, so the computer has to know about the temperature of what's entering the manifold in order to keep the mixture right, and that's the job of the MAT (Manifold Air Temperature) or charge temperature sensor. Although it's exposed to a gas instead of a liquid, it works in the same way as a CTS.

PSI and Hg

Pressure sensors contain a silicon diaphragm element that flexes when vacuum or pressure bears on it, which action varies its electrical resistance (the more it flexes, the higher the ohms). There are two types, absolute and differential. In the former, there's a trapped reference pressure against which the variable value acts. The latter has one side open to manifold vacuum and the other open to the atmosphere.

The MAP (Manifold Absolute Pressure) sensor sends data to the ECU on the amount of vacuum the engine is producing. This is an important input—in fact, some systems revert to it if the TPS fails. The BARO (for "barometric") sensor reports on atmospheric pressure as it changes with altitude and weather conditions. Commonly, these are combined in one differential-type sensor.

Airflow Sensors

Sensors that provide input on the volume of air entering the engine are especially important. The most common type is the vane-type airflow meter. Flaps inside a housing rotate against spring pressure as more and more air is taken in, and this rotation turns a variable resistor.

Mass Airflow—Some late models use a more sophisticated approach—the mass airflow sensor. The Bosch uses a heated wire, and the Delco uses a heated film, but the principle of both is to judge the actual mass (this differs from simple volume because density is taken into account) of intake air by measuring the cooling effect it has on the heated element.

The Karman-Vortex airflow meter, as

Measuring the actual mass (not just the volume) of airflow is more accurate, so we got the MAF (Mass airflow) sensor. This is the AC-Delco/GM hot film version.

The typical Bosch MAF uses a hot wire to measure the volume of intake air.

The Infiniti Q45 V8 is a good example of how computerized engine management systems provide top performance with low emissions.

found on some Toyota/Lexus and Mitsubishi models, optically senses the speed of vibration of a thin metal mirror in the intake vortex to generate a frequency signal that varies according to the volume of air entering the engine.

Knock Sensor

Contrary to oil company advertising, until the detonation sensor circuit first appeared on the 1978 Buick V6 turbo, a car wouldn't really get any more performance out of high-octane gasoline than it did from low-octane (as long as pinging wasn't present). With this cleverly conceived system, however, the anti-knock qualities of the fuel do have a direct effect on power output and efficiency.

The phenomenon that makes this possible is *piezoelectricity,* which is the generation of current in dielectric crystals subjected to mechanical stress. In a computerized engine control system, the crystal is commonly quartz (sandwiched between two electrodes), and the stress involved is in the form of the high-frequency vibrations caused by detonation.

Operation is simple. The knock sensor, which is threaded into the block, head, or intake manifold, "listens" for detonation. When it hears this distinctive sound, it sends a warning voltage signal to the computer, which responds by retarding the spark. As soon as this potentially damaging combustion condition subsides, the signal ceases, and timing is once again advanced to the optimum setting.

On GM ESC (Electronic Spark Control) systems, the detonation sensor signal doesn't go directly to the ECM, but instead is interpreted by a separate module, which, in turn, notifies the main computer by modifying a reference signal.

Switches

In some cases, instead of using a sensor that returns a varying signal to the ECU, simple on/off switches are used. For example, a temperature switch used in place of a sensor may close to ground at a particular coolant temperature, thus informing the computer that the engine is warm. ■

15
EMISSIONS TESTING

Nobody likes the hassle, inconvenience and expense of having to take a vehicle in for an emissions test. But like it or not, the annual or biennial emissions inspection has become yet another aspect of vehicle ownership in most major metropolitan areas today. Emissions testing is considered to be an essential element of making sure the vehicles that are on the road are running clean. After all, if they didn't actually test vehicle emissions, how could they enforce compliance with clean air standards? Some experts have argued that roadside monitors capable of detecting "gross polluters" (vehicles with serious emission problems) would be much more cost effective than the current method of detecting polluters. Others say "clunker laws" that permanently remove older, dirtier vehicles from the road by offering financial incentives to people who voluntarily scrap their vehicles is a better way to reduce overall pollution. Both alternatives have their merits. Even so, the EPA as well as most other state regulatory and enforcement agencies that are responsible for overseeing clean air programs are committed to mandatory emissions testing as a means of assuring general compliance.

If you live in an area that requires annual emissions testing, you're going to have problems if your vehicle is not emissions-legal. Flunking an emissions

Emissions testing is now required in most areas, so there's no getting around it. You can prepare yourself by following some of the tips for passing outlined in this chapter.

Race cars built "for off-road use only," as well as pre-67 and antique cars, are generally exempt from emissions testing. Photo by Michael Lutfy.

test didn't used to be that big of a deal. But it is now. With waiver limits now as high as $450 in some areas, and no waiver limit in many instances on vehicles that have missing or inoperative emissions controls, you may find yourself having to make some rather expensive changes or repairs to pass the inspection.

For example, if you've replaced your stock intake manifold and carburetor with aftermarket parts that are not emissions-legal, you may be forced to reinstall the stock parts to pass the test. The same could apply to a non-certified exhaust system, camshaft or other such part.

Many states enforce compliance with their emissions testing programs by refusing to renew or issue license plates to vehicles that fail to pass the test. Some will also suspend your driver's license if you ignore a notice of inspection.

When you receive a notice of inspection, or your inspection sticker is due to expire, you must have your vehicle inspected by the deadline—or suffer the consequences. If you don't know what the consequences are in your area, call someone and find out. In most cases the penalty is tough enough to make most people comply with the law.

Sometimes you'll receive an inspection notice for a vehicle you no longer own. If you've recently sold your vehicle and the paperwork hasn't filtered its way through the state bureaucracy yet, be sure to fill out the postcard indicating you've sold the vehicle and return it to the state so they don't think you're ignoring them and suspend your license. Many a driver has been stopped in a routine traffic check or for some other infraction only to find that their license has been suspended for ignoring an emissions notice!

EXEMPTIONS

Many states with emissions inspection programs exempt certain vehicles from testing. Diesels and vehicles powered by alternative fuels such as propane or compressed natural gas are often exempt and do not have to be tested. Trucks over 8000 GVW may be exempt. Government vehicles (police cars, ambulances, fire trucks, school buses, city buses, military vehicles, etc.) are often exempt.

Agricultural and off-road vehicles are typically exempt as are race cars, antique cars or pre-emissions-control vehicles (1967 and earlier). Some states require all vehicles from 1968 and up to be inspected while others exempt vehicles that are 20 or more years old.

People can be quite creative when it comes to circumventing the rules. Re-registering a vehicle at a relative's address that is outside the boundaries where emissions inspections are required (such as in the next county, township or across a state line) is one ploy that's sometimes used. But you can get into trouble with this trick if you're caught because many communities have ordinances requiring you to register your vehicle at your current address within a certain time period (typically 6 months or less). This is done so you'll have to buy a city or village sticker. Your insurance company may also not take kindly to such deceptions since your address often affects your insurance rates (city dwellers pay more than rural folks).

Another way to avoid emissions inspection hassles is to sell a vehicle.

During an emissions test, a technician will sample the gases coming out of your tailpipe with a "sniffer."

Transferring ownership to another family member causes a delay in the official paperwork. But sooner or later, they catch up with you and you have to make the trip to the dreaded inspection lane.

Some states also have laws against selling "problem" vehicles. These would be illegally modified cars that the seller knows won't pass an emissions test. If you pass off such a vehicle on an unsuspecting buyer, they may have the legal right to come back against you for any expenses they may have incurred trying to get the vehicle inspected. The state may also come after you for violating their laws.

By the same token, if you've bought a vehicle since the new Clean Air Act rules went into effect and it turns out that the vehicle has non-certified parts on it that cause you to fail an emissions test, you too may have legal recourse against the previous owner.

It's a complicated world we live in these days, so the best advice is play by the rules and hope that everybody else does, too.

TESTING OVERVIEW

Given the fact that there's no way around periodic emissions testing (short of moving to someplace where it isn't required or giving up the private ownership of a motor vehicle for public transportation or bipedal locomotion), you're faced with the responsibility of keeping your vehicle's emissions within acceptable limits. The emissions test tells you whether you're in compliance or not. If your engine's emissions are within the cut points, your vehicle passes the test and life goes on as usual. You get your smog certificate or inspection sticker, and the "right" to renew your vehicle registration (license plates). But if your vehicle doesn't pass the test, then you're in "emissions limbo" until you get your emissions problem fixed.

Test Facilities

In states that have "decentralized" inspection programs, the state or municipalities license independent repair facilities such as repair garages, service stations, car dealers, tire stores, retailers, etc., to perform the actual tests using special state-approved test equipment. These same shops may also be qualified to perform any emissions repairs that might be needed—which creates an obvious conflict of interest in the eyes of many consumers. Safeguards built into such testing program such as tamper-resistant computerized test equipment, electronic data recording and monitoring, spot checks and so on supposedly keep everyone honest. But where there's a will, there's a way, so buyer beware. Even so, most emissions testing programs are policed fairly well and abuses are not a serious problem.

The other approach to checking emissions are the "centralized" test programs where all inspections are performed at special test lane facilities operated by a state-licensed single contractor. Because no repairs are performed at these facilities, a centralized testing program eliminates any temptation on the part of the emissions tester to take unfair advantage of a motorist. In other words, they've nothing to gain by flunking your vehicle because they can't do the repair work. The

TYPICAL EMISSIONS CUT POINTS			WELL-TUNED ENGINE EMISSIONS	
Model Year	CO%	HC ppm	CO%	HC ppm
pre-1968	7.5-12.5	750-2000	2.0-3.0	250-500
1969-70	7.0-11.0	650-1250	1.5-2.5	200-300
1971-74	5.0-9.0	425-1200	1.0-1.5	100-200
1975-79	3.0-6.5	300-650	0.5-1.0	50-100
1980	1.5-3.5	275-600	0.3-1.0	50-100
1981-93	1.0-2.5	200-300	0.0-0.5	10-50
1994 & up	1.0-1.5	50-100	0.0-0.2	02-20

Here are the cutoff points for carbon monoxide and hydrocarbon emissions, and the emissions generally given by a well-tuned engine. As you can see, there is quite a bit of leeway between them. The bottom line is to keep your car well tuned.

disadvantage of the centralized test approach is that you then have to take your vehicle someplace else to get it fixed, and then make a second trip back to the central test facility for a re-inspection. This sometimes results in a "yo-yo" effect if your emissions problem isn't diagnosed properly and repaired. You end up bouncing back and forth between repair shop and test facility trying to get the problem resolved. To prevent this from happening, some states authorize repair facilities to do the retest. Others do not, fearing the potential for abuse.

THE ACTUAL TEST

Outside of California, most emissions test programs check for two pollutants: unburned hydrocarbons (HC) and carbon monoxide (CO). Most also measure carbon dioxide (CO_2) but for diagnostic purposes only since CO_2 is not a pollutant (though it is a "greenhouse gas" that may contribute to global warming). In California and some states, vehicles are also inspected for emissions tampering. This includes:

•Checking the restrictor in the fuel tank filler neck to make sure it hasn't been knocked out or enlarged to accept regular leaded gasoline.

•Inspecting the gas cap to make sure it is the correct type for the application and seals tight.

•Checking under the hood to make sure the engine has all the required emission control components.

•Looking under the car to see that the catalytic converter is in place.

•Checking the instrument panel to see if a "check engine" or "Sensor" warning light is illuminated, possibly indicating a performance problem that could affect emissions.

•In California, they also check to make sure that any non-stock aftermarket parts on the engine are emissions-certified. If the engine has been swapped or replaced, they'll also check to make sure it has all the required emissions equipment for the original model year and application.

Some states also include various "safety checks" as part of their emissions test. These include checking the brakes, tires, glass, exhaust system, lights, horn, wiper blades, etc. Such items don't have anything to do with emissions, but they can affect the safe operation of your vehicle (which is a good reason to check them yourself even if these items are not part of an emissions inspection).

I/M 240 Test—The new I/M 240 emissions test program adds oxides of nitrogen (NOx) to the tailpipe checks, "loaded mode" dyno testing to simulate actual driving conditions, and a performance check of your vehicle's evaporative emission control system (charcoal canister & purge valve) to make sure evaporative emissions are being controlled. The I/M 240 test is also a longer and more involved test (see sidebar on p. 116), and one that's harder to pass.

Test Standards

The test standards for acceptable HC and CO levels are based on the emission standards that the vehicle was required to meet when it was new. Older vehicles do not have to meet the same standards as newer ones, which means the test requirements will vary depending on the model year of your vehicle. The actual cut points for the emissions test are determined by the state or municipality, and are generally more lenient than the original new vehicle limits. Cut points are typically set 20% to 30% higher than the same year new car standards—primarily to minimize public dissatisfaction with the emissions testing program. If emissions cut points are set too low, too many vehicles would fail and there would be a lot of unhappy motorists (voters) clamoring for a change. Consequently, the cut points established for most emissions testing programs are designed to catch the majority of vehicles with the worst emission problems.

How clean do your emissions have to be? It all depends on the model year of your vehicle. Generally speaking, tailpipe emissions must be less than the levels indicated in the chart above.

Notice that the actual emissions produced by the average well-tuned engine are substantially less than the typical cut points required for an official emissions test. As we said, the goal of emissions testing is to identify vehicles with serious emission problems so the problems can be corrected.

Test Procedures & Equipment

The primary tool for measuring emissions testing is an infrared exhaust analyzer. It operates on the principle that different gases absorb different wavelengths of infrared light. Using these values, the analyzer can precisely measure the relative amounts of unburned hydrocarbons and carbon monoxide in the exhaust.

The infrared analyzer reads HC and CO content by splitting a beam of infrared light with a mirror and passing half the beam through a sample of the exhaust and the other half through a reference gas. The beam is chopped on and off by a spinning wheel. The beam then passes through a series of optical filters and strikes a light detector. The detector senses the difference between the two beams and generates an electrical signal that is processed electronically to produce the meter readings. Unburned hydrocarbons are displayed on one meter in parts per million (ppm) and carbon monoxide is displayed on a second meter in percent.

Four-Gas Analyzers—Most analyzers also have the ability to measure carbon dioxide (CO_2) and/or oxygen (O_2) in the exhaust for diagnostic purposes. Neither gas is a pollutant, but are good indicators of what's going on inside the engine. Using HC and CO alone to measure combustion efficiency is difficult because the catalytic converter "masks" many problems by significantly lowering HC and CO emissions. In other words, the converter cleans up the exhaust pollutants so well that some

---

OBD II & I/M 240

In an attempt to "push" diagnostic technology to a higher plateau so consumers can improve their chances of having complex emissions problems fixed properly, California passed a law that requires all vehicle manufacturers who sell cars in the state to offer a standardized onboard vehicle diagnostic system that conforms to its "Onboard Diagnostic II" (OBD II) requirements. Some 1994 California cars have the new OBD II diagnostic system but most received it in 1995. Federal clean air rules will also require OBD II in the remaining 49 states. In essence, the OBD II rules require a standard 16-pin diagnostic connector (9 of which provide OBD II diagnostic data while 7 are used for "other" data by the automaker).

The OBD II diagnostic rules require the ability to detect various kinds of emissions problems while the vehicle is being driven, such as exhaust gas recirculation (EGR) faults that increase oxides of nitrogen (NOx) emissions and transient misfiring. The fault codes and data stream is standardized so technicians can read out the information with a standard scan tool, thus eliminating dedicated cartridges and diagnostic connector adapters.

The OBD II regulations tie in with the new I/M 240 emissions testing program which is being adopted in cities with the worst air quality. The new "enhanced" inspection/maintenance program, which is called "I/M 240" because it involves a 240-second tailpipe emissions "transient loaded mode" test, is designed to catch various emission problems that cannot be detected by ordinary tailpipe idle tests.

Unlike a simple tailpipe emissions test, which only checks hydrocarbon (HC) and carbon monoxide (CO) emissions at idle, the I/M 240 test procedure simulates a typical urban driving cycle by monitoring tailpipe emissions while the vehicle is run on a dynamometer (actually more of a road simulator). During the I/M 240 test procedure, HC, CO and oxides of nitrogen (NOx) emissions are all sampled by the special test equipment and calculated in "grams per mile" (gpm) rather than the more familiar "parts per million" (ppm) for HC or percentage of concentration (%) for CO. Carbon dioxide (CO_2) is also monitored for diagnostic purposes. At the same time, a "purge" test is conducted to test the integrity and function of the charcoal canister and sealed fuel system.

During the initial phase-in period of I/M 240 testing, the failure cut points for the I/M 240 test are being set on par with those of existing emissions inspection programs (which generally allow a vehicle to exceed the applicable model year new car standards by as much as 20 to 30%). But as time goes on, the cut points will get tougher says the EPA.

One plus for consumers under the new I/M 240 test program is that emissions-related repairs must be performed by "emissions-certified" technicians. Technicians have to take special training courses and pass a difficult test. California has licensed smog repair shops and technicians for some time, but the new I/M 240 program takes the concept of certifying emissions technicians nationwide. Licensing technicians and shops allows the government to keep tabs on who's doing a good job at repairing emission problems and who isn't—a step that will hopefully result in better service for the public and cleaner air for the nation.

An exhaust analyzer is a great help in troubleshooting — and performance tuning. This four-gas unit measures CO, CO_2, HC, and O_2.

Exhaust analyzers are equipped with filters that must be kept clean.

alternative means of measuring combustion efficiency must be used. That's where the ability to read oxygen and/or carbon dioxide helps. The relative proportions of these two gases in the exhaust can reveal whether the air/fuel ratio is correct or not as well as other problems that affect engine performance and emissions. As combustion efficiency decreases, the oxygen content in the exhaust rises and carbon dioxide falls. An engine that is running at a nearly ideal air/fuel ratio of 14.5:1 will show about 14.5% carbon dioxide and 2.5% oxygen in the exhaust. Carbon dioxide readings of less than about 13% and oxygen readings greater than about 4% or 5% indicate poor combustion efficiency. This translates to an over-rich or over-lean air/fuel ratio, poor compression or an ignition problem.

High Readings

Hydrocarbon—High HC readings result when a lot of unburned gasoline and/or oil passes through the engine. This could be due to a misfiring plug due to fouling or a bad secondary ignition component(which can increase the HC reading 10 times over normal), worn piston rings or valve guides and valve stem seals (which allow oil to be sucked into the combustion chamber, see Chapter 13), or a super-rich air/fuel mixture. Other causes may be an idle mixture that's either too rich or so lean that it causes misfiring, a leaky needle and seat or a float problem, trouble in the power or main metering circuit, a malfunctioning air injection pump or other emissions-control device.

Carbon Monoxide—CO is a by-product of combustion in the presence of too little oxygen—if there were sufficient O_2, harmless CO_2 would be generated—so any condition that restricts air intake or causes extra fuel to be drawn into the engine can produce excessive carbon monoxide. These conditions include a rich idle mixture or choke setting, too much timing advance or low idle speed (both of which keep the throttle plates closed more than they should be, hence cutting down on air ingestion), a leaking power valve or accelerator pump, a clogged PCV system, an inoperative air injection pump or other smog-control device, and a clogged air filter. CO, by the way, is measured in percentage by volume.

Carbon Dioxide—Regardless of whether or not the vehicle is equipped with a catalytic converter, CO_2 will be about the same, the desired range being between 12% and 15%. Generally, an engine is operating most efficiently when CO_2 is at its highest. So, if you restrict the engine's intake and the reading rises, something is interfering with combustion. The most common causes of low CO_2 are misfiring and an improper mixture.

Oxygen—High oxygen levels indicate

ILLINOIS AIR TEAM

0813639

VEHICLE EMISSION TEST REPORT

YOUR VEHICLE'S EXHAUST WAS TESTED FOR HYDROCARBONS (HC) AND CARBON MONOXIDE (CO). HYDROCARBONS ARE UNBURNT GASOLINE WHICH CAUSE SMOG. CARBON MONOXIDE IS A COLORLESS, ODORLESS, AND TOXIC GAS. EXCESSIVE LEVELS OF THESE POLLUTANTS ARE CAUSED BY ENGINE MALFUNCTIONS WHICH CAUSE POOR GAS MILEAGE AND SHORTEN ENGINE LIFE. THE RESULTS OF THE TEST ARE SHOWN BELOW.

IF YOUR RESULTS ARE **PASS**, REMOVE THE OLD STICKER FROM YOUR WINDSHIELD AND REPLACE IT WITH THE NEW STICKER ON THE BACK OF THIS REPORT.

TEST RESULTS

FAIL

---->

FOR EXPLANATION OF **TEST RESULTS**, SEE COMMENT SECTION BELOW.

IF YOUR RESULTS ARE **FAIL**, DO NOT REMOVE YOUR OLD WINDSHIELD STICKER. SEE REVERSE SIDE FOR INSTRUCTIONS AND DIAGNOSTIC CODE EXPLANATION

DIAGNOSTIC CODES
2

FOR A MODEL YEAR 1981 OR NEWER LIGHT DUTY VEHICLE, YOUR **TEST RESULTS** ARE BASED ON BOTH HIGH AND LOW IDLE READINGS. FOR ALL OTHER VEHICLES THE RESULTS ARE BASED ON LOW IDLE READINGS ONLY.

POLLUTION LEVELS

HIGH IDLE				LOW IDLE		
HC (PPM)	CO (%)	CO+CO$_2$(%)		HC (PPM)	CO (%)	CO+CO$_2$(%)
0220	01.20	06.00	STATE STANDARDS	0220	01.20	06.00
0103	(02.21)	09.70	TEST DATE: 02/05/91	0175	00.80	14.15
			TEST DATE:			
			TEST DATE:			

VEHICLE INFORMATION

LICENSE #: VE2310
MAKE: MERC
YEAR: 83
ODOMETER: 62,000
VIN #: 1MEBP6520DW612251
CYLINDERS: 04
WEIGHT: 1

TEST INFORMATION

DATE: 02-05-1991
STATION: 10
OPER ID: 009
02/91 (05/91) 8.0
STICKER #: 4013936948
TIME: 08:56:59
LANE: 4
TEST #: 1
RPM: 2522

COMMENTS

IF YOU FAILED:
YOU ARE ELIGIBLE FOR TWO RETESTS.

SEE REVERSE SIDE OR STATE INSPECTOR FOR INSTRUCTIONS.

STEP 1 ➡ Clean the inside area of glass to which you are going to affix the sticker.

STEP 2 ➡ BEFORE REMOVING STICKER, READ THE INSTRUCTIONS ON THE BACK OF THIS FORM.

THIS LABEL MUST STAY WITH STICKER

V O I D
V O I D
V O I D
V O I D
V O I D

THIS LABEL MUST STAY WITH STICKER

THIS TEST IS AUTHORIZED BY THE VEHICLE EMISSION INSPECTION LAW, ILLINOIS REVISED STATUTES CHAPTER 95½ SECTION 13A-101 ET. SEQ., AND WAS PERFORMED IN ACCORDANCE WITH U.S.E.P.A. REQUIREMENTS AS SET FORTH IN 40 CFR, PART 85, SUBPART W.

Here's what bad news looks like. A failure means some type of repair or adjustment is necessary to meet the applicable emission standards for your model year of vehicle.

Forms vary from state to state, but may provide diagnostic information on the back to help you or a technician figure out what's causing an emissions problem. Most also require you to document any repairs that are made before a retest is given.

lean running, and a dramatic rise in the reading can be caused by lean misfiring. A typical engine with the air injection pump disabled will give an O_2 reading of between 0.5% and 4%, so anything within that range will be of questionable diagnostic value. Above 4% means a lean ratio and misfire. Below 0.5% means gasoline is being wasted.

NOx—What about NOx? Well, there's no practical way to measure NOx in an inspection or service situation—a laboratory would be required. So, inspection of the EGR system and the three-way catalyst is used to determine indirectly whether or not NOx is being produced in unacceptable amounts.

To get accurate readings with an exhaust analyzer, the analyzer must be properly calibrated (which is automatically done and self-checked on most equipment today). The vehicle must also be at normal operating temperature, and the exhaust system must be leak-free.

If you fail the test because of a loose or leaking vacuum hose or a clogged PCV valve, the repairs are normally done on-site for little or no charge. However, if a major item is missing (such as a catalytic converter) or has obviously been modified (such as inserting an H-pipe), this is considered a violation. If this is proven to be intentional, you may be fined. In addition, if your engine compartment looks like this, you may be required to prove that the performance aftermarket items added are street legal, so bring documentation such as EO numbers. Believe it or not, this engine puts out nearly 300 horsepower and is completely emissions-legal.

FAILING A TEST

A vehicle that fails an emissions test is essentially a vehicle that's in trouble. Unless the emissions problem is corrected, you won't receive the required certificate, sticker or registration. Consequently, you won't be able to legally drive your vehicle after the current certificate, sticker or registration expires. The options are: (1) fix the problem, or (2) get rid of the vehicle. Laws vary from state to state, and in some states selling or trading a vehicle in "as is" condition may not relieve you of your responsibility to have an emissions problem corrected.

What if you just ignore the rules? Every emissions program has some means of enforcing the rules. Penalties range from fines to suspension of driving privileges. The bottom line is this: You must have your vehicle's emissions tested when required to do so, and you must make an attempt to have an emissions problem repaired if your vehicle fails to pass the test.

On your inspection form, you'll generally find instructions that tell you what to do if you failed to pass the test. Don't throw the test form away because it contains valuable diagnostic information that can help you or your technician correct your emissions problem, and you may be required to return the form when repairs have been completed and/or for reinspection.

Every state has their own emissions test report format, so we'll have to stick to generalities in describing what's on the form itself. As a rule, most reports are computer printed and contain the following:

• Vehicle identification (your year, make, model, license number and/or VIN number).

• Test information (date, time, test location, lane number, operator identification, etc.).

• Emissions test results (your vehicle's tailpipe readings).

• A checklist of visually inspected items (if applicable).

• The applicable emissions standard cut points that your vehicle must meet to pass the test.

• A statement or notation indicating that you either passed or failed the test.

• If you failed, a possible explanation as to why your vehicle failed (diagnostic information).

• Instructions on what you must do to correct an emissions problem, including requirements for waivers and reinspection.

EMISSIONS VIOLATIONS

Emission violations are more serious for three reasons:

1. An emissions violation for tampering may also result in a fine if there's obvious evidence that you intentionally removed, disconnected or otherwise rendered inoperative any emissions-control device.

2. Missing emissions-control equipment must be replaced regardless of cost. Either the parts are replaced or the vehicle will not be considered legal for street use.

3. The cost of replacing missing emissions-control parts cannot be applied toward a waiver limit for emissions repairs—which means you may end up spending more money on any additional repairs that might be needed to bring emissions down to acceptable levels.

If your vehicle failed an inspection because some minor piece of emissions-control equipment was missing or incorrect (like the heat riser duct hose to the air cleaner or the wrong gas tank filler cap), correcting the violation should be no big deal. On the other hand, if the violation occurred because a major emission component was missing (such as the catalytic converter, air pump, no EGR system, etc.), replacing the missing part or parts can get mighty expensive. A original equipment replacement converter can cost several hundred dollars. Even an aftermarket "universal" type of replacement converter can cost well over a hundred dollars by the time it's installed. For a list of what is and isn't okay, see Chapter 3. With an older vehicle, it may be very difficult if not impossible to replace missing emissions-control components if the components are no longer available either through a car dealer or automotive parts stores. The only source for such parts may be a salvage yard, which means your chances of finding all the correct components may be zilch if the vehicle is not a "collector" car or popular model. Most salvage yards don't keep vehicles for more than a few years, and once most of the valuable parts have been stripped off they're crushed for scrap.

In some instances, the cost to repair or replace missing emissions-control components may be more than the vehicle is worth. It makes no sense to spend a lot of money replacing missing or defective emissions-control parts on a vehicle that itself may only be worth a few hundred dollars.

If your vehicle failed the test because of excessive emissions and not some other type of violation, reading the explanation as to why your vehicle failed, along with any diagnostic information that's provided, can help guide you in making your repair decisions. If the problem seems minor, a few simple adjustments may be all that's necessary to bring the emissions down to acceptable levels. If you're competent under the hood, you can probably do these adjustments yourself. Most inspection programs allow you to attempt the repairs yourself but may or may not allow you to apply the cost of do-it-yourself repairs toward the waiver limit. The other option is to take your vehicle to a professional service facility of your choice for repairs.

Warranty Coverage

Newer vehicles are covered by an emissions performance warranty and

---

DIRTY OLD CARS

It's no secret that older cars are dirtier than newer ones. The tailpipe emissions from late-model cars with computerized engine controls and three-way catalytic converters are only a fraction of older "pre-emissions" cars: 1981 and later cars produce 96% less carbon monoxide (CO) and unburned hydrocarbons (HC), and 76% less nitrogen oxides (NOx).

Studies have found that cars and light trucks five or more years old represent 54% of the vehicle population—yet account for 75% of all automotive pollution!

Increased pollution from older cars and trucks is often due to wear and lack of proper maintenance. And even when they're properly tuned, older vehicles typically produce much higher levels of pollutants in the exhaust because they lack the emission controls of their later model counterparts.

Pre-1968 vehicles are the worst because most lack any emission controls whatsoever. Cars from the early 1970s aren't much better. Those with catalytic converters (most 1975 and later) are better, but the cleanest ones are those with computerized engine controls built since 1981 (1980 in California).

In 1994, the federal emissions limits for new cars and light trucks were lowered even more. The previous federal standards allowed no more than 0.41 grams/mile (g/m) of HC, 3.4 g/m of CO and 1.0 g/m of NOx for the first 50,000 miles. The new standards lower HC to 0.25 g/m HC and NOx to 0.4 g/m. California has gone even further, requiring "ultra-low emission vehicles" and even "zero emission vehicles" by the latter part of the 1990s. So what do all these numbers really mean? It means a 1975 to 1979 model year car puts out roughly as much HC and CO pollution as four late-model cars; a 1972 to 1975 model year car produces as much of these pollutants as seven late-model cars; a 1968 to 1971 model year car produces pollution equivalent to 10 late-model cars; and a '67 or earlier pre-emissions controlled car pumps as much crud into the atmosphere as 25 late-model cars!

Every time one of these older cars or trucks is scrapped, it offsets the pollution produced by anywhere from 4 to 25 new cars. Most, in fact, are actually much dirtier, producing the pollution equivalent of as many as 75 new cars!

Time and natural attrition removes millions of older cars and trucks from the vehicle population every year. Even so, there are still a lot of these older, dirtier vehicles that are still on the road and will be for some time to come.

emissions defect warranty that covers 100% of the cost of parts and labor for repairing any emissions related part or system during the covered period. The coverage is for 5 years or 50,000 miles, whichever comes first.

To get free repairs under this warranty, you must take your vehicle back to an authorized new car dealer. It doesn't have to be the same one where the vehicle was purchased new, but it does have to be a dealer who sells the same make (Ford, Chevy, Toyota, etc.). Most dealers will not honor warranty claims for work done by anyone except in the most unusual circumstances (your car broke down in Nowhere, Utah and the nearest dealer was 800 miles away).

Every vehicle that's been built to federal emission standards since 1981 has had this 5 year/50,000 mile emissions warranty. The list of items covered is specific to each vehicle, and varies somewhat from one model to another and from one make to another. Your dealer should be able to help you determine what's covered and what isn't. Before you authorize major repairs on your nearly new vehicle, you should check the warranty coverage. For more details on warranty coverage, see Chapter 3.

Retesting

Once the cause of the emissions failure has been accurately identified and the necessary adjustments or repairs made, all that remains is taking the retest. You should bring along all proof of repairs (parts and/or labor receipts, a repair order, diagnostic printout, etc.) as required to qualify for a retest. Many states will not allow you to have your vehicle retested unless you can provide proof that an attempt has been made to repair it. Such rules are considered necessary to prevent "gamblers" from taking repeated tests in hopes of getting lucky and hitting an "easy" lane or test facility. If an inspection program is working correctly, all the test equipment should be calibrated to the same standard and read more or less the same. Realistically, this doesn't always happen. If you know of a lane or test facility that's easier than others, keep it a secret.

If your vehicle has a serious emissions problem (like this one obviously does), you may be able to qualify for a waiver. But you'll first have to spend the required dollar amount before they'll give you a waiver.

Waivers

The one "loophole" in every emissions inspection program with which you should be familiar is the "waiver." A waiver is a way around the emissions test if your vehicle can't meet the required emission standards and you've made an attempt to repair your engine's emission problem. To qualify for a waiver, however, you first have to spend a certain amount of money on attempted repairs. Furthermore, you must document what you've spent on repairs by providing receipts or repair orders. As mentioned earlier, some programs won't count any repairs you've made yourself towards the waiver limit—only those made by a professional repair facility.

Limits—Waiver limits vary from state to state, but generally range from $50 to $300 or more. The waiver limit may be fixed (one amount for all vehicles), or it may vary, depending on the model year of your vehicle. In some programs, the waiver limit for older vehicles is set lower than that for newer ones. The logic is that people who own older vehicles are probably less able to afford expensive repairs than those who own newer vehicles. Unfortunately, it's often the older vehicles that need the most attention. (See sidebar "Dirty Old Cars" on p. 121.)

In some states (California, for example), you may be able to qualify for a waiver if the cost of "anticipated" repairs will obviously exceed the waiver limit. If the waiver limit happens to be $175 for your model year and your engine flunked the emissions test because of a serious oil consumption problem, the cost of an overhaul would obviously be a lot more than the $175 waiver limit.

One thing to keep in mind about waiver limits is that they're going up. Most states have raised their waiver limits in recent years to reflect a more serious commitment to clean air. The truth of the matter is the EPA has put the pressure on to establish waiver limits that are high enough to produce meaningful results. Waivers are a political hot potato because the cost must be borne by motorists (voters). If waivers are set too high, people may be forced to make repairs they can't afford. If set too low, the repairs may be ineffective in reducing emissions. So states have generally tried to set waiver limits that realistically reflect both the public's ability to pay and their willingness to reduce pollution.

In states that have adopted the new federal I/M 240 emissions testing program (see sidebar p. 116), waiver limits of up to $450 have been imposed! The EPA says the higher figure is necessary to address the more serious emissions problems. The $450 figure also means a lot of older vehicles that have become too expensive to fix will end up going into early retirement and be replaced by newer, cleaner vehicles.

TIPS FOR PASSING

Here are some suggestions for improving your chances of passing an emissions test. Generally speaking, the following items should be taken care of before you take the test, especially if you think there's a possibility your vehicle might not pass. In addition to these steps, you should tune your vehicle according to the guidelines in Chapter 17.

•Keep your vehicle properly maintained. A well-maintained vehicle is generally a good running vehicle with low emissions.

•Fix any known problems before you go for the test. Why risk failing if you know there's something that needs attention.

Before you go to an emissions test, you should follow some of the tips below to help increase your chances of passing. It is especially important to get your car tuned prior to a test, particularly if it is an older model that hasn't been serviced in a while.

Missing converter? Replace it. Punched out filler restrictor? Replace or repair it. Misfire? Diagnose and fix the problem. Problem with the choke? Fix it. Lousy gas mileage? It may indicate a fuel, ignition or compression problem. Check it out.

•Replace the spark plugs (or clean and regap the old ones), replace the air filter and check and adjust the idle speed to specs. Fouled spark plugs, a dirty air filter or an idle speed that's too high are some of the most common reasons for excessive emissions.

•Change the oil. The oil in the crankcase can become badly contaminated with gasoline if you've been doing a lot of short trip driving, especially during cold weather. These vapors can siphon back through the PCV system and cause elevated CO readings.

•Fill up with a tank of reformulated gasoline or oxygenated gasoline that contains alcohol. It burns slightly leaner and produces lower CO emissions. That may help you make the cut if your CO readings are marginal.

•Make sure the engine is at normal operating temperature before it is tested. Drive the car an adequate distance to bring it up to normal operating temperature, preferably at highway speeds briefly to blow out the cobs.

•Go to the test lane at a time when they're not busy. Excessive idling while waiting in a test lane may cause your catalytic converter and/or oxygen sensor to cool down enough where they won't control emissions properly, causing higher than normal readings. ■

16
TROUBLESHOOTING

Whether from neglected maintenance or a part failure, problems with emissions-control systems are pretty common. Of course, that's to be expected when a mind-boggling amount of extra plumbing and hardware is added to a vehicle. But don't start wishing for the old days because they're not coming back. Instead, learn how to deal with these modern systems. You may even find them to be scientifically interesting and challenging.

PCV

A common test of a typical system is to remove the PCV valve from its grommet in the valve cover and shake it to make sure it rattles. This will indicate whether or not the valve is clogged or jammed, but it can't tell you if the spring has broken or lost its tension.

Another check is to feel for strong vacuum at the end of the valve while the engine is idling. If there isn't any, the valve, hose or passages are blocked. Even if vacuum is present, you still won't know if the valve's spring is weak or broken.

Testers are available that attach to the oil filler hole and measure the vacuum or air flow the system produces.

A PCV valve is an inexpensive part, especially when you consider the value of the engine it's largely responsible for protecting. So, it only makes sense to replace it at the intervals the manufacturer recommends, which range from every 12,000 to every 30,000 miles.

A manual vacuum pump is a big help in diagnosing emissions-control systems.

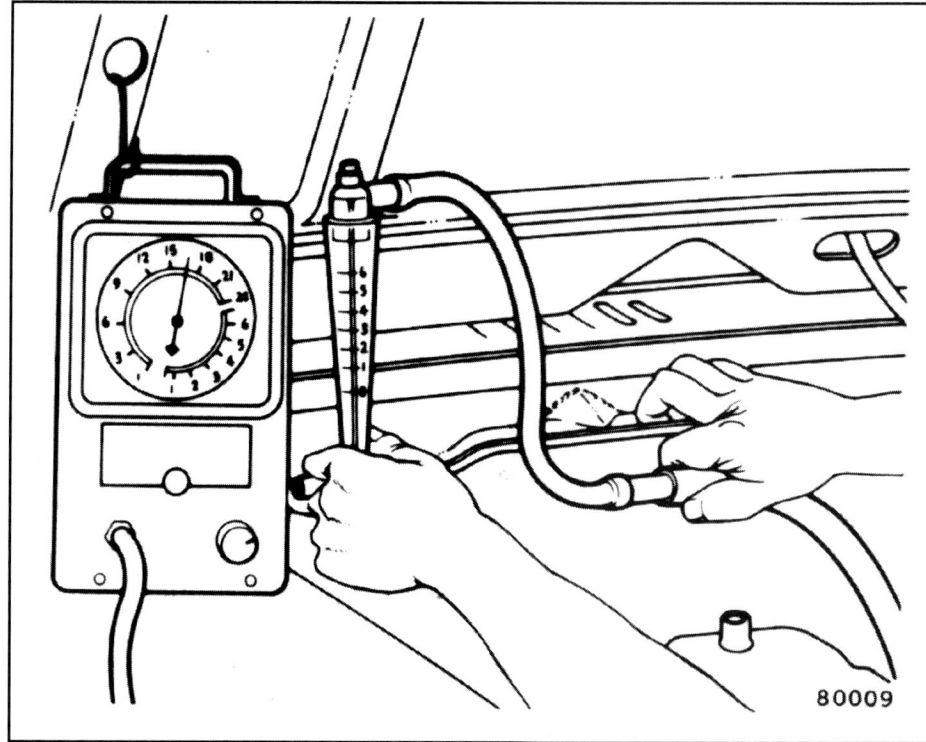

Way back when PCV systems were considered high-tech, set-ups like this were used to check for adequate flow. But all you really have to do is shake the valve to see that it rattles, and feel for vacuum (American Motors).

EVAP

One of the most common problems with evaporative emissions control systems is the smell of gasoline inside the vehicle, which is usually due to a disconnected or improperly routed vapor line. Other possibilities are fuel starvation, and even a collapsed tank caused by a restriction in the system or the installation of the wrong gas cap.

The liquid/vapor separator is mostly trouble-free, although it's possible for the liquid return to become plugged with debris, such as rust or scale, from inside the tank.

Any blockage between the tank and the charcoal canister can cause trouble. Check tank venting by removing the gas cap, disconnecting the vent line from the canister, then blowing into the line—it should be clear.

Charcoal canisters are usually good for the life of the car. Problems with purge control valves and clogged filters do occur, however. You can check a typical purge valve with a vacuum pump. It should hold vacuum for 15-20 seconds.

Also, there is always the chance that somebody may have mixed up the vent, purge, and control vacuum lines.

On units equipped with computer-controlled solenoid purge valves, you'll have to refer to the manufacturer's service literature for information on when the solenoid is supposed to open.

HEATED AIR INTAKE

For some inexplicable reason, the heated air intake system is frequently ignored as the cause of driveability complaints. Many supposedly incurable hesitation problems are cured simply by replacing a missing hot air duct, reconnecting a vacuum hose, or installing a new bleed valve.

The symptoms associated with a malfunction in the heated air intake system are exactly what you'd expect: hesitation, stalling and rough idling if too little hot air is ducted to the carburetor or throttle body, and the opposite; reduced power output and fuel efficiency if the flapper valve keeps the unheated air passage closed.

Since a leak or other interruption in the vacuum signal to the vacuum motor, or a perforated diaphragm in the vacuum motor itself, will normally result in the flapper valve closing off heated air (in other words, always open to unheated air), the first set of problems just mentioned is the most common. Only if there's some sort of binding condition or if the bleed valve jams closed will the flapper valve stay up and close off unheated air.

Diagnosis

Diagnosis is essentially a process of making sure the valve closes below the specified temperature, and opens above it, and that the vacuum motor raises the door when you hook it to a manual vacuum pump.

First, give it a visual examination. Make sure the vacuum lines are not cracked, crushed or broken, and that they're properly connected. Next, see that the heat tube that runs from the exhaust manifold stove to the air cleaner snorkel is intact. Look inside the snorkel (you may have to remove the air cleaner or use a mirror) to see if the flapper valve is in the down position. Or, you can insert a pencil or screwdriver to find out the position of the valve.

Start the engine. Providing the temperature is below that at which the bleed valve opens, the flapper valve should rise to the heat-on position, then drop gradually as the engine warms up.

If you don't get these results, there are several ways to proceed to uncover the fault. One logical progression is as follows:

1. Attach a spare line to a known source of manifold vacuum or to a manual vacuum pump.

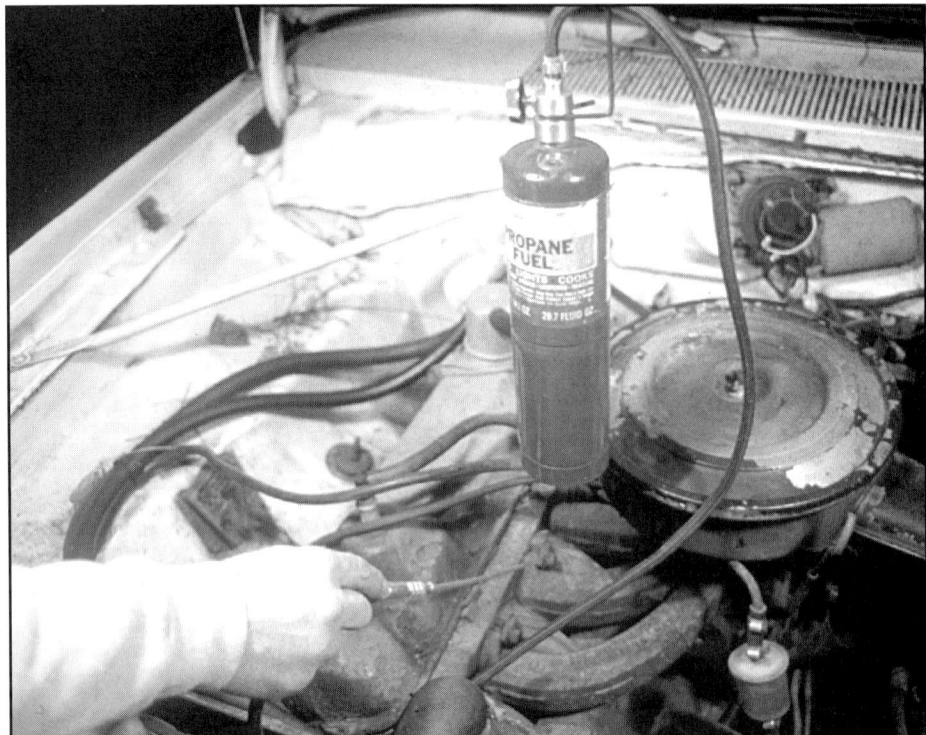

A good way to find a vacuum leak is to introduce propane to likely joints and see if the idle changes.

2. Remove the standard vacuum line from the nipple of the vacuum motor.

3. Connect your source of vacuum to the vacuum motor.

4. If the flapper valve rises now, the trouble is probably in the bleed valve or the line that feeds it vacuum.

5. If the flapper valve doesn't rise when given direct vacuum, there's most likely a leak in the vacuum motor diaphragm or possibly the flapper valve or linkage is jammed.

6. If you're using a manual vacuum pump and the flapper valve does rise when vacuum is applied to the vacuum motor, see that the motor holds vacuum for a reasonable length of time (a common limit is less than a 10 in. Hg. drop in 5 minutes). If not, the diaphragm is leaking.

7. To check the bleed valve, first look up the temperatures at which it should close and open. A common Chrysler system, for example, will raise the flapper valve with a cold engine and the ambient temperature below 50 degrees F.

8. Remove the air cleaner and cool the bleed valve below the closing temperature. This can be done by covering it with ice cubes.

9. With the hand pump, apply vacuum to the inlet side of the bleed valve (make sure the outlet side is properly connected to the vacuum motor).

10. If the flapper valve doesn't rise, the bleed valve is probably defective. It's relatively inexpensive and very easy to replace.

Non-Vacuum Type

The normal procedure for checking the non-vacuum type of thermostatically controlled air cleaner is, first, to see that the flapper valve and its linkage move without binding, then to immerse the thermostatic bulb in cold water to see that it raises the flapper valve, and hot water to see that it lowers it. The specific steps for a typical example, the old AMC six-cylinder, are as follows:

1. Remove the top of the air cleaner and immerse the entire snorkel in a large pan of cold water, making certain that the thermostatic bulb is covered.

2. Place a thermometer in the pan and observe it while heating the water slowly.

3. The flapper valve should remain closed below 105 degrees F At 130 degrees F, the valve should be open completely.

4. If this doesn't happen, and the valve mechanism isn't binding, the thermostatic bulb is defective.

EFE

Testing mechanical EFE's is just a matter of applying vacuum to the motor to find out if it closes the heat riser (on a garden-variety GM system, 10 in. Hg. shouldn't bleed off in less than 20 seconds), and feeling for vacuum at the line below the specified temperature. Of course, make sure the heat riser shaft isn't frozen in the manifold.

On the electrical variety, current must be present at the heating element's lead when the engine is cold. If not, check the fuse, the thermostatic switch and the wiring and connections. On the computer-controlled type, check for a trouble code. The element itself should have the specified resistance between its hot lead and ground, typically no more than three ohms.

AIR INJECTION

The following is a list of common problems and diagnostic tips associated with pump-type air injection systems:
• Pump noise: Remove the belt and spin the pulley by hand. Don't expect silence, but reject loud squealing and heavy

turning resistance. Sometimes the cause of pump failure is a leaking check valve that allows exhaust to enter the pump.

- Backfiring: With an ordinary diverter valve, see if vacuum from a hand-operated pump applied to the valve's nipple causes air to flow from the muffler. If not, either the valve isn't functioning, or the pump isn't producing flow.

- Failing an emissions test: Disconnect the pump outlet hose at the check valve to make sure there's sufficient pressure and volume at idle.

- Exhaust leakage noise: Especially on GM vehicles, the air injection manifold tends to rust through, causing a noisy exhaust leak under the hood.

With aspirator valves, all you can do is listen for exhaust noise and see if there's pressure at the intake instead of vacuum.

CATALYTIC CONVERTERS

Even skeptics will have to admit that catalyzing the exhaust stream has turned out to be a terrific idea—it allows us to have high performance and cleaner cars. That doesn't mean, however, that there aren't any problems. Fouling, clogging, melt-down, breakage of the ceramic substrate, etc., can cause a converter to stop doing its job, and/or plug it and raise backpressure to the point where performance, driveability, and fuel mileage are seriously affected, or even of stalling and no-starts.

Both monolithic and pellet-type converters are subject to plugging from a melt-down (overheating can occur if there's too much gasoline in the exhaust from a rich mix or a misfiring cylinder) or disintegration of the ceramics (GM dual-bed units are prone to this).

Causes

The consensus of opinion among experts is that the most common cause of failure is an engine that pumps out too much unburned fuel, which can overheat or carbon-clog the catalyst. The excess gasoline in the exhaust may be due to a bad spark plug or valve, but an overly rich air/fuel mixture is certainly a good possibility. With older cars, you might attribute it to something like a heavy carburetor float, but on later models with computerized engine controls, most authorities point to continuous running in open loop. In other words, cases where the ECU (Electronic Control Unit) doesn't have the opportunity to tailor the air/fuel mixture to conditions, so it falls back on the limited operating strategies in its PROM (Programmed Read-Only Memory). These emergency calibrations are usually on the rich side.

Furthermore, most auto repair technicians say a failed oxygen sensor is usually the reason closed loop can't be achieved (of course, there are many possible causes, ranging from some other faulty sensor to a thermostat that doesn't close). It makes sense to check out oxygen sensor activity any time you're presented with a converter problem, and not to replace a catalyst until you have corrected whatever condition has caused its failure.

Another cause of clogged and contaminated catalysts is oil. A set of bad valve seals can cause a great deal of carbon formation, and metals present in the oil will coat the catalytic agents.

Diagnosis

The easiest test for plugging is done with a vacuum gauge. Note the reading at idle, then hold rpm at 2500. The needle will drop when you first open the throttle, then stabilize. If the reading then starts to fall (for example, Chrysler says down to 10 in. Hg. from 15), excessive backpressure is the probable cause, so suspect a blockage somewhere in the

If you want to get fancy, you can cobble up something like this to measure air injection pump pressure. Or, just feel for sufficient flow (Ford).

A falling manifold vacuum reading at a steady 2500 rpm indicates a restriction in the exhaust system, probably from a clogged cat.

exhaust system.

The next step is to check backpressure directly. If the car has air injection, disconnect the check valve from the distribution manifold, and plug in a pressure gauge with a low scale. Or, remove the oxygen sensor and take your reading at its hole in the manifold or headpipe. There's some dispute over how many psi is normal. Check the specifications for the vehicle at hand, but if you see over 1.25 psi at idle, or more than 3 psi at 2000 rpm, there's a restriction in the exhaust system.

Another direct kind of test is to open a connection to relieve pressure, then see how the engine runs. Take another vacuum reading under these conditions.

In cases where you still aren't sure whether or not the converter is restricting exhaust flow, you can resort to one of the inexpensive kits now available that allows you to check pressure both fore and aft of the unit. Typically, these comprise a hole punch for your air chisel, a self-tapping hollow nipple, and some screw-in plugs for the holes you'll be making. If there's any noticeable difference between the readings, you've found a restriction.

Thump Test—The "thump" test can be very useful. Raise the car securely, then put on a heavy glove or use a rubber mallet to give the converter a few solid blows. With the pellet type, you should hear some rattling. If not, the pellet bed is probably clogged or heavily contaminated. On the monolithic type, a rattle means the ceramic honeycomb substrate is at least partially broken, although the situation may not be bad enough to require replacement.

Heat Test—Temperature comparison can be used to find out if the catalyst is working. Using a pyrometer, check the temperature of the pipe just ahead of the converter with the engine fully warmed up and running. Then take a reading on the outlet pipe. If the outlet temperature is at least 100 deg. F higher than the inlet temperature, the catalyst is operating. If both readings are nearly the same, or the outlet is cooler than the inlet, the rapid oxidation reaction is not occurring. This, however, may be due to lack of air injection rather than to catalyst contamination.

Check Air Injection—If emissions are high, or there's little temperature differential between the inlet and outlet pipes, you still can't be sure if the catalyst itself is bad, or if it's just not getting enough air to support oxidation. So, check the air injection system or aspirator valve, either of which has the same basic effect as a blacksmith's bellows (see Chapter 7). See that the air pump drive belt is okay and properly adjusted, and examine the check valve to see if it's corroded through. You may hear an exhaust leak, or the pump may start making excessive noise. If you suspect that the check valve is allowing exhaust to flow backwards, remove it and blow through both ends. It should let air pass in one direction, but not in the other.

Examine the air injection manifold, another component that tends to rust out. Find out if the diverter valve is dumping pump output when the throttle is opened, then allowed to snap closed. You should feel and hear air escaping from the little muffler on the valve. If not, backfiring may occur, which can shatter the catalyst. Before condemning the diverter valve itself, make sure its vacuum line is intact and properly connected.

With aspirator valves, you should be able to hear and/or feel the fluttering of the internal flapper.

EGR PROBLEMS

If you grasp the fundamentals of how EGR is supposed to work (see Chapter 10), and you're able to figure out what type of EGR system is on your engine (call a dealer or check a service manual), then you ought to be able to determine whether or not your EGR system is functioning properly. Troubles associated with a malfunctioning EGR system can be reduced to recirculation occurring when it shouldn't, or not occurring when it should. Typical symptoms include:

- Since EGR's cooling effect keeps the charge from exploding, detonation is probable if the valve fails, the vacuum signal is interrupted or the exhaust passages become plugged.

- Stalling or roughness at idle if the valve jams open or the signal is constant due to a hose mix-up.

- Poor cold driveability if the signal reaches the valve too soon after startup.

It would be dumb not to take EGR into account when troubleshooting any of these symptoms. However, before focusing on the EGR system, you should check other areas first. Does your engine have a detonation (spark knock) problem when accelerating under load? Refer to the timing specs for your engine and check ignition timing. Your timing may be over-advance. If the timing is within specs, check the engine's operating temperature. A cooling problem may be causing your engine to detonate. If the temperature is within its normal range and there are no apparent cooling problems, other possibilities to investigate include spark plugs that are too hot for the engine application, a lean air/fuel mixture, low octane fuel or too much compression (due to a buildup of carbon in the combustion chambers or because of pistons or heads that have too much compression for the fuel you're using).

As a preliminary, see that all the vacuum lines involved are intact and properly routed (use the diagram under the hood, or look it up in a suitable manual). You may have seen cars with hoses so scrambled that nothing was working properly, not to mention those

A common cause of heavy detonation/pinging is an intentionally disabled EGR valve.

on which they've been plugged intentionally to disable the EGR valve or other emissions-control devices.

A related point: On vehicles which have the automatic transmission modulator hose teed into the EGR vacuum line, anything that disrupts vacuum to the EGR will cause shifting problems and eventually a ruined transmission.

There are several ways to proceed. You can follow the EGR troubleshooting procedure that's listed in a service manual for your engine (which may be very specific depending on the application). On late-model computer-controlled engines, there may be trouble codes that relate to the EGR system. On such an application, the first step would be to read out the code or codes using a special diagnostic procedure (a special "scan tool" may be required to read the codes if the vehicle doesn't provide manual flash codes), then to refer to the specific diagnostic charts in a service manual that tell you what to do next.

GM—On late-model GM applications, for example, a code 32 indicates an EGR problem. The logic by which the onboard diagnostics detects trouble follows one of two routes. On some applications, a code 32 is set when the computer detects a richer fuel mixture off idle (indicating no EGR). On others, a code is set if the computer energizes the EGR vacuum solenoid but does not detect a corresponding drop in intake vacuum.

Ford—On late-model Fords, a code 31 indicates a problem with the EGR valve position sensor (EVP). It works like a throttle position sensor, going from high resistance (5500 ohms) when the EGR valve is closed to low resistance (100 ohms) when it is open. You'll find these EVP sensors mostly on Ford EEC-IV V-6 and V-8 engines. Other codes include a code 32, which indicates the EGR circuit is not controlling. A code 33 means the EVP sensor is not closing, and a code 34 indicates no EGR flow. Any of these codes may indicate a faulty EGR valve as well as a problem in the EGRC or EGRV vacuum solenoids. Other codes include a code 83 (EGRC circuit fault) and code 84 (EGRV circuit fault). Both indicate an electrical problem in one of the solenoid circuits. The solenoids should have between 30 and 70 ohms resistance.

Diagnosis

If you don't have OBD or the equipment to take advantage of it, then the following procedures will uncover

If opening the EGR valve by hand doesn't cause the engine to idle roughly or stall, the passages are plugged.

many common, generic EGR problems:

1. Use a vacuum gauge to check the EGR valve vacuum supply hose for vacuum at 2000-2500 rpm. There should be vacuum if the engine is at normal operating temperature. No vacuum would indicate a problem such as a loose or misrouted hose, a blocked or inoperative ported vacuum switch or solenoid, or a faulty vacuum amplifier (or vacuum pump in the case of a diesel engine). Sometimes loss of EGR can be caused by a failed vacuum solenoid in the EGR's vacuum supply line. Refer to a vacuum hose routing diagram in a service manual or the hose routing information on your vehicle's emissions decal for the location of the solenoid. If the solenoid fails to open when energized, jams shut or open, or fails to function because of a corroded electrical connection, loose wire, bad ground, or other electrical problem, it will obviously affect the operation of the EGR valve. Depending on the nature of the problem, you may have no EGR, EGR all the time, or insufficient EGR. If bypassing the suspicious solenoid with a section of vacuum tubing causes the EGR valve to operate, find out why the solenoid isn't responding before you replace it. The problem may be nothing more than a loose or corroded wiring connector.

2. Inspect the EGR valve itself. If the EGR valve stem is accessible, push it against spring pressure. It should move freely and return fully. Because of the valve's location, it may be difficult to see whether or not the valve stem moves when the engine is revved to 1500 to 2000 rpm by slowing opening and closing the throttle. The EGR valve stem should move if the valve is functioning correctly. A hand mirror may make it easier to watch the valve stem. Be careful not to touch the valve because it will be hot! If the valve stem doesn't move when the engine is revved, remove the hose and feel for vacuum as you speed up the engine again. If you find vacuum, the valve is at fault. If there's no vacuum present, check out the controls. Note that if you're dealing with a Ford pressure-operated unit, you should feel pressure instead. there's probably something wrong with the EGR valve.

Another way to "test" the EGR valve on some engines is to apply vacuum directly to the EGR valve. Depending on the type of EGR valve that's used, vacuum should pull the valve open, creating the equivalent of a large vacuum leak. This should cause a sudden change in the engine's idle quality, causing a noticeable increase in roughness and/or a drop of at least 100 rpm. But this test doesn't work on many engines with backpressure EGR valves.

Backpressure type EGR valves are more difficult to check because there must be sufficient backpressure in the exhaust before the valve will open when vacuum is applied. One trick that's sometimes used is to create an artificial restriction by inserting a large socket into the tailpipe, then applying vacuum to the valve to see if it opens. Don't forget to remove the restriction afterwards. Remember, don't condemn a backpressure-type valve until you're sure the exhaust system is stock, has no leaks, and isn't clogged. Also, don't try to get the positive backpressure type to hold vacuum with the engine off or idling. Backpressure EGR valves sometimes fail if the hollow valve stem becomes clogged with carbon or debris. This you can see for yourself.

Vacuum Amplifier Diagnosis

On EGR systems with vacuum amplifiers, there are two types of amplifier design. Early-model units typically use a single connector, while late-model amplifiers have two connectors. To test the system:

Throttle Test—With the engine at operating temperature, perform the throttle test. Slowly open and close the throttle while keeping the engine under

2500 rpm. If the EGR valve opens and closes, everything is okay. If the valve fails to open, proceed to the next step.

Check Vacuum—Pull the vacuum hose off the EGR valve and connect the hose to a vacuum gauge. Repeat the throttle test. If a strong vacuum signal (10 or more inches) is reaching the EGR valve, the amplifier is okay and the problem is a bad EGR valve diaphragm. If no vacuum or a weak vacuum is detected, move on to the next step.

Check TVS—If the system has a temperature vacuum switch or solenoid between the amplifier and EGR valve, pull the hose off the TVS that leads to the amplifier and connect your vacuum gauge. Repeat the throttle test. If vacuum is reaching the TVS, the amplifier is okay and the problem is in the switch.

Reconnect the TVS Hose—Pull the vacuum hose from the carburetor venturi signal port. Temporarily plug the port and connect a hand vacuum pump to the signal hose. Start the engine and send a 1- to 3-inch vacuum signal through the line to the amplifier using your hand pump. If the engine starts to idle rough, indicating that the EGR valve is opening, the venturi vacuum port may be clogged or obstructed inside the carburetor. A thorough cleaning will be necessary to clear the path. If nothing happens and the vacuum supply line to the amplifier is okay, the amplifier is defective and should be replaced.

It should be noted that on some cars, the vacuum amplifier is calibrated to generate a small vacuum (2 inches or so) at all times, even when there is no carburetor venturi vacuum signal. This is to maintain vacuum in the system for quicker response. This small amount of vacuum is not enough to cause the EGR valve to open (most require at least 10 inches of vacuum), so do not think that the amplifier is defective. If you discover that full manifold vacuum is reaching the EGR valve at idle, the amplifier is leaking vacuum internally and should be replaced.

Unclogging

GM says never to use a solvent to dissolve deposits in an EGR valve, but Chrysler tells you it's okay, providing you're careful not to get any on the diaphragm. With most specimens, you'll be cleaning the pintle and valve seat with a dull scraper or wire brush, and knocking out loose carbon by tapping the pintle. But these parts are expensive, so a more satisfactory means of cleaning them is desirable. Some GM versions, for instance, can be disassembled for this purpose. Just make sure you don't damage the seating surface in the process, and that you scribe marks so you can get it back together in the proper alignment.

EGR manifold passages often need cleaning, too. Frequently, the cause of a detonation condition can be traced to carbon build-up in these passages. If you saw that the EGR valve is indeed opening, remove the valve and look into the passages (use a mirror if necessary). Deposits can usually be broken loose with a stiff wire or an awl.

The thing to remember when troubleshooting EGR systems is that the basic principle of operation is the same regardless of how many components and gizmos are incorporated into the system. There should be no EGR at idle or when the engine is cold. And there should always be EGR in a warm engine under part-throttle operation. If the EGR valve is not working, start at the valve and trace backward to see why vacuum is not getting through. The same applies to those situations where vacuum is reaching the valve when it is not supposed to. The most likely cause here is a temperature vacuum switch or solenoid that is open all the time, or a defective vacuum amplifier.

One other thing that should be checked when inspecting any EGR system is to make sure that all the vacuum plumbing is correctly routed. With all the vacuum

Hard starting with CIS may mean the cold start injector isn't doing its job. You can check its action directly like this.

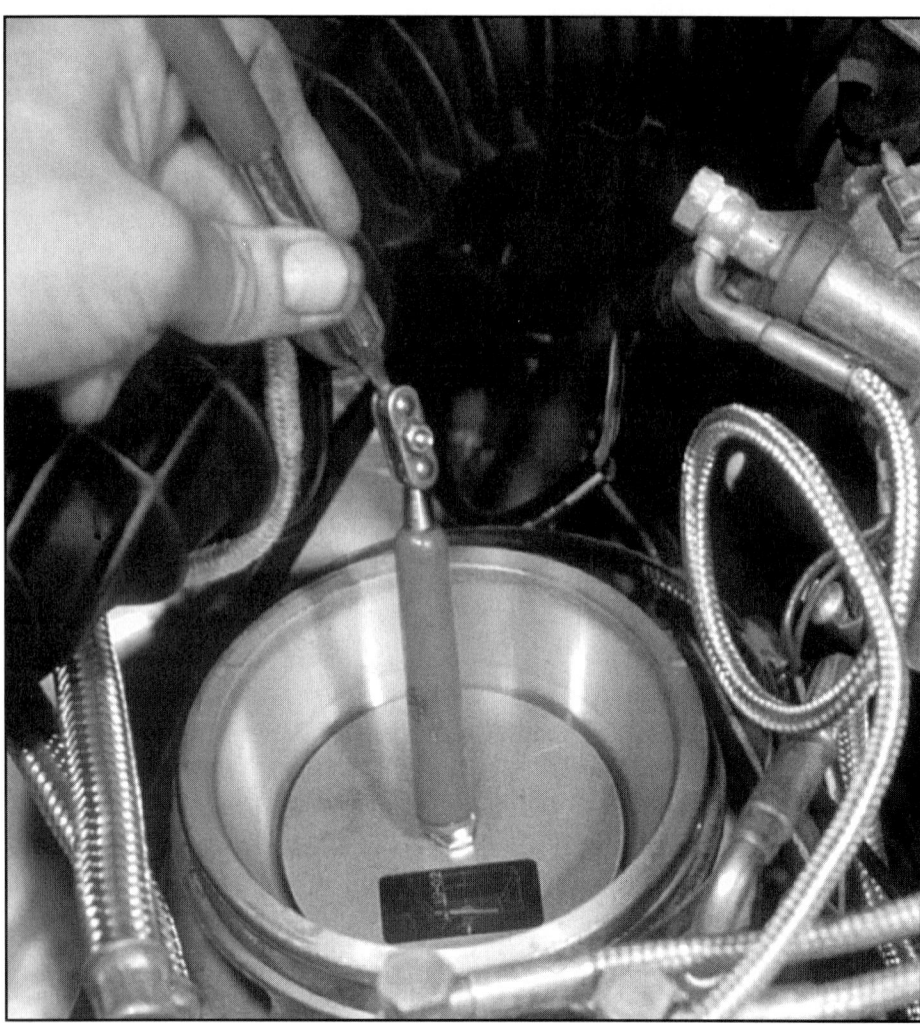

Make sure the air sensor plate moves freely, and that lifting it turns on the fuel pump. The injectors should spray fuel.

hoses under the hood, it is easy to get things mixed up if you have to disconnect something, say, during a routine tune-up.

CONTINUOUS INJECTION

A list of the most common CIS symptoms and their probable causes is a logical beginning for troubleshooting:

• Difficult cold starting may be due to a malfunctioning cold start valve or thermo-time switch.

• Rough running or stalling during warm up should cause you to suspect the auxiliary air regulator, system pressure, and the idle speed and mixture adjustments.

• Surging or missing when the engine is warm points to a problem in system pressure, warm control pressure, the fuel distributor and injectors, and the idle speed and mixture settings.

• Hard starting when the engine is hot may be caused by a faulty hot start relay (if present), improper air sensor plate height, incorrect idle speed and mixture settings, or faulty leakdown/cut-off pressure.

Diagnosis

A fuel pressure gauge of sufficient capacity, connecting fittings and specific service information for the vehicle at hand, will be necessary for checking fuel pressure at various points in the system. If you find a deviation from specifications, first be certain the pump is delivering the proper pressure, then you can make adjustments by changing shims in the pressure relief valve.

Ideally, this system supplies all the cylinders with the same amount of fuel. If not, poor idle quality and part-throttle performance will be the symptoms. To find out if all the injectors are providing equal amounts of gasoline, the car makers say that you're supposed to use a set of graduated measuring tubes and a special fixture that depresses the air sensor plate. You probably don't have this type of equipment, but you can make a similar check by using any type of graduated container for each injector, and moving the air sensor by hand.

Adjustments

The two emissions-related adjustments you can perform on K-Jetronic are those of the idle speed and the air/fuel mixture. To set speed, turn the screw next to the throttle plate linkage, which varies the size of a bypass drilling. Counter-clockwise equals faster. Mixture is adjusted by turning a screw which bears on the air sensor lever. You'll find a rubber plug between the fuel distributor and the air sensor cone, which must be removed for access to the screw. Use a long, 3mm Allen wrench to adjust, and clockwise richens the mixture.

ELECTRONIC FUEL INJECTION

Whenever you have a problem with an EFI-equipped car, heed this list of symptoms and their most probable causes before jumping to any unfortunate conclusions:

•No start: Lack of spark, fuel pump circuit, fuel pump, fuel filter, control unit
•Hard starting: Cold start valve, thermo-time switch, intake air sensor, fuel

One of the basic checks of an EFI system is that of available fuel pressure. This may uncover a weak pump, a clogged fuel filter, etc.

pressure regulator, temperature sensor

• Stalls during warm-up: Auxiliary air regulator, temperature sensor

• Runs poorly when cold: Auxiliary air regulator, temperature sensor

• Too lean: Fuel pressure regulator (pressure too low), restricted injectors, vacuum leak, clogged fuel filter

• Too rich: Fuel pressure regulator (pressure too high), leaking cold start valve

• Flooding: Thermo-time switch, cold start valve

• Low power: Fuel pressure regulator (pressure too low), restricted injectors

• Poor fuel economy: Intake air sensor, fuel pressure regulator (pressure too high)

• Erratic performance: Vacuum leak, intake air sensor

• Rough idle when warm: Vacuum leak, fuel pressure regulator, leaking cold start valve, bad injector

Diagnosis

There are some good quick checks that'll tell you quite a bit about the condition of the system. First, listen to each individual injector using a mechanic's stethoscope or a long rod held to your ear. They should all make an identical clicking sound while the engine is idling. If one sounds different from the others, or makes no noise at all, suspect it as the cause of missing, rough running, or other performance problems.

In cases where one or more injectors are silent, or if the engine won't start, disconnect the injector harness plug and check for 12 volts across the plug terminals using a test light. If you get no flashing while the engine is being cranked, the trouble is in the electrical circuit, not the injector itself.

You should bear one important point in mind: Just because the car has EFI, it's a serious diagnostic blunder to automatically blame it for any problem that may be present. Always check the basics, including ignition, compression, vacuum tightness, camshaft drive components, etc., before condemning the highly dependable parts that supply and meter gasoline.

Rochester CPI

The following diagnostic notes pertain to the AC Rochester Central Point Injection (CPI) system:

• The top of the manifold will have to come off for any service, but that's just a

When you get a no-start with EFI, hook a test light or a little "noid" light like this across the leads to an injector. If it doesn't flash while you crank, the injectors aren't getting a complete circuit.

matter of removing 10 Torx bolts and stud nuts. At reinstallation, torque them to 124 in-lbs. starting with the one at the front of the driver's side, then continue clockwise around the perimeter. This joint is sealed with a reusable gasket, not a chemical compound as you might have expected.

• It's easy to remove the poppets. Squeeze the two plastic prongs together as you might on an electrical connector.

On an EFI with an air meter or mass air flow sensor, any air leaks between the meter/sensor and the throttle will cause hesitation and possibly a shift into the "limp-in" mode.

• Resistance across the injector should be 1.5 ohms.

• The TPS is non-adjustable.

• The O_2 sensor heating element gets current through the IGN/GAUGES fuse.

• Key on/engine off fuel pressure should be 54-62 psi.

• Just as with many other GM systems, long cranking time may mean the fuel pump relay isn't working, so the pump has to get its complete circuit by means of the oil pressure switch.

• There's just one nationwide emissions calibration.

ELECTRONIC ENGINE CONTROL SYSTEMS

This is definitely a case where the specific diagnostic information for the model at hand will be necessary if you're going to nail down a subtle problem. You'll need the proper test procedures and voltage or resistance specifications in order to be sure your efforts will result in an accurate diagnosis. Also, if a self-diagnostic mode is included, you'll have to find out how to generate, read and interpret fault codes.

So, all that's appropriate to include here are the most important general rules that apply to all cases, and which may keep you from making embarrassing and expensive mistakes, and some generic testing procedures for common sensors.

Test Equipment

Before we tackle overall system testing procedures, we'd better discuss what you need to perform them. Of course, you could use a scan tool, engine analyzer or oscilloscope, but we're not going to try to tell you how to operate all the different versions of those pieces of equipment. Instead, we'll talk about generic testers.

The logical first choice is a high-quality DVOM (Digital Volt Ohm Meter, also called a Digital Multi-Meter or DMM) that'll give you minimum, maximum, and average readings at the push of a button, and also report on cross-counts (that is, the frequency with which the sensor crosses the 450 mV rich/lean line) with a Hertz function. What about a no-frills

Resistance across the injectors solenoid coil is important, too.

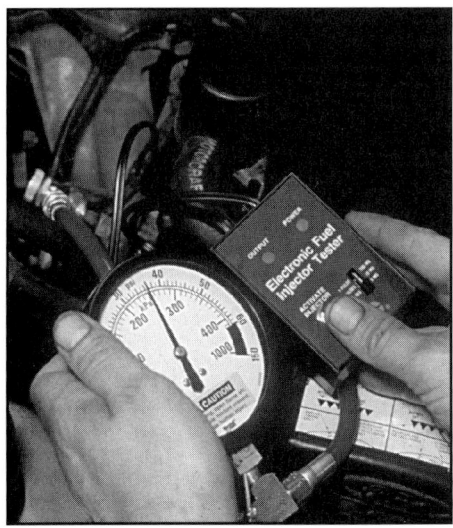
To find clogged injectors, you can use a "buzz box" to pulse them, then compare pressure drop.

digital meter? Its display won't be able to keep up with the changing readings, and neither will you.

Then there are job-specific O_2 sensor testers. We're especially fond of one we've been using that has 13 LEDs representing various voltages, and a rotary switch that allows you to stay in closed loop while monitoring, isolate the sensor from the computer for open loop checks, or send the ECM either a rich or lean signal.

Maybe you'd like to use your old analog volt-ohmmeter because you figure the swinging of the needle on the two-volt scale would be easy to interpret. Maybe you'd better reconsider. Every automaker cautions us to use only meters with high impedance for testing O2 sensors. One reason for this is the possibility of current flow breaking the bond between the zirconium dioxide and the platinum, or blowing a channel through the sandwich. And an extra ground may skew the signal. If you still like the idea of watching a needle, analog voltmeters with 10 megohms per volt input impedance are available from electronics supply outfits.

Regardless of your choice of tester, you need a method of tapping into the sensor lead. A jumper between the lead and the harness will work as a connecting point, or you can use one of those convenient insulation-piercing hooks available at most auto supply stores.

Overall System Checks

First and foremost, make sure all the traditional basics are okay. Just because a computer system is present, don't jump to the conclusion that it's the cause of the trouble. Check compression, ignition and coolant temperature, and look for vacuum leaks. In many cases, the fault will lie in one of these rather than in the electronics.

Next, when symptoms or trouble codes lead you to suspect a particular sensor, make sure you check its circuit before replacing it. Look for frayed or broken wires and loose or corroded connectors. Then, check the sensor itself with a digital volt/ohmmeter to see if it's doing what it's supposed to.

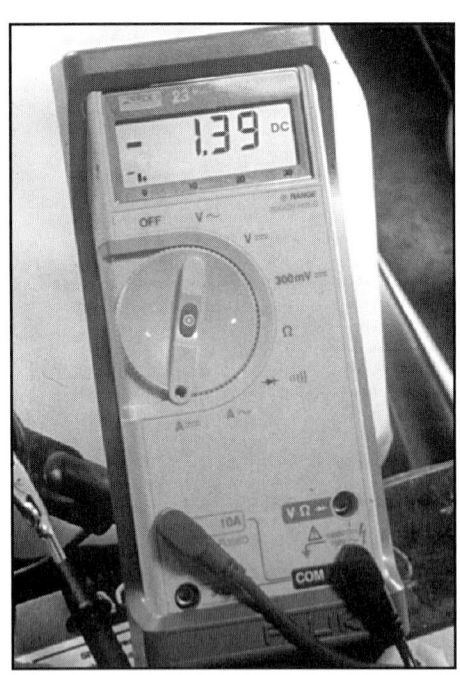

A quality DMM (Digital Multi-Meter, also called a DVOM for "Digital Volt-Ohm Meter") is the most essential piece of equipment for checking out electronic engine management systems. But you'll need the specific factory diagnostic info.

The most basic check of both the sensor and feedback system operation begins with running the engine at fast idle until everything's hot, then tapping your tester into the lead (with electrically heated units, make sure you use the output wire). You should see rapidly changing readings as the computer keeps adjusting the blend. Deciding whether or not response is slow enough to justify replacement requires some judgment. A common rule of thumb for minimum activity is eight trips across the rich/lean line in ten seconds, and sometimes you can find specs for this under the heading "cross-counts."

Next, try pulling the vacuum hose off the brake booster. This should send the reading to zero, then it'll start working its way back up. Don't expect it to go very high, because the feedback carb or injectors probably won't be able to supply enough fuel to make a rich mix with that much extra air. Thumbing the hose should bump the voltage to about .9, then it'll gradually come down to the midpoint. Some technicians like to blow a generous amount of acetylene or propane into the air intake to see what happens at the rich end—a rise in voltage is normal. Shutting off the gas suddenly should produce a momentary zero reading, then the ECM will start to compensate.

O_2 Sensor Checks—On early feedback/closed loop systems, most automakers recommended replacing the oxygen sensor every 30,000 miles, but no interval is given in the owner's manuals of later versions.

That, however, doesn't mean failure can't occur anymore. Several things can wreck an oxygen sensor's ability to provide accurate info. Mechanical damage in the form of a broken element or wire happens, but the most common killer is contamination. Lead, carbon, silicon (from RTV or anti-freeze—expect O_2 sensor problems whenever you replace a blown head gasket) can all coat that precious platinum and make the unit sluggish or altogether inoperative.

Contamination isn't always the kiss of death. When you get poor readings, try running the engine at 3000 rpm for a few minutes, then retest. You may have burned off whatever was interfering, plus you'll be sure the sensor's hot enough to make juice.

To test the O_2 sensor all by itself, disconnect its pigtail and attach the meter's positive lead to it (the other lead to ground, of course) so that the computer is cut off from this source of data. Cause an artificial lean condition (pull a vacuum hose or the PCV valve) and voltage should drop. Then, cause a rich condition (with a carb, close the choke partially) and you should see the reading rise. If the O_2 sensor doesn't respond, it's either too cold, or it's dead.

Knock Sensor Checks—The most obvious trouble you can have with a knock/detonation sensor system is pinging. There shouldn't be anything beyond a trace of this noise. A more subtle malady is less-than-sparkling performance and fuel mileage, which may be the result of the computer believing that detonation is present when it really isn't.

The most straightforward means of testing the knock sensor circuit is to simulate the sound of detonation by tapping on the area nearest the sensor with a wrench or extension while you hold rpm at 2000 or so (the system's disabled at idle) and seeing what happens. If you're using a scan tool, you'll read degrees of retard, get a yes/no display, and/or see a continuous 0-255 counter loop (no change in the number means no knock signal).

But you can observe the action without a scanner. Just hook up your timing light, shine it at the marks, hold fast idle and start tapping. If you don't see retard occurring, first change to a different tool and modify the force you're using. If that

Special hand-held testers, called "scan tools," can give you a lot of important data on sensor activity, etc., but they're too expensive for most do-it-yourselfers.

doesn't produce a change in ignition timing, check out the wiring and connections before condemning either the detonation sensor itself or the computer.

But maybe you're getting retard all the time. In that case, you should suspect an overly sensitive sensor, or an internal engine noise that mimics spark knock. Although the crystal assembly is calibrated to respond only to the exact frequency that detonation causes, it can be fooled. After all, that's what you do to test it.

TPS Sensor Checks—When the TPS fails, as it often does, driveability problems can be expected, especially tip-in hesitation. If it's hanging up, say on a typical Chrysler product, it can cause high idle speed because there'll be more voltage than the specified minimum in the return wire. Combined with high vacuum, this will make the computer think you're in decel so it'll back out the AIS (Automatic Idle Stabilizer) motor. Another possible glitch if the sensor doesn't return completely is hard starting. On our Chrysler example, if the signal is 2.4 volts above the closed throttle value while cranking, the ECM will read it as full throttle and engage the clear-flood mode—the injectors won't get the necessary pulses and fuel won't flow.

As we said, the TPS is a variable resistor or potentiometer—the same device that adjusts the volume of your stereo. The direct DVOM test involves pulling the sensor's plug, then connecting all three terminals to the harness again using jumpers with contact taps. Hook your meter between the output and ground wires, switch on the ignition, then look for the minimum voltage spec. (about .26V on a typical feedback carb-equipped GM car) with the throttle at curb idle position. Open the throttle slowly to do the sweep test—you want to see an even rise to about 5V. Voltage should go up smoothly. If it drops back or jumps ahead at any point in its travel, get a new part. Of course, you can also use your scan tool to do this check. If you want to back up your findings, repeat the sweep test using an ohmmeter across the sensor's supply and output terminals.

Many versions are adjustable, and we have it on good authority that a setting that's off just .2 volt can cause trouble.

Coolant Temp Sensor Checks—There are several ways of checking a CTS. With a scan tool, for instance, you can read coolant temp directly. A cold engine should obviously show about ambient, and a hot engine should produce a display of between 190 and 220 deg. F.

If those basics are off, you can usually find specs in factory diagnostic info for the resistance the sensor is supposed to have at various temperatures. Check the actual coolant temperature with a thermometer or digital pyrometer, switch your DVOM to "Ohms," then take a reading across the sensor's terminals (don't mistake this component for the one that controls that electric radiator fan, which also has two terminals—if you can't be sure from looking at the locator drawings or wiring diagram, trace wires back from the fan motor to its relay, then from the relay to the fan thermoswitch).

Next, start the engine and watch the reading. If you don't see at least a 200 ohm change within a minute, unscrew the

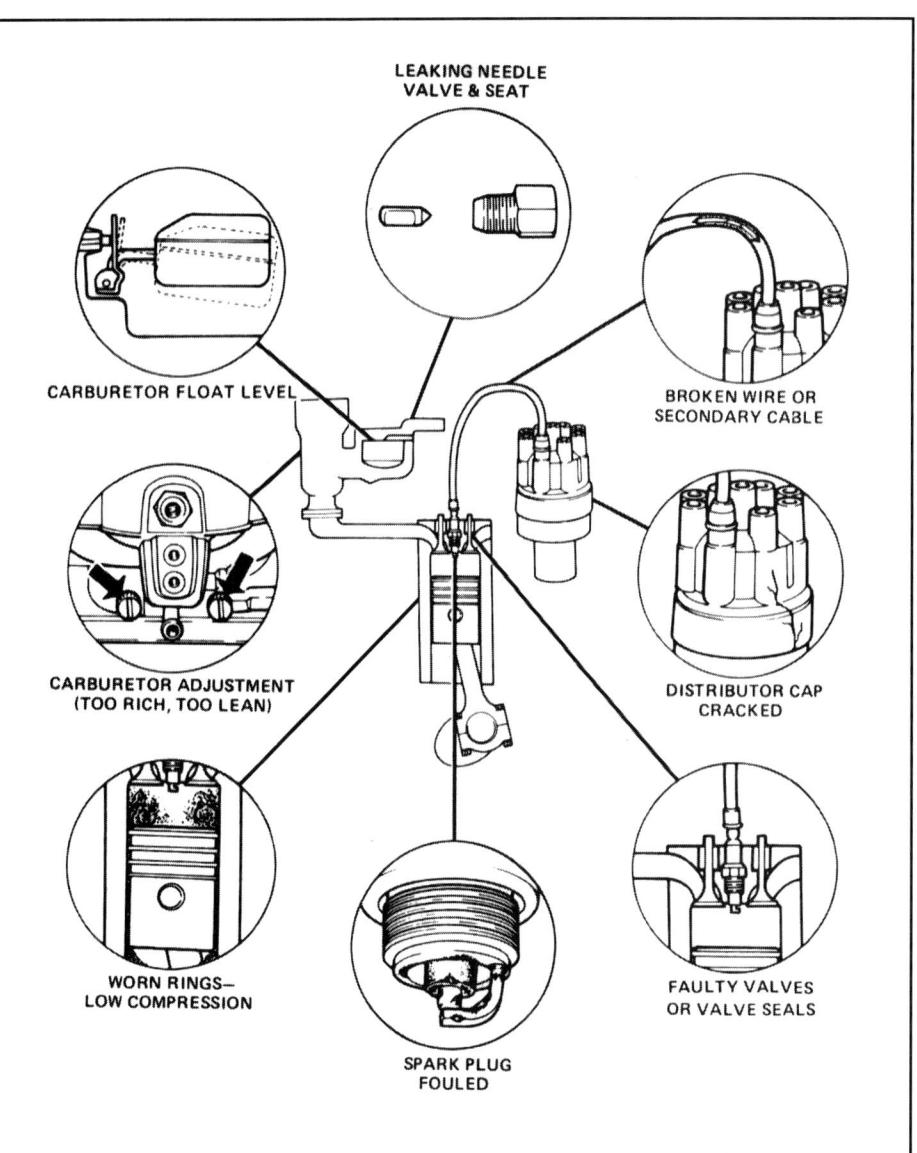

Before you blame the electronic engine controls for a problem, make sure you check all the traditional basics (Chrysler).

137

To check the coolant temperature sensor, check the actual coolant temperature with a thermometer or digital pyrometer, switch your DVOM to ohms and take a reading across the sensors terminals. Next, start the engine and watch the reading. If you don't see a 200 ohm change within a minute, then check the sensor to see if it is caked with crud. If the readings still don't come up to spec, then replace the sensor.

sensor to see if it's coated with crud. Clean it and try again. If you don't get that rapid resistance change, or your readings don't match specs, replacement is justified.

There are several ways of finding out how the system is responding to CTS info. Using a scan tool on an NTS unit, unplug the sensor's lead. This open circuit will simulate the high resistance of super low temperatures, so you should see the minimum temp reading, typically -40 deg. F. Go to the other extreme by jumping the two connector terminals, which should display the max temperature (250 deg. F and up).

Another procedure involves setting up to read pulse width (a scanner or scope maybe, or you could even use the six-cylinder scale of a digital dwell meter tapped into an injector's output wire), then putting a 50K ohm pot and a 100 ohm resistor (to provide some resistance when the pot's turned to full hot) in series in the sensor wire. If the dwell/pulse width varies as you change resistance (you'll have to keep adjusting idle speed to keep it constant), the computer is indeed using input on coolant temp in its calculations. We have a hand-held sensor simulator that makes such substitution tests easy on most types of sensors.

MAP Sensor Checks—The biggest clue is a lean condition. Using your scan tool, look at BARO (barometric pressure). It should be about what the weatherman tells you. Get both BARO and MAP (manifold absolute pressure) readings, which should be the same with the engine off. With the engine running, subtract MAP from BARO to get actual in. Hg, which you can compare to a mechanical gauge reading.

As with the TPS, most MAP sensors can be checked using jumpers with contact taps so the component can stay in the system. On a garden-variety GM car with a feedback carb, terminal A is ground, B is sensor output voltage, and C is reference voltage. With the key on, first see that you've got pretty close to five volts between A and C. Then, look for 4.6V between A and B. Using a hand pump, apply vacuum to the sensor's nipple, and make sure 2.3V is available at 10 in. Hg, and 1.0V at 20 in. Hg.

For a dynamic test, use your favorite method of reading pulse width and apply vacuum to the sensor using a hand pump. You should see pulse width decrease as vacuum is increased. Commonly with EFI, 23 in. Hg or so at idle will kill the engine because it simulates decel, a mode in which the computer shuts down the injectors. Techs have told us they've seen engines that cut out intermittently while idling because of this action. This happens when somebody has advanced timing beyond specs, which results in a smaller-than-normal throttle opening for proper idle speed. That, in turn, produces enough vacuum at idle to trigger the injector cut function.

A possible source of trouble you'd probably never think of is frozen condensation in the MAP's vacuum line (reroute it to eliminate low spots).

MAF Sensor Checks—When a mass air flow sensor or its wiring or ducting goes awry, what kind of symptoms can you expect? The answer depends on whom you ask. GM guys will tell you the engine will crank up and die. Bosch talks about starting problems both hot and cold, hesitation, stalling (especially under load), rough idle and low power output. Nissan gives stalling, poor idle, black smoke and switching to the fail-safe mode as evidence of air flow meter problems (in some models, this mode will be manifested by the inability to exceed 2000 rpm). Generally, contamination of a hot wire or film sensing element, which slows response, will result in stumble.

That's all fine, but the most prominent logical effect of a bad signal or lack of any signal is trouble at transient throttle—stalling, sagging, missing. If the signal is far enough out of range to cause the electronics to shift to LOS (Limited Operating Strategy), overall performance and driveability will be lousy.

Unfortunately, other things can cause many of those same symptoms, some of them very basic. Before you jump to conclusions, you've got to think about ignition, compression, fuel supply, etc. And a simple problem that's commonly overlooked is a hole or rip in the duct between the sensor and the throttle body, which admits unmeasured or "false" air and leans out the mixture. An open PCV can do the same thing, and a plugged air filter can mean trouble, too. So, at the risk of repeating ourselves, don't automatically blame the engine management electronics or fuel injection setup just because it's present and perhaps not that familiar to you. In fact, chances are the trouble is elsewhere, because these systems are generally quite dependable.

Which is not to say you'll never run into a bad air flow/mass sensor. For instance, the common GM MAF has a pretty poor reliability record (the higher-frequency 10kHz Hitachi unit used on late-model GM cars has a much lower failure rate).

Trouble Codes—So if the basics check out and you suspect the EFI, how do you find out if the air flow/mass meter is the culprit and not some other sensor?

Well, most late-models have at least some onboard diagnostics, so you might as well use them first. That's why they're there, after all, and with OBD II regulations, they'll become more and more important.

As just mentioned, fault codes vary, so you'll need the specific service information for the car at hand, but we'll offer a couple of examples for illustration. On a garden-variety Ford EEC-IV, Code 26 tells you that the VAF (Vane Air Flow) or MAF is out of self-test range, Code 56 means the signal is above max voltage, and Code 66 indicates signal voltage below minimum. But Hyundai has just one code for the air flow sensor—12—and the manual says that if you see that code and the harness and connector are both okay, go ahead and replace the sensor.

But codes can get you into trouble. All they should really be used for is aiming your diagnostic efforts in a general direction, not as the final word on what's wrong. You'll need to do some more troubleshooting and rely on your experience before you can be sure.

A scan tool works well here. Errors in sensor calibration are magnified as air flow increases, so being able to test during road load can be helpful. In some cases, both the air flow value the ECM is using and the actual signal from the sensor can be displayed. If these two numbers don't match, the computer is probably reverting to a substitute value from memory because the MAF info is faulty. By the way, GM units should produce a signal that results in a reading of 4-7 grams per second at idle, or 100-240 gps at W.O.T. (naturally, depending on the displacement and horsepower of the particular engine).

Vane-Type Air Meter Checks—

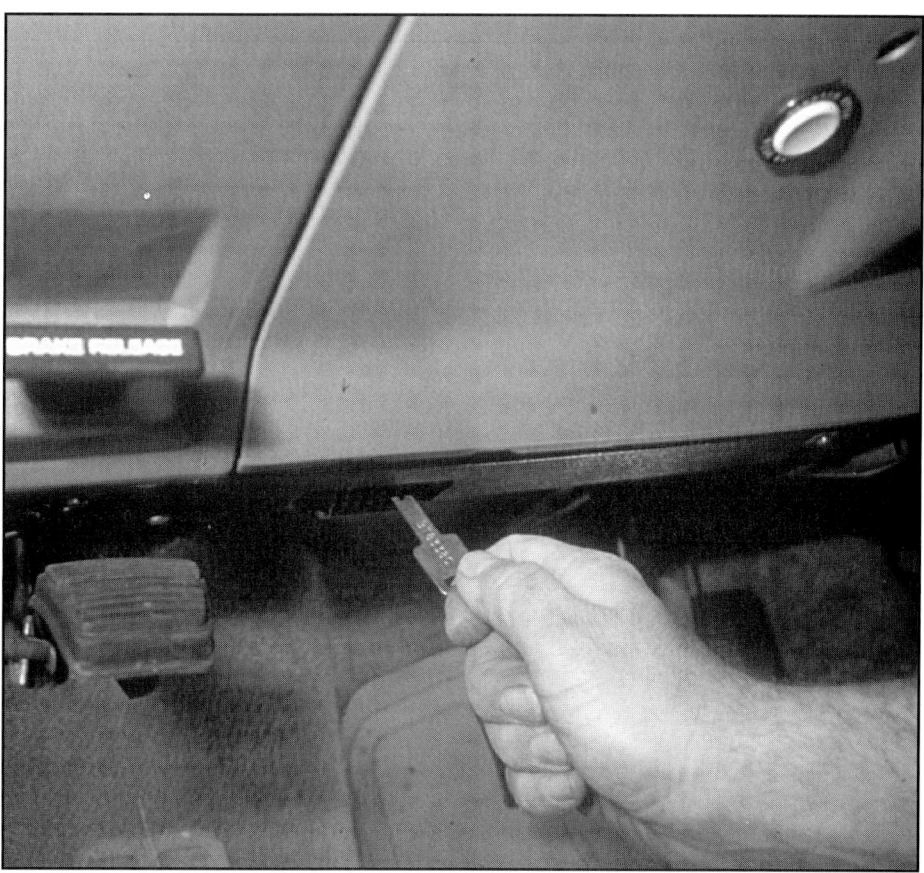

On GM C3 (Computer Command Control) systems, you can get the dash light to flash out trouble codes by jumping two terminals of the ALDL (Assembly Line Diagnostic Link, which you'll find under the dash in most cases) with a little tool like this, or just with a bent paper clip.

On the rotating vane type air meter, the first thing to do is reach inside the air box and move the flap through its range by hand. You should feel no binding or roughness. If it's a version that incorporates a fuel pump switch, make sure you hear the pump start when you push the vane (key on).

Using your favorite connection method (backprobing, jumpers, wire piercing, or break-out box) and a digital volt-ohmmeter, look for reference voltage input (that's 5V). Switch to the output contact and you should see the reading change smoothly as you push the flap to the fully-open position (in most cases, this will be a falling reading, but on Ford EEC-IV units it will rise—look for .25V closed and 4.50V open). Some automakers give you direct resistance specs, too. For instance, on a Mazda 323 the reading between sensor terminals E2 and VS should be 20 ohms with the vane closed, and 1,000 ohms with it open. Just as with a TPS (Throttle Position Sensor), you're checking the condition of the resistive strip or track, and any jumps in either voltage or ohms readings mean it's time for a new sensor.

Bosch—To check out a Bosch hot wire air mass sensor, first look for battery voltage at the appropriate terminal, then measure output. A typical unit should read about 2V at idle, rising to almost 3V at 3,500 rpm. To give you some more ballpark references, common output specs for Ford hot wire units are 0- 0.5V key on/engine off, 0.5-1.0V at hot idle, 1.5-2.5V at hot cruise, and 3.0-4.7V at W.O.T. Also, you should see the voltage change when you blow air through the sensor.

Delco MAF Sensor Checks—We'll bet many of you have heard of the basic GM MAF test: When tapped with a tool (the handle of a screwdriver, not a two-pound ballpeen) at idle, a bad sensor will not only produce a dramatic change in frequency, it may also cause the engine to stumble or stall. This is certainly a convenient check, and it's almost 100% accurate. You might even want to try it before pulling codes.

But there's another quick check that's almost as fast. With the key off, unplug the MAF's harness connector, then start the engine. If it runs appreciably better now, it's time for a new sensor.

If you have a digital multi-meter that can measure frequency, you can use that mode to check AC-Delco, Hitachi and any other unit you run into that produces a frequency signal. Set the meter to read Hz or kHz, and connect its leads to the sensor's signal and ground wires. An ordinary AC-Delco MAF, as found on a 2.8L Chevy V6, should show you about 45 Hz at 1000 rpm and 72 Hz at 3500 rpm, whereas the high-frequency type of a late-model 3800 will read 2.9 kHz and 5.0kHz at those same speeds. Record the readings at various rpm and compare them to specs. You should see a linear frequency rise with no dips or jumps as speed increases.

Finally, remember that there's a much better chance that a sensor, its wiring, or something else altogether is causing the problem than that there's a fault in the computer itself. It's an undisputed fact that most computers returned to their manufacturers as defective are actually okay. You certainly don't want to make that kind of big-bucks diagnostic mistake. ■

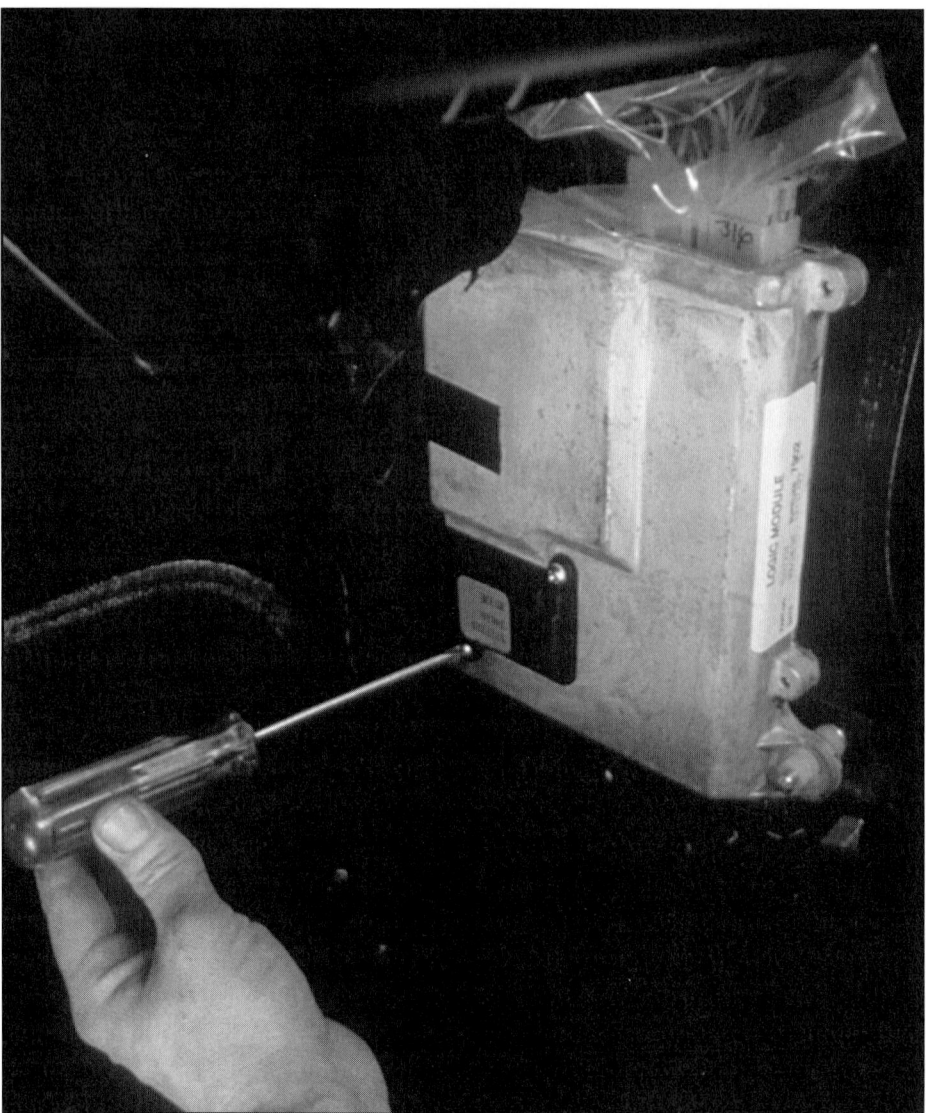

The computer itself is terrifically dependable—and expensive. It's the last thing to blame for a problem. Most supposedly faulty ECU's returned to manufacturers are really okay.

17
TUNING EMISSIONS-CONTROLLED ENGINES

In the days before anti-pollution equipment appeared on cars, a tune-up was basically a matter of replacing the spark plugs, points and maybe the condenser, then setting the ignition timing, the idle speed and mixture, and, if the car had one, adjusting the automatic choke and fast idle. The secondary ignition system—rotor, distributor cap, and spark plug cables—was thoroughly inspected if the mechanic was conscientious. If the engine had mechanical, non-hydraulic valve lifters, the valve lash might also be set to specifications. Replacement of the air filter element, or a thorough cleaning of the mesh in oil-bath-type air filters, and fuel filter service were usually included.

HOW THINGS CHANGE

With the advent of emissions controls, that routine started to change. Some items were added to the list, and some were deleted. On pre-1975 domestics and many later imports, ignition service was the same as it had always been. But with the appearance of catalytic converters and breakerless electronic ignition in 1975, it was no longer necessary to routinely replace anything but spark plugs, and timing could be set once at the factory and never needed to be adjusted again, because there was no point rubbing block wear to make it vary.

Despite the advent of 100,000 mile spark plugs, they are still a top priority. You should always check the gap—it's commonly quite a bit off (Autolite).

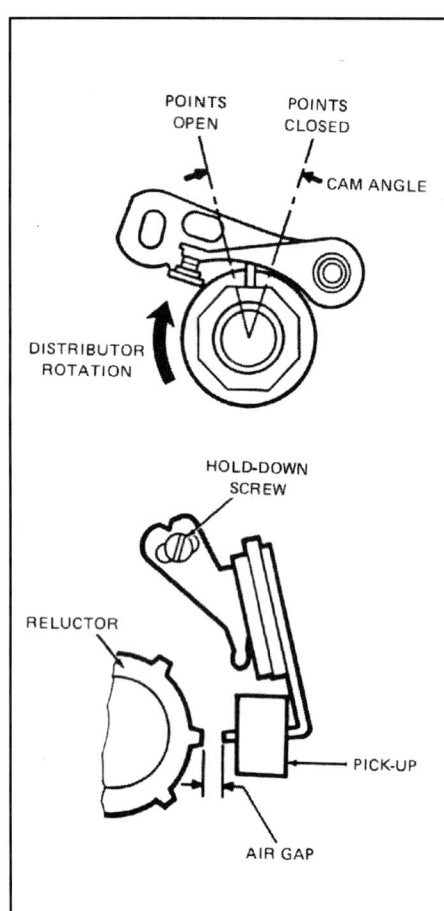

Here's the real reason we got electronic ignition: Point rubbing block wear causes spark timing to change gradually, whereas primary ignition control stays put forever with electronic pick-ups (Chrysler).

Limiter caps on the idle mixture screws that reduced the amount of adjustment allowed appeared quite early in the fight against air pollution. Then came plugged or locked mixture adjustments that were pretty well tamper-proof, electronically controlled spark advance and idle speed, and closed-loop/feedback carburetors and fuel injection.

Today's Tune-Up

So, on today's cars the term "tune-up" is really just a sacred cow—everybody has been hearing it all their lives, and they stick with it out of habit even though it's not appropriate anymore. Think about it: To tune means to adjust, doesn't it? Well, you won't find much to adjust on late models.

A new definition of this job is definitely needed, and we'll start on that by saying what a tune-up is NOT, and never has been: a repair. It's always been a maintenance operation meant to ensure against problems such as hard starting, rough idling, etc. If a problem is fixed in the process, it's more by accident than by design.

But that doesn't necessarily mean there's less work to do. Performing a new-style tune-up requires inspecting hardware and plumbing and perhaps running functional tests. Anybody can just replace spark plugs (although there's a right way and a wrong way to do it), but we believe that the definition of a tune-up should include more than that, especially since driveability complaints are still common.

PRIMARY IGNITION

We'll begin with the system that produces the spark. On pre-electronic ignition cars (1974 and older domestics; most imports used points until later years), replacement of the points and condenser is basically the same as it's been for decades. It's timing that became more critical. Many distributors incorporated vacuum units that both advanced and retarded the spark. For example, a common design has separate chambers on either side of a diaphragm that is linked to the plate on which the points are mounted. The outer chamber is connected to a carburetor port above the throttle plates so that it advances the timing when the throttle is opened, as in most ordinary configurations. But the inner chamber is hooked up to a source of manifold vacuum so that it retards the spark at idle and during deceleration in order to help control emissions. Similar configurations have been used on some later electronic ignition distributors.

This makes it absolutely necessary that you know for sure if the initial timing setting should be done with the advance or retard hoses connected or disconnected on the particular model at hand. The specifications label in the engine compartment will give you this information. Also, some makes that have only a single-action vacuum unit (many Hondas, for instance) require that timing be adjusted with the hose connected.

Stability

The main reason the automakers adopted electronic ignition was not to produce a stronger spark (in fact, some systems, Chrysler's for example, produce relatively low voltage), but to take advantage of the fact that timing never varies. In a point system, spark is produced when the points open, so timing is affected by the dwell angle or gap width. The longer the dwell or narrower the gap, the later the spark. So, as the rubbing block on the movable point wears, timing becomes retarded. Obviously, without points this situation is eliminated.

Of course, just because timing doesn't change over time with electronic ignition doesn't mean it was adjusted properly in the first place, so it's necessary to at least check this setting with a timing light when performing a tune-up.

Because an engine's spark advance curve is so critical to the amount of all three pollutants that it pumps out, and due to the limitations of mechanical and vacuum devices, automotive engineers have created various types of electronic advance systems. These make use of electronic logic and various sensors to tailor the spark much more accurately to conditions than could be done with mechanical/vacuum setups. Chrysler's Electronic Lean Burn (ELB) was the first example of this, and it allowed engines to use less exhaust gas recirculation (EGR) and still achieve low levels of NOx emissions. The other makers followed suit shortly thereafter, so many cars on the road today have electronic spark advance control.

It's beyond the scope of this book to provide you with the checkout procedures

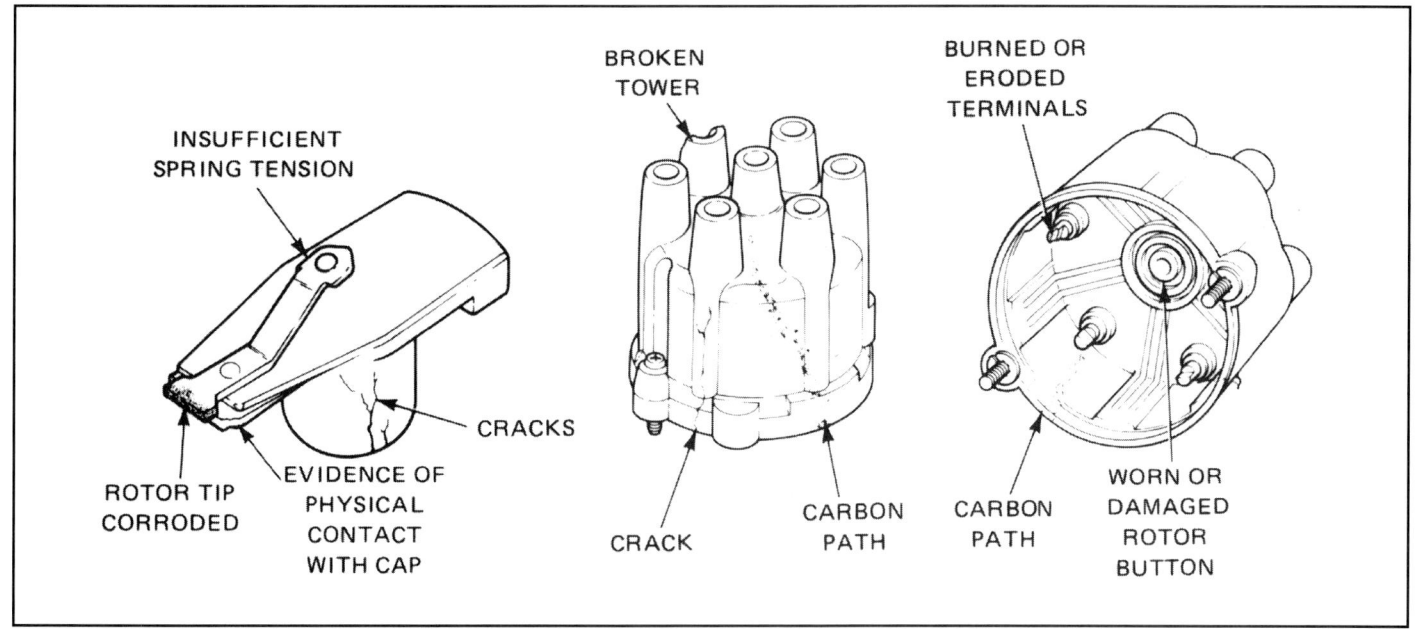

The ignition system's primary circuit is prone problems such as these.

for all these systems. Suffice it to say that if there's a performance or fuel mileage problem, the proper factory service troubleshooting information should be consulted before expensive electronic components are replaced. Often, the problem is due to nothing more than a bad electrical connection or a cracked vacuum line.

SECONDARY IGNITION

The secondary ignition circuit comprises the rotor, distributor cap, spark plug and coil wires, the plugs themselves, and half of the coil. On pre-electronic-ignition cars with emissions controls, service of these components is basically the same as it had been for many decades, except that the higher temperatures found under the hood relative to those of uncontrolled models can cause the wires to deteriorate faster.

But with electronic ignition systems, things are different. The higher voltage produced by most versions is naturally that much more prone to escaping. In other words, the strong electrical pressure will "leak" out at any opportunity. Contaminated or cracked insulation, accumulations of dirt on the cap, and/or dampness can all contribute to this situation. Certainly, secondary ignition components have been improved to cope with this extra burden—GM's 8mm silicone wires, for example—but the potential for problems is still there.

To illustrate: The authors encountered a 1974 model car that had points but was equipped with an aftermarket capacitive discharge ignition system that produced voltage on a par with that of, say, GM's HEI (High-Energy Ignition). The engine would absolutely not start in wet weather, but as soon as the CD was taken out of the circuit, the powerplant fired up and ran perfectly. It was a textbook case of the secondary components not being capable of handling the voltage the system generated.

Spark Plugs

Spark plugs are the business end of the secondary system. Until the advent of

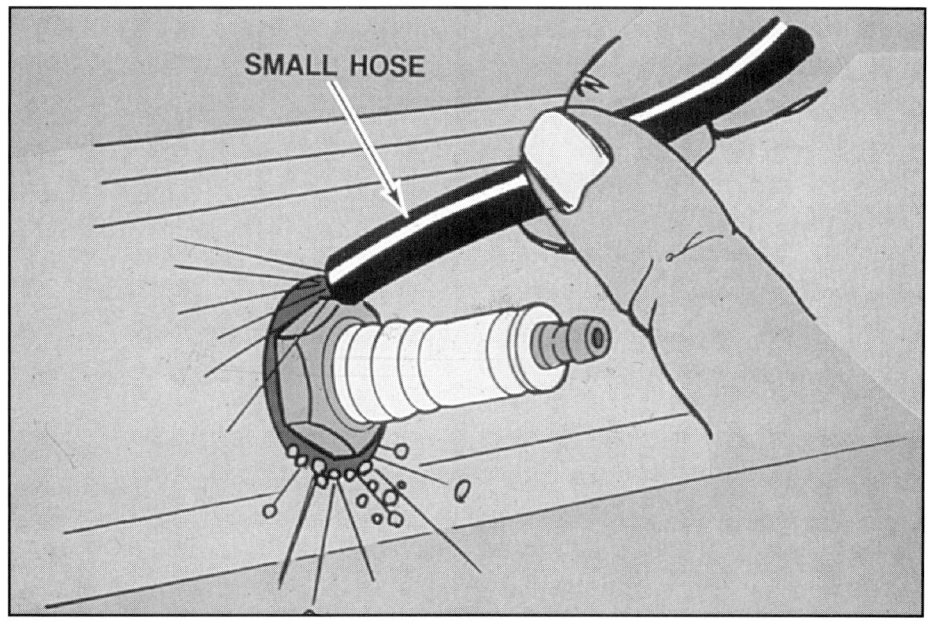

Before removing the old plugs, blow out the wells to keep flotsam and jetsam out of the cylinders (Champion).

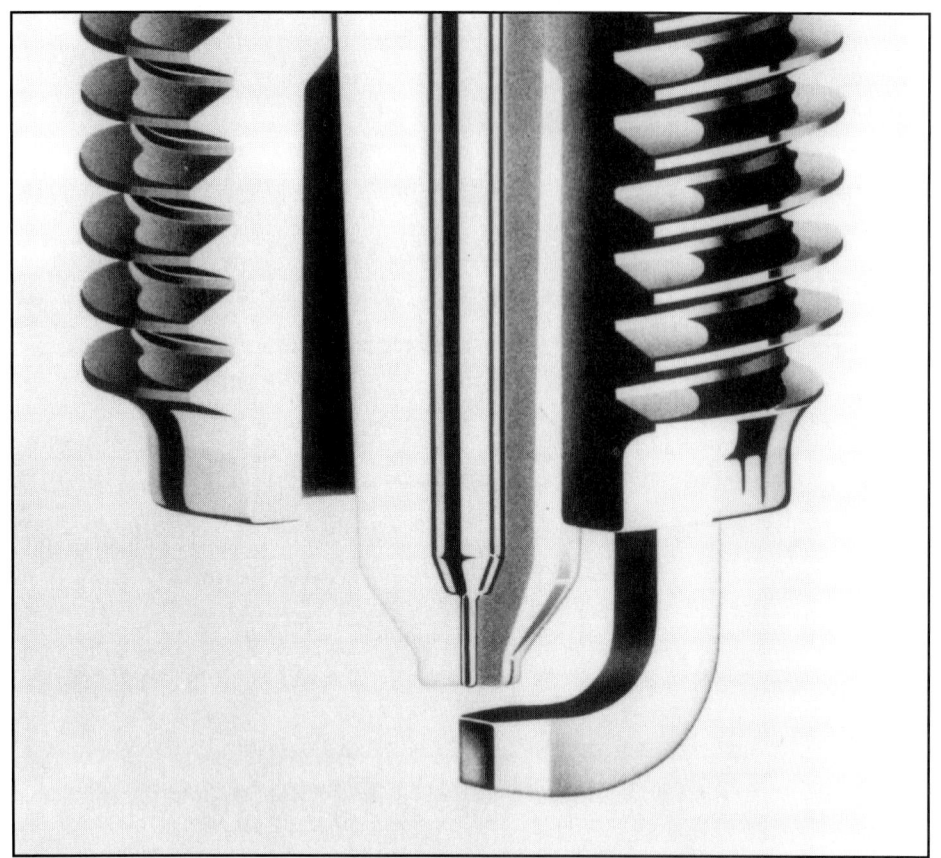

Platinum plugs use just a tiny pin of that precious metal to cut erosion and gap growth (Bosch).

electronic ignition, plug service was basically the same as it had always been: a matter of periodic replacement (every 12,000 miles), or occasional cleaning, filing, and regapping. Designs and heat ranges were altered to accommodate emissions-controlled engines, but that didn't affect procedures or maintenance intervals.

With electronic ignition and unleaded gas, however, the life expectancy of a typical spark plug rose dramatically (30,000 or more miles between change intervals). Deposits on the insulator nose, which act as shunts that bleed voltage away from the gap and cause missing, were much less likely to form. This, combined with the fast rise-time of the voltage surge with electronic primary circuit switching, allowed plugs to keep firing dependably for a much longer time. Now some automakers are actually recommending plug replacement only every 100,000 miles (platinum tipped spark plugs).

The wide-gap plug (up to 0.080 inch) was another development. This helped assure that the lean anti-pollution mixture in the cylinder was indeed ignited. Never, by the way, attempt to set ordinary plugs to wide-gap specs because the electrodes won't be square and thus will erode unevenly.

In the days when cars were permitted to spew pollutants into the atmosphere without limit, the insulator nose of a plug from a cylinder that was firing properly was supposed to be light brown. As mixtures became leaner, the color went to tan, and now it should be almost white.

Cleaning Plugs—A word about cleaning spark plugs is appropriate here. The main goal of the process is to eliminate the deposits that can act like shunts on the insulator nose. Solvent and a brush or sandblasting are the usual methods. Then, the electrodes should be filed square so they have sharp edges that promote proper spark formation. Finally, the gap is reset to specifications.

Spark plugs should ideally be tightened with a torque wrench. If that isn't practical for you, tighten plugs with gaskets one-half turn after the gasket touches the head, and those with conical seats one-sixteenth turn after contact.

Aluminum heads are becoming more and more popular, and the thing to remember about them is always to use a small dab of suitable anti-seize

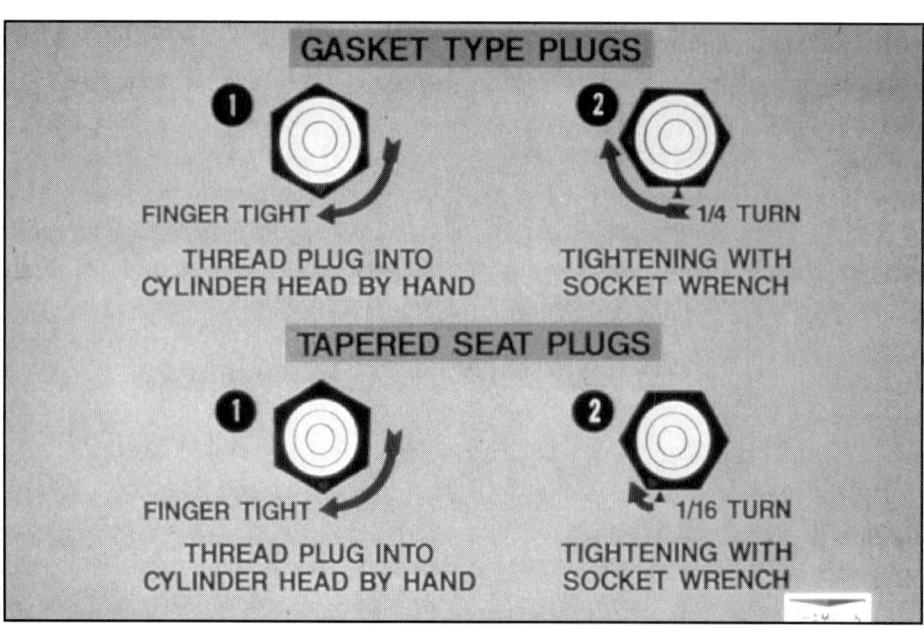

If you don't use a torque wrench when installing plugs, at least follow this procedure (Champion).

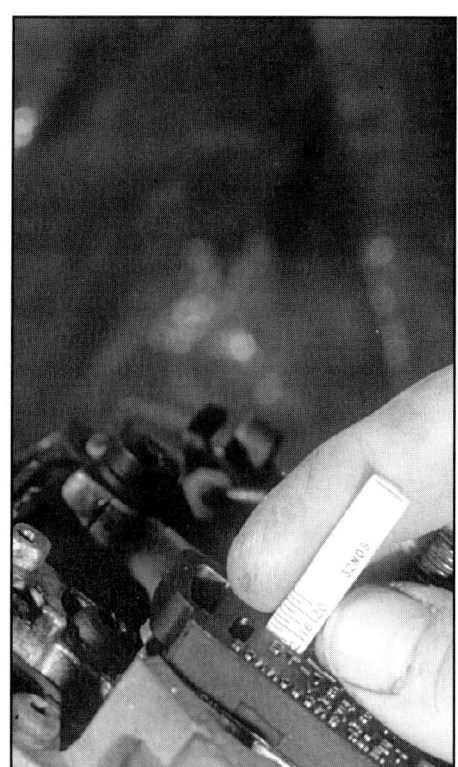

Feedback carbs have numerous complex adjustments. Make sure you have the exact factory info at hand.

compound or graphite on the spark plug threads.

CARBURETORS

Carburetors changed considerably early on in the anti-pollution war because they're so critical to the composition of what comes out of the tailpipe. Calibrations became leaner in every mode of operation, chokes were made to open faster, and adjustments were made differently. All this resulted in driveability problems and service misunderstandings.

One of the first changes found on almost all early emissions-controlled carburetors is the presence of limiter caps on the idle mixture screws. These reduce the amount of adjustment possible to keep the mixture from being set too rich or too lean. Of course, they were frequently removed by people who believed they could achieve a smoother idle by enriching the mixture more than the caps allow. That usually didn't work, because the idle passages had restrictions that didn't allow a grossly rich mix.

Normally, acceptable idle smoothness could be attained within the limiter cap's range. Very specific procedures were published by the automakers for setting the mixture in cases where the caps were removed, and new caps were supposed to be installed.

Propane

Propane enrichment idle mixture setting procedures were adopted in the 1970s to allow the idle mixture to be adjusted properly without the need for an infrared exhaust analyzer. All that's required is a propane tank, valve and hose setup. Although this idea is now pretty much obsolete, we'll give you a typical procedure just for historical perspective. Who knows? You might find this information useful someday.

1. Warm up the engine and disconnect the PCV hose and the charcoal canister purge hose from the air cleaner, and plug the air cleaner nipples. Or, remove the air cleaner.

2. With the engine idling, hook up a tachometer and insert the propane hose into the air cleaner snorkel or attach it to one of the air cleaner nipples. If the air cleaner is removed, connect the propane hose to the choke pull-off vacuum nipple.

3. Open the propane valve slowly until maximum rpm is reached (keep the propane bottle vertical).

4. Turn the carb idle speed screw or solenoid plunger until you have achieved the specified propane-enriched rpm.

5. Continue to open the propane valve. If the speed rises, let it do so until it levels off. Then reset the curb idle speed screw to the enriched rpm.

6. Shut off the propane valve and turn the idle mixture screw until the smoothest idle at the specified hot idle speed is attained.

Curing Dieseling

The idle stop solenoid (see Chapter 11) has been around since the late 1960s but is still widely misunderstood, which is unfortunate because it's the most powerful deterrent to that annoying automotive phenomenon known as *dieseling*.

This little device is attached to the carburetor throttle linkage. It consists of an electromagnetic coil with a core or plunger that can move in and out. When energized, the plunger projects out from the coil and holds the throttle open to a certain position. When de-energized, the plunger drops back inside the coil, allowing the throttle to close as much as the ordinary throttle stop screw permits.

The solenoid is connected to the ignition switch so that it's energized when the key is on, keeping the engine at normal idle speed. When the key is switched off, the plunger retracts, allowing the throttle to close more. Since this reduces the amount of air that can enter the intake manifold, and the engine can't run without air, it stops the engine

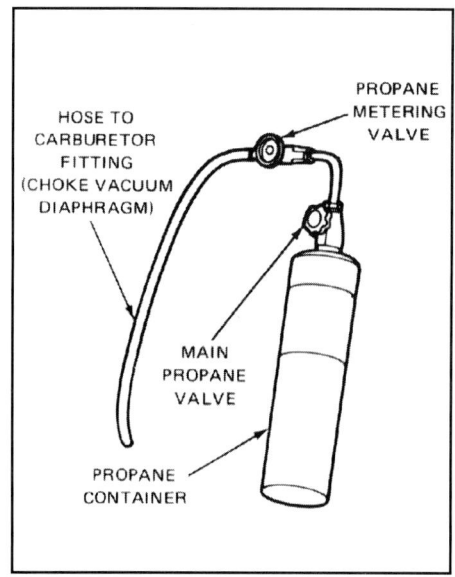

To eliminate the need for an exhaust analyzer, the automakers once recommended the propane enrichment method of setting idle mixture (Chrysler).

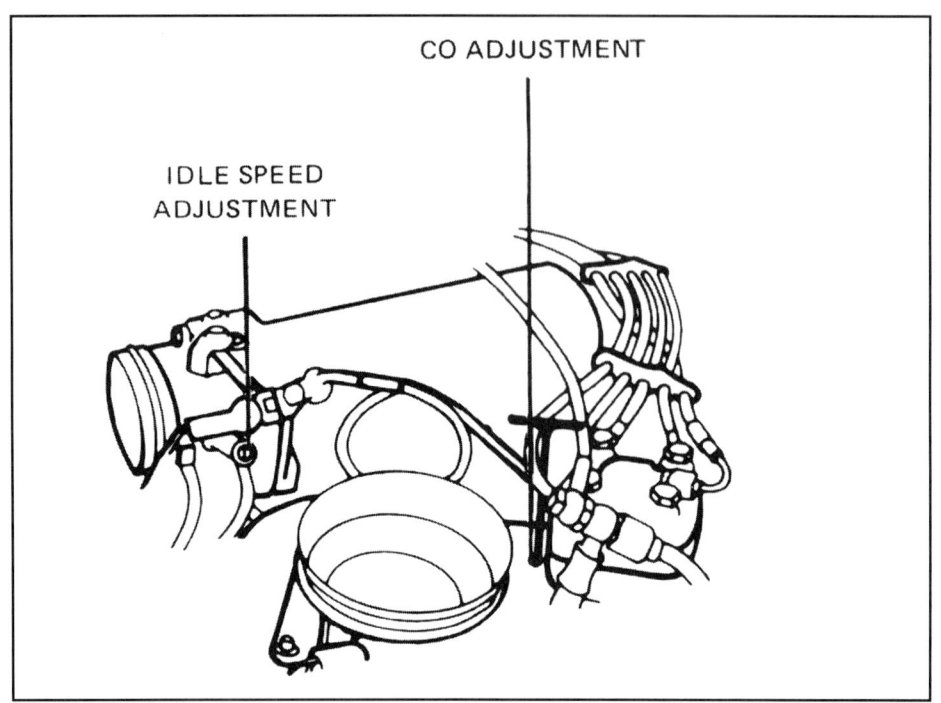

While a typical EFI allows no adjustments, you can set both the air/fuel ratio and the idle mixture on older versions of K-Jetronic/CIS (Volkswagen).

In spite of improvements in fuel additive packages, many experienced technicians still say injector cleaning is helpful. This can be done by means of an aerosol . . .

dead even if self-ignition is occurring.

It's easy to set an idle stop solenoid properly. First, find out both idle speed specifications—here will be one with the solenoid energized and another, lower one, with it de-energized. Set the former by either screwing the plunger in or out of the solenoid or by moving the solenoid itself. Then unplug the solenoid's electrical connector (which will cause the plunger to retract) and set the latter by means of the ordinary throttle stop screw.

Remember that the solenoid is not strong enough to overcome the throttle return spring, so when the key is switched off, then back on, or when the electrical connector is plugged back in, the throttle will have to be opened manually to allow the plunger to move out to its energized position.

Kickers

Throttle kickers are devices that open the throttle a specified amount above normal idle under certain conditions, as when the air conditioner is switched on. They may operate electrically or through a combination of electricity and vacuum. Setting them properly requires only that you have the right specifications for the vehicle.

A variation on this theme is the idle stabilizer found on Volkswagens with electronic ignition and a few other cars. The idea here is to maintain constant idle

. . . or with a pressurized tank of cleaner. Either method is acceptable and recommended for clogged injectors or intake deposits.

rpm, regardless of the accessories that are on, by advancing the timing. This, of course, is handled by a program in the ignition control module.

FUEL INJECTION

Then there's fuel injection, which has taken over completely—you can't buy a new car today that has a carburetor. Except for changing the fuel filter every 60,000 miles or so, most systems require no periodic maintenance whatsoever. For a while back in the late 1980's it looked like regular injector cleaning was going to become a standard procedure in every modern tune-up. But that didn't happen for two reasons: One, the oil companies got with it and started providing excellent detergent packages in all their gasolines. And two, late-model injector designs are much less susceptible to deposit build-up. So, injector clogging became more and more unusual. It does still happen sometimes, though, but now it falls into the category of a repair.

EMISSIONS-CONTROL DEVICES

Finally, there are the emissions-control devices themselves. The PCV system should certainly be examined at every tune-up, and the valve replaced at least at the intervals specified.

There's always a good chance that some of those important vacuum hoses under the hood have gone hard, cracked or have become disconnected or incorrectly routed, so check them during a tune-up.

An EGR valve that's stuck open will destroy that idle, and one that's stuck closed or a system with clogged passages will promote detonation. Clean the valve and the passages at the mileage suggested. Troubleshooting and repairing emissions-control systems are covered in detail in Chapter 16. ■

TUNE-UP CHECKLIST

The following list will complete our new definition of the tune-up, which in most cases should be done every 30,000 miles or two years, or prior to taking an emissions test.

- In cases of a driveability or idle roughness in an EFI-equipped vehicle, listen to each injector separately using a mechanic's stethoscope. If one sounds different from the others, none of the following steps will help. Injector cleaning or replacement is needed.
- If low power or poor mileage leads you to suspect a plugged catalytic converter, do the tests in Chapter 9.
- Replace and gap the spark plugs.
- While the plugs are out, do a compression test to tip you off to any emerging internal engine problems.
- If the engine hasn't got hydraulic valve lifters/cam followers, check valve lash and adjust it to specs if necessary.
- While you've got the valve cover off, retorque the head (do it cold).
- Check the distributor cap, rotor, and spark plug and coil wires. Spraying the ignition secondary system with water will help you find voltage leaks.
- Check choke (carbureted engines)
- Make sure the spark timing is right and that the marks don't wave excessively under the light (an indication of worn distributor drive gears or shaft bushings).
- Check idle speed against specifications.
- Examine the vacuum lines and air intake ducts.
- Check manifold intake vacuum (reveals exhaust restrictions and vacuum leaks)
- See that the PCV system is operating properly.
- Open the EGR valve to be certain it roughens or kills the idle.
- Check oxygen sensor activity (see Chapters 14 and 17).
- If the "Check Engine," "Service Engine," or "Power Loss" dash light is on, pull any trouble codes that have been set in the self-diagnostic memory.
- Replace the air and fuel filters unless the recommended intervals haven't been reached.
- Inspect and adjust the accessory drive belts.
- Scrutinize the radiator and heater hoses.
- If the radiator fan is electric, see that it comes on when the engine gets hot. If there's a clutch on a belt-driven fan, feel for excessive play or insufficient drag.
- Look for leaks at the water pump shaft seal or vent hole.
- Service the battery terminals. Check charging and voltage.
- If the engine has an OHC (OverHead Cam) timing belt, replace it at every other tune-up. This is especially important with designs that aren't free-wheeling—in other words, where the valves and pistons collide if the belt snaps.
- Check all vital fluid levels (engine oil, transmission fluid, coolant, brakes, power steering).

GLOSSARY

acid rain—Corrosive rain formed when sulfur emissions from motor vehicles, industrial plants, or electric generating stations combine with hydrogen and oxygen in the atmosphere to form sulfuric acid (H2SO4). This is not only generally corrosive to anything it may come into contact with, it also raises the acidity of lakes and ponds, often to the point that fish and other aquatic life cannot survive.

air—The combination of gases that makes up the earth's atmosphere. Composed of nitrogen (76 to 78%), oxygen (18 to 21 %), carbon dioxide, argon, and other gases. When air is drawn into an engine, the oxygen combines with the fuel during combustion, producing carbon dioxide and water vapor. If there is too much fuel for the available oxygen, carbon monoxide and unburned hydrocarbons are produced in the exhaust. What is more, at temperatures above 2500 deg. F, oxygen can combine with nitrogen to form NOx another harmful pollutant.

AIR—See *Air Injection Reactor*.

air/fuel ratio—The relative proportions of air and fuel entering an engine's cylinders as produced by the carburetor or fuel injection system. The ideal or "stoichiometric" ratio for gasoline is 14.7:1 air to fuel by weight. A higher ratio would contain more air and less fuel, and would be considered a "lean" mixture. A lower ratio with more fuel and less air would be a "rich" mixture. The air/fuel ratio is determined by the orifice size of the main jets inside a carburetor, the dwell duration of the mixture control solenoid inside a feedback carburetor, or the orifice opening and fuel pulse duration of a fuel injector.

Air Injection Reactor (AIR)—The GM name for the air injection system, which comprises a vane pump, a diverter valve, and a check valve.

air injection system—Supplies fresh air to the exhaust stream, which helps oxidize HC and CO, and, on models equipped with a catalytic converter, gives the catalyst the extra air it needs to oxidize those pollutants.

air pump—A belt-driven vane pump that supplies the flow of air needed for most air injection systems.

altitude—The distance of a point from sea level. Important to automotive emission control because the higher the altitude, the fewer oxygen molecules per given volume of air, which alters the effective compression and air/fuel ratios.

ambient temperature—The temperature of the air surrounding a vehicle.

anti-percolation valve—A carburetion system device used to prevent fuel evaporation from the fuel bowl while the engine is running. It is connected to the throttle linkage so it is closed when the throttle is open, and open when the throttle is closed. With the engine off, hot fuel vapors boil out through the vent line and into the charcoal canister.

ASE—An abbreviation for the National Institute for Automotive Service Excellence. This organization certifies professional mechanics in emissions repair and other specialties.

aspirator—A one-way valve attached to the exhaust system of an engine that admits air during periods of vacuum between exhaust pressure pulses. Used to help oxidize HC and CO, and to supply additional air which the catalytic converter may require. Can be used instead of a belt-driven air injection pump in some applications. Called "Pulsair" in GM systems.

backfire—1. An explosion in the exhaust system of a motor vehicle caused when unburned air/fuel mixture is ignited, usually upon deceleration. 2. An explosion of the air/fuel mixture in the intake manifold, which is evident at the carburetor or throttle body, and may be caused by improper ignition timing, crossed spark plug wires, an intake valve that is stuck open, etc.

back pressure—Resistance of an exhaust system to the passage of exhaust gases. This can have an adverse effect on performance, fuel economy and emissions. Excessive back pressure may be caused by a clogged catalytic converter, or a dented or crimped pipe.

back pressure EGR—Some emissions-control systems use a back-pressure sensor or diaphragm to monitor backpressure so that exhaust gas recirculation (EGR) flow can be

GLOSSARY

increased when the engine is under maximum load (and producing maximum back pressure).

barometric pressure—The pressure exerted by the weight of the earth's atmosphere, equal to one bar, 100 kilopascals, or 14.7 psi (often rounded off to 15 psi) at sea level. Barometric pressure changes with the weather and with altitude. Since it affects the density of the air entering the engine and ultimately the air/fuel ratio, some computerized emissions control systems use a barometric pressure sensor so that the spark advance and EGR flow can be regulated to control emissions more precisely.

blowby—Byproducts of combustion that leak out of the combustion chamber past the piston rings into the crankcase. Mostly hydrocarbons, this represents approximately 20% of the air pollution a pre—emission controlled engine produces. In modern vehicles, the blowby vapors are drawn into the intake stream through the positive crankcase ventilation system (PVC) to be burned in the cylinders.

British Thermal Unit (Btu)—The amount of heat required to raise one pound of water one degree Fahrenheit. The heat value of various motor fuels are often compared in Btu's per gallon or per pound.

California Air Resources Board (CARB)—A state agency responsible for regulations intended to reduce air pollution, especially created by motor vehicles.

carbon dioxide (CO_2)— A harmless, odorless gas composed of carbon and oxygen. In automotive science, the product of complete combustion.

carbon monoxide (CO)—An odorless gas composed of carbon and hydrogen. A major air pollutant and potentially lethal if breathed in small doses. CO is formed by the incomplete combustion of any fuel containing carbon (gasoline, diesel fuel, alcohol, coal, wood, etc.). If additional oxygen is provided (as by an air pump) and allowed to combine with the CO (as happens inside a catalytic converter), carbon monoxide is transformed into harmless carbon dioxide (CO_2). Excessive CO emissions in vehicles are caused by overly rich air/fuel ratios. Possible causes include restrictions in the air intake, a clogged air filter, a partially closed choke, a fuel-saturated carburetor float, too high a float level, leaky needle valve and seat, oversized carburetor jets, internal carburetor fuel leaks, defective oxygen sensor, or leaky injectors.

catalysis—The action of a catalyst.

catalyst—A substance that accelerates or enhances a chemical reaction without being changed itself. When used in a catalytic converter, they can reduce the level of harmful pollutants in the exhaust. Catalysts which are commonly used include platinum, palladium and rhodium, all of which are quite expensive. To minimize costs, a thin layer of catalyst is all that's needed on a ceramic honeycomb, ceramic pellets or a metallic substrate inside the converter. Catalysts are quite sensitive to lead, however. If leaded gasoline is used, the lead coats the catalyst and renders it useless.

catalytic converter—An automotive exhaust system component containing a catalytic element to reduce NOx and/or HC and CO tailpipe emissions. When hot exhaust gases pass through the converter, the catalyst allows oxygen from the air pump or aspirator to "reburn" the pollutants lowering the concentrations of HC and CO to almost zero. This is called an "oxidation" reaction. It creates a great deal of heat (1200 to 1600 deg. F) so the converter is made of stainless steel surrounded with a heat shield. In "Three-Way" converters, a separate "reduction" chamber mounted just ahead of the oxidation chamber contains a different mixture of catalysts to reduce NOx emissions.

Central Port Injection (CPI)—An AC Rochester fuel injection system originally installed on the 4.3L Chevrolet Vortec V6 that produces port injection performance at a low price by using one TBI—style injector to pulse fuel directly to individual nozzles at the intake ports.

centrifugal advance—A mechanical means of advancing spark timing with flyweights and springs. These weights are located inside the distributor (except on engines with computerized engine controls), under the rotor (GM window-type distributors), or under the point or pickup base plate. The size of the

GLOSSARY

weights, the amount of spring tension, and the engine rpm rating determines the rate of advance.

charcoal canister—The basic component of evaporative emission control systems, this is a small cylindrical or rectangular container that contains activated charcoal particles. The charcoal traps gasoline vapors from a vehicle's sealed fuel system.

charge temperature sensor—On computer-controlled engines, a sensor which sends a signal to the computer that varies with the temperature of the intake stream.

check valve—A valve which permits the passage of a gas or fluid in one direction, but not in the other. For example, the check valve between the air pump and exhaust manifold in an air injection system allows air to flow to the manifold, but stops exhaust gas from entering the air pump in the event that the pump belt breaks.

choke—A manually or thermostatically actuated plate mounted on a shaft at the mouth of a carburetor which is closed when the engine is cold to greatly increase the amount of gasoline in the air/fuel mixture, thus aiding starting when fuel evaporation is low. Choke problems are the primary cause of hard starting. If the choke is not closed when the engine is started, the air/fuel ratio will be too lean while the engine is cold. This results in a slow idle and stalling. If the choke does not open up at the correct rate as the engine warms up, the air/fuel ratio becomes too rich. This causes a rough idle, poor fuel economy, and excessive HC and CO emissions. The choke is controlled by a coiled bimetallic spring that reacts to temperature changes. An electric heating element inside the choke housing, or warm exhaust gases or engine coolant routed nearby, are used to speed up the rate at which the choke opens.

choke heater—A device which warms the thermostatic coil of an automatic choke, causing it to open quickly. A typical late model version comprises an electrical heating element and a timer circuit.

chlorofluorocarbons (CFCs)—A family of chemicals that includes R-12 automotive air conditioning refrigerant. CFCs have been blamed for a deterioration of the Earth's protective ozone layer. CFCs are being phased out of production by international agreement.

Clean Air Act—Originally passed in 1970 by the U.S. Congress, and updated in 1990, this legislation created today's auto emissions laws and established the Environmental Protection Agency as the watchdog over our nation's air quality.

closed loop—The basic principle of electronic engine management in which input from an oxygen sensor allows the engine control computer to determine and maintain a nearly perfect air/fuel ratio.

Cold Weather Modulator—A component of certain Ford heated air intake systems, this is a thermostatically controlled check valve that traps vacuum in the vacuum motor when the car is accelerated hard at temperatures below 55 deg. F. This eliminates hesitation by allowing heated air to enter the engine in spite of the drop in vacuum that naturally occurs when the throttle is opened wide.

Combined Emission Control system (CEC)—An early GM transmission controlled spark system that uses the solenoid valve's plunger as an auxiliary throttle stop. During high gear deceleration, it holds the throttle open a predetermined amount, leaning the mixture in that normally rich, high-vacuum mode.

compression ratio—The relationship between the piston cylinder volume from bottom dead center to top dead center. Higher compression ratios improve combustion efficiency but also require higher-octane fuels. Pre-emission control engines often had compression ratios as high as 11.5:1 whereas most of today's engines are between 8.5:1 and 9.5:1. Diesel engines have very high compression ratios, from 18:1 to 22:1.

computerized engine controls—A microprocessor-based engine management system that use sensor inputs to regulate spark timing, fuel mixure, emissions and other functions. Most have a certain amount of self-diagnostic capability, and store or generate fault codes to help a technician diagnose system problems.

GLOSSARY

coolant temperature override switch (CTO)— A vacuum controlling device that prevents the overheating associated with late ignition. At normal operating temperature, the CTO allows ported vacuum to reach the distributor. But when overheating begins to develop at idle, it routes the strong manifold signal to the advance unit, which makes the spark occur earlier, cooling combustion and increasing idle speed.

coolant temperature sensor—In computerized engine control systems, a thermistor which informs the ECU as to the temperature of the coolant. In the PTC (Positive Temperature Coefficient) type, ohms go up with temperature. In the more common NTC (Negative Temperature Coefficient) type, resistance goes down as heat goes up.

crankcase emissions—See *blowby*.

decel valve—A device which reduces exhaust emissions during vehicle deceleration by keeping rpm up and vacuum down. Found on some Ford engines such as the Pinto 2.0L, it works by opening an extra intake air passage under high vacuum conditions.

detonation —A phenomenon of internal combustion wherein the compressed air/fuel charge explodes violently instead of burning smoothly, usually due to the creation of a second flame front in the chamber away from the spark plug, which collides with the spark-ignited flame front, causing a noise known as "pinging" or "spark knock," and potentially damaging to the engine. Detonation can be caused by excess spark advance, low-octane fuel, lean air/fuel mixtures, and/or overheating. Carbon buildup inside the combustion chambers and on the piston face can also increase compression sufficiently to cause detonation. Mild detonation is not harmful, but heavy detonation can damage pistons, rings and rod bearings. It can also cause elevated HC emissions.

detonation sensor—An electrical device mounted on an engine block, cylinder head, or intake manifold that generates a small voltage signal when it encounters the vibration frequency associated with detonation. This signal is sent to an electronic control unit that retards ignition timing until detonation stops.

dieseling—A condition in which a carbureted engine continues to run after the ignition is shut off. Caused by hot spots inside the combustion chambers, spontaneous fuel ignition keeps the engine running (see also preignition). This does not occur in fuel injected engines, however, because the injectors stop the flow of fuel when the ignition is shut off. To prevent dieseling in carbureted engines, an idle stop solenoid or fuel cut-off solenoid may be used.

Digital Electronic Fuel Injection (DEFI)—An early Cadillac electronic fuel injection system.

direct-acting thermostatic air cleaner—The basic component of certain heated air intake systems, which uses a thermostatic bulb that is connected to the rod that operates the flapper valve in the air cleaner snorkel.

diverter valve—In an air injection system, a vacuum-operated valve that directs air pump output to the atmosphere during high-vacuum deceleration to eliminate backfiring.

dual diaphragm distributor—A distributor incorporating a vacuum advance unit which contains two separate diaphragms and vacuum chambers. Usually, one is connected to ported vacuum, which advances the spark during acceleration and cruising, and the other to intake manifold vacuum, which retards the spark at idle and deceleration.

duty solenoid—On a feedback carburetor, a solenoid that cycles many times per second to control a metering rod, hence the air/fuel mixture.

dynamometer—A machine used to simulate loaded driving conditions for emissions and diagnostic purposes. A vehicle's drive wheels are placed on a pair of rollers so the vehicle can be driven in place at various speeds while being subjected to changing loads. Engine performance and emissions are monitored with additional test equipment to determine emissions compliance and/or to diagnose emissions or performance problems.

Early Fuel Evaporation (EFE)—A

GLOSSARY

GM system that promotes the evaporation of gasoline in the intake manifold through the application of heat from exhaust or an electrical heating element, thus improving the engine's cold running characteristics and reducing emissions.

EEPROM—Electronically Ereaseable Program Read Only Memory, a special type of computer calibration chip that can be reprogrammed electronically by a service technician using the proper equipment and codes.

EGR valve—The exhaust gas recirculation valve meters a small amount of exhaust gas into the intake manifold to dilute the air/fuel mixture. This keeps combustion temperatures below 2500 degrees F to reduce the formation of NOx. The amount of exhaust gas recirculated into the engine is only a few percent. The EGR valve is mounted on the intake manifold. A vacuum-operated diaphragm or small electric motor lifts a small plunger-type valve that uncovers an opening to the intake manifold. Exhaust gases then flow through plumbing from the exhaust manifold past the EGR valve and into the engine.

EGR valve position sensor—A potentiometer that keeps the engine control computer informed as to how far the EGR valve is open.

electronic control unit (ECU)—A digital computer, especially one that controls engine and sometimes transmission functions. Also known as an electronic control module (ECM), or electronic control assembly (ECA).

electronic fuel injection (EFI)—Any type of fuel injection system that uses electronic rather than mechanical controls. See also Central Port Injection (CPI), multiport fuel injection (MFI), K-Jetronic, L-Jetronic, sequential fuel injection (SFI), throttle body injection (TBI) and Tuned Port Injection (TPI).

electronic spark control (ESC)—The process whereby spark advance is controlled electronically rather than by mechanical means. A microprocessor calculates the necessary amount of spark advance based on engine operating conditions and what has been programmed into its memory. The traditional vacuum and centrifugal advance mechanisms are not used in the distributor.

emission controls—The components directly or indirectly responsible for reducing air pollution, including crankcase emissions, evaporative emissions and tailpipe emissions.

emissions—Unwanted, harmful chemicals and chemical compounds that are released into the atmosphere from a vehicle, especially from the tailpipe, crankcase, and fuel tank. These include unburned hydrocarbons, carbon monoxide, oxides of nitrogen, particulates (soot), and sulfur.

emissions warranty—A federally mandated 5 year/50,000 mile performance and defect warranty that covers all emissions related components on all new vehicles built since 1981. Parts covered include the PCV system, EGR system, catalytic converter, exhaust manifolds, head pipe, air pump and related plumbing, charcoal canister and fuel vapor recovery system, heated air intake system, fuel injection system (except pump & filter), engine sensors and computer. Spark plugs and other normal maintenance items are not covered after 2 years or 24,000 miles. On 1995 and later vehicles, the emissions warranty is reduced to two years/24,000 miles on all emissions related components except the computer and catalytic converter which are extended to 8 years/80,000 miles. If a covered component fails during the warranty period, you can return to a new car dealer for free replacement of the faulty component — but you are under no obligation to do so.

Environmental Protection Agency (EPA)—A regulatory agency of the U.S. federal government responsible for creating and enforcing regulations concerning the protection of the environment from various forms of pollution, including that which is generated by motor vehicles.

evaporative emissions—Hydrocarbons from fuel which evaporate from a vehicle's fuel tank and carburetor. They are eliminated by sealing the fuel system and using a charcoal canister to trap vapors from the fuel tank and carburetor. Some states also require that filling stations

GLOSSARY

include evaporative emission recovery devices on gasoline pumps to catch gasoline fumes when the fuel tanks are being filled.

exhaust analyzer—An automotive test and service device which uses a process involving infrared energy to determine and display the composition of an engine's exhaust gases. The two-gas type measures hydrocarbons and carbon monoxide content, while the four-gas type also measures oxygen and carbon dioxide content.

exhaust gas oxygen (EGO)—The amount of oxygen present in the exhaust stream, as measured by an oxygen sensor and reported to the computer in closed-loop, feedback systems. The computer uses this information along with signals from other sensors to control the air/fuel mixture.

exhaust gas recirculation (EGR)—An emissions-control system which reduces an engine's production of oxides of nitrogen by diluting the air/fuel mixture with exhaust gas so that peak combustion temperatures (those which cause nitrogen and oxygen to form these harmful compounds) in the cylinders are lowered. A malfunction which prevents EGR will cause elevated NOx emissions and possibly detonation under load.

exhaust emissions—Pollutants identified by clean air legislation as being harmful or undesirable. These include lead, unburned hydrocarbons (HC), carbon monoxide (CO), oxides of nitrogen (NOx). Permissible levels for these pollutants are specified in grams per mile or percent of volume.

fast idle—The higher speed at which an engine idles during warm-up. When first started, a cold engine needs more throttle opening to idle properly. On carbureted engines without computer idle speed control, a set of cam lobes on the choke linkage provides a fast idle speed of 1100 to 1500 rpm during engine warm-up.

feedback—1. A principle of fuel system design wherein a signal from an oxygen sensor in the exhaust system is used to give a computer the input it needs to properly regulate the carburetor or fuel injection system in order to maintain a nearly perfect air/fuel ratio. 2. A signal to a computer that reports on the position of a component, as an EGR valve. Typically, the feedback device is a variable resistor.

feedback carburetor—A carburetor that controls the air/fuel mixture according to commands from the engine control computer, typically through the operation of a duty solenoid.

fuel injection—A system that uses no carburetor but sprays fuel under pressure into the intake manifold or directly into the cylinder intake ports. The advantage of fuel injection over carburetion is that it allows more precise control of the air/fuel mixture for improved performance, fuel economy, and reduced emissions.

grams per mile (GPM)—A measurement of the amount of emissions a vehicle produces.

gulp valve—In an air injection system, a valve that opens to admit extra air into the intake manifold upon deceleration, thus leaning out the mixture to prevent backfiring.

Hall Effect—A phenomenon in which voltage is generated by the action of a magnetic field acting on a thin conducting material, commonly used to control the primary circuit of an electronic ignition system. Named for the American scientist Edwin Hall (1855-1938).

HC-CO meter—Uses an infrared sensing device to measure the amount of HC and CO in vehicle exhaust (see also infrared analyzer). A probe is inserted into the tailpipe of the vehicle being checked so that samples of the exhaust gases can be drawn into the machine. HC is read in parts per million and CO is read in percent.

heated air intake system—A system that maintains intake air at a more or less constant temperature by blending outside or underhood air with heated air picked up from a shroud over the exhaust manifold. A typical version uses a vacuum motor to power a door in the air cleaner snorkel, and a thermostatic bleed valve to control the signal to the vacuum motor. Also called Thermostatic Air Cleaner (TAC). A system malfunction that prevents the door from closing can cause hesitation and stumbling when the engine is cold. An air temperature

GLOSSARY

control flap stuck shut will overheat the air/fuel mixture, possibly causing detonation and elevated CO levels (due to a rich air/fuel ratio, as warm air is less dense than cold air).

heated exhaust gas oxygen sensor (HEGO)—An oxygen sensor which is heated electrically as well as by engine exhaust so that it warms up to normal operating temperature more quickly, thus allowing the engine to enter closed-loop operation sooner than with a non-heated sensor.

heat riser—A channel in an intake manifold through which exhaust gas flows in order to heat the manifold, thus aiding in fuel vaporization.

heat riser valve—A control valve between the exhaust manifold and exhaust pipe on one side of a V8 or V6 engine that restricts the flow of exhaust causing it to flow back through the heat riser channel under the intake manifold. This aids fuel evaporation and speeds engine warm-up. A heat riser valve stuck open will slow engine warm-up and may cause hesitation and stalling when the engine is cold. A valve stuck in the closed position will greatly restrict the exhaust system and cause a noticeable lack of power and drop in fuel economy.

High-Swirl Combustion (HSC)—Ford's name for a cylinder head and valve design that promotes turbulence in the combustion chamber during the power stroke, thus contributing to complete, efficient burning of the air/fuel charge.

Honda CVCC—(Controlled Vortex Combustion Chamber). An efficient combustion chamber design that uses a small auxiliary combustion chamber (containing the spark plug) that receives a rich mixture to get an overall lean mixture in the cylinder to fire dependably.

hot idle compensator—A temperature-sensitive carburetor valve that opens when the inlet air temperature exceeds a certain level. This allows additional air to enter the intake manifold to prevent overly rich air/fuel ratios.

humidity—The amount of water vapor in the air. The amount of water air can hold before it becomes saturated depends on temperature. Warm air can hold more moisture than cold air. At 100 percent humidity, the air is completely saturated with moisture. Humidity affects engine performance because it tends to boost the effective octane rating of the air/fuel mixture. Engines can therefore tolerate more spark advance during humid weather than during dry weather.

hydrocarbon (HC)—Any chemical compound composed chiefly of hydrogen and carbon, especially petroleum. As an automotive pollutant, HC is simply unburned fuel and lubricating oil, and may be found in the crankcase as blowby, evaporating from the gas tank, and escaping from the tailpipe.

hydrogen—A flammable gas with the symbol H, often occurring as H_2. The most common element in nature.

Idle Air Control valve (IAC)—The GM name for an electrically-operated valve which, according to commands from the engine control computer, varies the size of an air passage that bypasses the throttle plate of an electronic fuel injection system, thus controlling idle speed.

idle adjustment—Refers to either the idle speed adjustment or the idle mixture adjustment.

idle limiter cap—A plastic device pressed over a carburetor's idle mixture screw which limits the amount of adjustment available during service. Adopted to help eliminate excessive air pollution that may be caused if the idle mixture is set too rich.

idle mixture—The air/fuel ratio at idle.

idle mixture adjustment—The idle air/fuel mixture adjustment is usually sealed to prevent tampering. It is adjusted on carbureted engines by removing the anti-tampering plugs (which may require carburetor removal first) and turning the idle mixture screws until the proper idle mixture is achieved. Turning the screw out richens the idle mixture, while turning it in leans the mixture. Using the idle drop technique, the mixture is set by adjusting the mixture for smoothest idle at the slowest rpm rating. Setting the idle mixture on an emissions controlled engine is tricky and may require monitoring exhaust

GLOSSARY

CO or using a special procedure called "propane enrichment."

idle mixture screw—A screw with a tapered point which is threaded into the body of a carburetor and can be turned to vary the amount of fuel which can pass through the idle circuit.

idle speed—The speed at which an engine idles specified in revolutions per minute (rpm). Idle speed is an important value because it can affect emissions, idle quality and driveability. For most engines, the idle speed is usually between 600 and 850 rpm. The exact idle speed is specified on the underhood emissions decal. If the idle speed is too low, the engine may stall, especially when the transmission is in drive and/or electrical accessories or the air conditioner is on. If too high, idle emissions will be increased, the vehicle may be hard to hold at a stop when in drive, and the jolt created when shifting into gear may damage the transmission or driveshaft joints.

idle speed adjustment—The idle speed on carbureted engines without computer idle speed control is set by turning a screw that opens the throttle plates. The screw is turned until the desired idle rpm level is achieved. On fuel injected engines without computer idle speed control, idle speed is set by turning an idle air bypass screw that allows air to bypass the throttle plates. Idle speed is not adjustable on engines with computer idle speed control.

idle stop solenoid—An electromagnetic device mounted on carburetor linkage that maintains the proper throttle opening for specified idle speed while the ignition is on, but allows the throttle to close farther when the ignition is switched off, thus reducing the amount of air that can enter the engine and reducing the likelihood of dieseling.

inspection/maintenance (I/M)—The periodic inspection and maintenance of a vehicle's ignition, fuel and emission control systems. By maintaining the engine's various subsystems, emissions are kept to a minimum while fuel economy and performance are enhanced. A well maintained engine is an efficient running and clean engine.

I/M 240—An "enhanced" emissions inspection program that uses loaded mode testing to check vehicle emissions. It is called I/M 240 because it is a 240 second test that simulates the federal urban test cycle for certifying new vehicle emissions performance. Loaded mode testing requires a dyno and special emissions analysis equipment that can determine tailpipe emissions in grams per mile and measure oxides of nitrogen (NOx) emissions.

inches of mercury (in. Hg.)—A measurement of vacuum related to the height to which atmospheric pressure can push a column of mercury within a tube. Standard atmospheric pressure at sea level is 14.7 in. Hg, commonly rounded off to 15.

integrated circuit (IC)—A miniaturized electronic circuit having all necessary components, such as transistors, resistors, capacitors, etc., integrated into a silicon chip.

K-Jetronic—A Robert Bosch mechanical port fuel injection system which injects gasoline continuously (the "K" stands for the German word for "continuous"). Also known as CIS (Continuous Injection System). Late-model variations include KE-Jetronic and K-Jetronic with Lambda, both of which employ oxygen and other sensors to keep the air/fuel mixture within a stricter range.

knock sensor—A sensor that signals the engine control computer when detonation is detected. When the computer receives a knock signal from the knock sensor, it momentarily retards ignition timing until detonation ceases. Knock sensors react to engine vibrations in a specific frequency band, but can sometimes be fooled by other engine sounds such as those produced by worn rod bearings.

lambda—The 11th letter of the Greek alphabet used by engineers to represent the air/fuel ratio. European auto makers typically refer to the oxygen sensor as the lambda sensor.

Lambda Sond—The first closed loop fuel injection system to appear in production, developed jointly by Robert Bosch and SAAB.

Lean Burn—A Chrysler Corporation electronic engine control system that

GLOSSARY

appeared in 1976. It used precise control of spark timing to allow a very lean mixture to burn properly, thus reducing emissions. Although it used an analog computer, which is relatively unsophisticated compared to a digital computer, it was the first mass production application of an engine control computer.

leaner and later—Refers to early calibration strategies for air/fuel mixture and ignition timing that reduced HC and CO formation.

lean misfire—A condition caused by an air/fuel mixture that is too lean to sustain combustion. Lean misfire causes one or more cylinders to pass unburned fuel into the exhaust system causing a big increase in hydrocarbon (HC) emissions. Symptoms include a rough idle and hesitation or stumble on acceleration. Lean misfire is often caused by vacuum leaks or an EGR valve that's stuck open.

light-off mini-oxidation catalytic converter—A smaller catalyst mounted just behind the exhaust manifold that gets hot very quickly after the engine is started, so it begins working in time to neutralize much of the extra pollution that is produced during cold running. Also called a "pre-cat" or "pup" converter.

Linear EGR—An AC Rochester EGR system using a linear motor to move the valve's pintle in small steps, which provides precise control of recirculation.

liquid/vapor separator—An evaporative emissions system component mounted above the fuel tank that prevents liquid gasoline from entering the vent lines.

L-Jetronic—A Robert Bosch port EFI system which uses input on the volume of intake air to calculate fuel delivery (the "L" stands for "Luftmengenmessung," German for "air flow management").

magnetic timing—A procedure for checking or adjusting an engine's ignition timing using a mag timing meter rather than a timing light. A magnetic probe is inserted into a receptacle near the crankshaft harmonic balancer or flywheel. The probe picks up a small notch in the balancer or flywheel with every revolution of the engine. An inductive pickup that clamps onto the number one plug wire tells the meter when the plug fires. The meter then displays the degrees of timing advance.

manifold absolute pressure sensor (MAP)—A variable resistor used as a sensor to inform an engine control computer as to the vacuum conditions in the intake manifold.

manifold air temperature (MAT)—The temperature of the intake stream in the intake manifold, as increased by a heat riser or EFE system, and/or converted to engine control computer input by a MAT sensor.

manifold vacuum—The vacuum available at an engine's intake manifold generated by the engine's pumping action, measured in inches or millimeters of mercury.

mass air flow sensor (MAF)—A device used in EFI systems which supplies the computer with input as to the volume of air entering the manifold. Found at the mouth of the intake manifold, it uses the temperature differential between a heated platinum wire or a plastic film and the passing air to generate a signal of varying voltage.

Mitsubishi Jet Valve—A tiny third valve that admits only air to churn up the air/fuel charge and promote lean running and a complete burn.

monolithic catalytic converter—A catalytic converter which has its catalytic materials coating a ceramic honeycomb, as distinguished from the pellet bed converter.

multipoint fuel injection (MFI)—A type of fuel injection system that has a separate fuel injector for each of the engine's cylinders. Multipoint injection systems deliver better performance and lower emissions than throttle body injection (TBI) systems, but are more costly and complex.

negative backpressure valve—An EGR valve incorporating a bleed hole that is normally closed. When backpressure drops, indicating reduced load, the bleed opens reducing vacuum above the diaphragm and cutting EGR flow.

nitrogen—A gaseous, nearly inert

GLOSSARY

element given the symbol N, which comprises approximately 80% of the earth's atmosphere, naturally occurring as N2.

noble metal—A relatively rare and very expensive metal, such as the platinum, palladium, and rhodium found in catalytic converters.

NOx—See *oxides of nitrogen*.

octane—Refers to gasoline's ability to resist detonation. The higher the octane number, the greater the fuel's resistance to detonation (spark knock or pinging under load). Most of today's gasolines have a pump octane rating of 87 to 92.

onboard diagnostics (OBD)—A term for special diagnostic software and hardware that detects performance problems that adversely affect emissions. OBD rules require a standardized diagnostic connector and fault codes for emissions troubleshooting.

open loop—In engines with a computer and oxygen sensor control system, a mode of operation during which the computer ignores the signal from the oxygen sensor, typically before the engine reaches normal operating temperature.

Orifice Spark Advance Control (OSAC)—A Chrysler Corporation emission control system which slows vacuum advance of ignition timing by means of an orifice in a component mounted on the air cleaner.

OSAC valve—An abbreviation for "orifice spark advance control valve," a device used on some older Chrysler engines to limit NOx formation. The valve delays vacuum to the distributor vacuum advance between idle and part throttle operation.

Otto cycle—The basic principle of operation of the common four-stroke piston engine, involving intake, compression, ignition/power, and exhaust. Named for Nikolaus Otto, German inventor (1832-1891).

oxidation—Any reaction in which a chemical joins with oxygen, as rusting or combustion.

oxidation catalyst—A two-way catalytic converter which promotes the oxidation of HC and CO in an engine's exhaust stream, as distinguished from a three-way or reduction catalyst

oxides of nitrogen—Harmful gaseous emissions of an engine composed of compounds of nitrogen and varying amounts of oxygen which are formed at the highest temperatures of combustion. With other gases and in the presence of sunlight, an ingredient of photochemical smog.

oxygen—A gaseous element given the chemical symbol O, and occurring as O2, which makes up approximately 20% of the earth's atmosphere.

oxygen sensor—A device, usually threaded into the exhaust manifold, which uses platinum inner and outer electrodes and a zirconium electrolyte to generate a small voltage, the strength of which is dependent upon the amount of oxygen present in the engine's exhaust stream. This voltage is used as a signal to the engine control computer, which takes the signal into consideration when determining the amount of fuel necessary to maintain a proper air/fuel ratio.

ozone—A molecule of oxygen chemically represented as O_3 which is formed by exposure of O_2 to an electrical discharge, and has a pungent odor and a strong oxidizing effect.

particulate emissions—Solid particles, such as carbon and lead, found in vehicle exhaust; soot. A problem especially in diesels.

particulate trap—An emission control device in the exhaust system of a diesel engine which is used to capture particulates before they can enter the atmosphere.

parts per million (PPM)—A measurement of the emissions of a motor vehicle given as the number of parts of a particular chemical within one million parts of exhaust gas.

PCV valve—A steel or plastic valve used in positive crankcase ventilation systems to meter blowby into the intake stream. It is commonly mounted in a grommet in the valve cover and connected to a hose that goes to a spacer under the carburetor, or to an intake manifold port with fuel injected engines. Inside the housing is a movable plunger and a light coil

GLOSSARY

spring that bears against it.

pellet bed catalytic converter —A GM converter design comprising a stainless steel shell and a bed of catalyst-coated ceramic pellets. Unlike monolithic converters, if the pellets become contaminated or otherwise rendered inoperative, they can be dumped and replaced with a new load, although this job (which requires special vibrator/aspirator equipment) is not actually done very often in the real world.

photochemical smog—A noxious, unhealthful gaseous compound in the atmosphere formed by the interaction of various chemicals such as the pollutants hydrocarbons and oxides of nitrogen in the presence of sunlight.

piezoelectricity—Voltage generated by a dielectric crystal under mechanical stress. The principle is used in knock sensors.

platinum—A rare, valuable metallic element given the symbol Pt, which is highly resistant to corrosion, and is used as a catalytic agent in automotive catalytic converters of the oxidizing type.

ported vacuum—Engine vacuum as available above the throttle plates of a carburetor, as used to advance ignition timing when the throttle is opened above its idle position.

ported vacuum switch—A valve which permits or stops the passage of ported vacuum to a vacuum-operated component, such as a distributor advance mechanism. It may be thermostatically operated, or controlled by electric current or the movement of a mechanical component.

positive crankcase ventilation (PCV)—An engine emission control system which picks up crankcase gases such as blowby and meters them into the intake stream to be burned. Clogging of the PCV system may cause oil leaks as blowby increases crankcase pressure. It can also lead to rapid sludge buildup in the crankcase and possible engine damage.

positive backpressure valve—A common type of EGR valve which uses exhaust system backpressure to sense engine load, thus more accurately metering the amount of exhaust recycled.

preignition—The ignition of the air/fuel mixture in the combustion chamber by means other than the spark; the same as dieseling. It is usually caused by hot spots in the combustion chamber (sharp edges, carbon accumulation or spark plugs with too hot a heat range). Preignition can burn holes in pistons and contribute to detonation.

Programmed Read Only Memory (PROM)—A computer component which contains values and programming which is not lost when the power supply to the computer is shut off or interrupted, its memory being non-volatile. Used to determine the basic parameters of operation in an engine control computer system.

propane enrichment—A service procedure common in the 1970's which was used to set idle mixture. A metered amount of propane gas is added to the intake stream and the resulting rpm increase is observed.

pulse width—In EFI systems, the length of time the injectors are energized and held open ("on" time), which determines the amount of fuel injected. Measured in milliseconds, a range of 1/1,000th to 7/1,000th of a second is common.

quench area—Any internal portion of a combustion chamber which causes combustion to cease because of the temperature drop in the air/fuel charge where it meets this area.

R-12—A type of refrigerant used in automotive air conditioning systems that contains ozone-damaging chlorofluorocarbons (CFCs). R-12 is being phased out and replaced with a new refrigerant, R-134a. Professional technicians are required by law to recover and recycle R-12 when performing any type of A/C service or repair work.

R-134a—An "ozone-safe" refrigerant that is used on most 1993 and later vehicles. It is not a direct drop-in replacement for R-12, and may require various modifications to a vehicle's A/C system before it can be substituted for R-12.

reduction catalyst—The section of a three-way catalytic converter that

G L O S S A R Y

breaks NOx down into harmless nitrogen and oxygen through a reduction reaction.

reference voltage—In computerized engine management systems, a five volt signal sent out from the ECU to a variable resistance sensor such as a TPS. The computer then reads the voltage value of the return signal. Called "V-ref" by GM.

remote backpressure transducer—In an EGR system, an exhaust backpressure sensing device mounted in the vacuum line leading to the EGR valve rather than on the valve itself. At idle or light loads, it bleeds off the vacuum signal to prevent recirculation.

resistance—The quality of reducing the flow of electrons in a circuit.

retard—To cause ignition spark to occur later in an engine's cycle.

rich mixture—An air/fuel mixture with insufficient air or excessive fuel. The ideal mixture for gasoline and air is 14.7:1 by weight. A richer mixture is needed during engine warm-up and during acceleration, but a rich mixture at other times causes elevated carbon monoxide (CO) emissions.

road-draft tube—A pre-emission control-era device for ventilating the crankcase, essentially a pipe routed under the chassis at an angle that produced a small amount of vacuum as the vehicle traveled forward. Fresh air was drawn through a mesh filter in the oil filler cap, circulated around inside the crankcase, and exhausted through the road-draft tube carrying blowby with it.

run-on—See *dieseling*.

self-diagnostics—In automotive computers, especially those for engine control, a program which assesses the condition of the system, including the sensors and the computer itself and communicates its findings to the technician by means of trouble codes.

sensor—An electrical device used to provide a computer with input as to temperature, rpm, vacuum, etc.

sequential fuel injection (SFI)—A type of multiport injection system where the individual fuel injectors are pulsed sequentially one after another in the same firing order as the spark plugs rather than being pulsed simultaneously. This allows more precise fuel control for lower emissions and better performance, but also requires more complex controls.

smog—A general term for air pollution, especially the photochemical variety. Smog forms when sunlight causes chemical reactions in air pollutants resulting in the formation of ozone and other compounds.

smog pump—A slang term for an air injection system pump.

spark advance curve—The rate at which ignition timing advances. If plotted on a graph, the line resembles a curve. It rises from some initial amount of advance and levels off at the maximum advance. On older ignition systems with mechanical and vacuum advance controls, the curve depended on engine rpm and intake vacuum. On later ignition systems with electronic spark advance, various sensor inputs are used to calculate the amount of advance needed.

spark decel valve—A vacuum valve located in the line between the distributor and carburetor. The valve advances spark advance during deceleration to reduce emissions.

spark delay valve—A vacuum valve used in the line between the distributor and carburetor to delay vacuum timing advance under certain driving conditions to reduce NOx emissions. The valve acts like a restriction in the vacuum line so that vacuum builds up and changes very slowly.

spark knock—See *detonation*.

stoichiometric—Referring to the ideal air/fuel ratio, 14.7:1 by weight, in which all the oxygen is consumed in the burning of all the fuel.

stoichiometry—The state of having a stoichiometric air/fuel mixture.

Sub-EGR Control Valve—A vacuum valve used on Chrysler and Mitsubishi 2.6L engines which is operated mechanically by means of the throttle linkage, so it varies the signal to the EGR valve according to the position of the accelerator pedal.

GLOSSARY

tetraethyl lead—A lead compound that increases gasoline's octane rating. Unfortunately, lead damages catalytic converters and oxygen sensors and therefore cannot be used in vehicles designed to operate on unleaded fuel.

Thermactor—The Ford name for an air pump or air aspirator air injection system.

thermal reactor—An emissions-control device comprising a large, heavy exhaust manifold in which hydrocarbons and carbon monoxide that escape from the cylinders are oxidized. Extra air is provided by a pump or an aspirator valve to promote this reaction. It is now obsolete as it is not as efficient as the catalytic converter.

thermistor—A resistor the value of which changes according to its temperature. Used as a sensor for a gauge or computer system.

Thermostatic Air Cleaner (TAC)—The GM name for an engine air cleaner assembly that controls the temperature of the intake air by blending relatively cool underhood or outside air with relatively hot air picked up from a shroud over the exhaust manifold.

thermostatic vacuum switch (TVS)—A valve that controls the passage of vacuum according to temperature. In a basic EGR system, it blocks vacuum until a certain coolant temperature is reached, at which point it opens, allowing vacuum to reach the EGR valve.

three-way catalyst (TWC)—A catalytic converter that oxidizes hydrocarbons and carbon monoxide, and also reduces oxides of nitrogen emissions. Usually, it has separate chambers, the one upstream handling reduction, and the one downstream handling oxidation. The noble metals used as the catalytic agents are platinum, palladium, and, for reduction, rhodium.

throttle body injection (TBI)—A fuel injection system that has fuel injectors located in a common throttle body. The throttle body resembles a carburetor from the outside, and sits in the usual position on the intake manifold, but instead of having a fuel bowl, float and venturis it has one or two fuel injectors. TBI provides many of the advantages of fuel injection (easier starting, lower emissions, etc.) without the cost and complexity of multiport injection.

throttle position sensor (TPS)—In computerized engine control systems, the variable resistor-type sensor which informs the ECU of throttle position. A strip of carbon provides resistance, and a brush or wiper moves along its face as the throttle is opened. It is a three-wire device with terminals for reference voltage, output back to the ECU, and ground.

transistor—An electronic component using a semiconductor to amplify or switch current.

transmission controlled spark (TCS)—An emissions-control system which prevents distributor vacuum advance from occurring at normal operating temperature until the transmission has shifted into high gear. An electrical switch mounted on the transmission controls a solenoid-actuated vacuum valve in the hose between the vacuum port above the throttle plate of the carburetor and the distributor vacuum advance unit.

trouble code—A number generated by a computer to indicate a failure in a sensor, circuit, or the computer itself. The number may be communicated to the technician by the flashing of a dash light when the diagnostic mode is entered.

Tuned Port Injection (TPI)—A General Motors multiport fuel injection system used on 5.0L and 5.7L V8 engines featuring tuned intake runners from a common plenum.

two-way catalyst—A catalytic converter that oxidizes hydrocarbons and carbon monoxide, but has little effect on oxides of nitrogen. The noble metals used to promote this reaction are platinum and palladium. Also called an "oxidation catalyst."

vacuum—A condition of pressure which is less than that of the atmosphere; negative pressure.

vacuum advance—The principle of using the vacuum generated by an engine to advance ignition timing, accomplished with a mechanism attached to the distributor which moves the breaker point or pickup coil plate when it receives vacuum.

GLOSSARY

vacuum delay valve (VDV)—An orifice-controlled valve which delays a vacuum signal to a diaphragm, such as in the distributor vacuum advance unit.

vacuum motor—A chamber containing a vacuum diaphragm used to create movement in a component, such as a heated air intake blend door, when engine vacuum is routed to the chamber.

vapor-liquid separator—Part of the evaporative emissions control system that prevents liquid gasoline from flowing through the vent line to the charcoal canister.

vapor lock—A condition in carbureted engines where excessive heat has caused the fuel in the fuel line or fuel pump to boil. The bubbles block the flow of fuel to the carburetor and prevent the engine from starting.

vapor recovery system—Part of the evaporative emissions control system that prevents gasoline vapors from escaping to the atmosphere. Vapors are trapped in the charcoal canister and then drawn into the engine and burned when the engine is started. Vapor recovery can also refer to preventing gasoline vapors from entering the atmosphere when a vehicle is being refueled.

variable resistor—An electrical component which reduces the flow of current in a circuit variably according to its position, such as a TPS. Also called a "potentiometer," or "pot."

venturi—The narrow part of the carburetor throat. When air passes this point, the restriction causes an increase in velocity and a drop in pressure that siphons fuel from the fuel bowl into the airstream.

venturi vacuum amplifier—In certain EGR systems, a device that uses the weak venturi vacuum signal to regulate the application of strong manifold vacuum to the EGR valve. It usually includes a reservoir that supplies sufficient vacuum when the engine is producing too little to operate the EGR valve.

ACRONYMS

EMISSIONS RELATED

AAV: anti-afterburn valve (Mazda)
AIR: Air Injection Reaction (GM)
AIS: Air Injection System (Chrysler)
CC: catalytic converter
CCP: controlled canister purge (GM)
CCV: canister control valve
CEC: Crankcase Emission Control System (Honda)
CO: carbon monoxide
CO_2: carbon dioxide
CP: canister purge (GM)
ECS: Evaporation Control System (Chrysler)
EEC: Evaporative Emission Controls (Ford)
EECS: Evaporative Emissions Control system (GM)
EFE: Early Fuel Evaporation system (GM)
EGR: exhaust gas recirculation
EGR-SV: EGR solenoid valve (Mazda)
EGRTV: EGR thermo valve (Chrysler)
EVRV: electronic vacuum regulator valve for EGR (GM)
HAIS: Heated Air Intake System (Chrysler)
HC: hydrocarbons
NOx: nitrogen oxides
OC: oxidation converter (GM)
ORC: oxidation reduction catalyst (GM)
PAFS: Pulse Air Feeder System (Chrysler)
PAIR: Pulsed Secondary Air Injection system (GM)
PCV: positive crankcase ventilation
PVS: ported vacuum switch
SO_2: sulfur dioxide
TAC: thermostatic air cleaner (GM)
TAD: Thermactor air diverter valve
TWC: three-way catalyst
TVS: thermal vacuum switch
TVV: thermal vacuum valve (GM)

COMPUTERIZED ENGINE CONTROL SYSTEMS

C3: Computer Command Control system (GM)
C4: Computer Controlled Catalytic Converter system (GM)
CAS: Clean Air System (Chrysler)
CCC: Computer Command Control system (GM)
CVCC: Compound Vortex Controlled Combustion system (Honda)
ECCS: Electronic Concentrated Control System (Nissan)
EEC: Electronic Engine Control (Ford)
ELB: Electronic Lean Burn (Chrysler)
MCU: Microprocessor Controlled Unit (Ford)
MISAR: Microprocessed Sensing and Automatic Regulation (GM)
PGM-FI: Programmed Gas Management Fuel Injection (Honda)
SCC: Spark Control Computer (Chrysler)
TCCS: Toyota Computer Controlled System

SENSORS

ACTS: air charge temperature sensor (Ford)
AFS: airflow sensor (Mitsubishi)
AFM: airflow meter
APS: absolute pressure sensor (GM)
APS: atmospheric pressure sensor (Mazda)
ATS: air temperature sensor (Chrysler)
BARO: barometric pressure sensor (GM)
BMAP: barometric/manifold absolute pressure sensor (Ford)
BP: backpressure sensor (Ford)
BPS: barometric pressure sensor (Ford & Nissan)
BPT: back-pressure transducer
CAS: crank angle sensor
CESS: cold engine sensor switch
CID: cylinder identification sensor (Ford)
CMP: camshaft position sensor (GM)
CP: crankshaft position sensor (Ford)
CTS: charge temperature switch (Chrysler)
CTS: coolant temperature sensor (GM)
CTVS: choke thermal vacuum switch
ECT: engine coolant temperature (Ford & GM)
EGO: exhaust gas oxygen sensor (Ford)
EOS: exhaust oxygen sensor
EPOS: EGR valve position sensor (Ford)
EVP: EGR valve position sensor (Ford)
FLS: fluid level sensor (GM)
HEGO: heated exhaust gas oxygen sensor
IAT: inlet air temperature sensor (Ford)
IATS: intake air temperature sensor (Mazda)
KS: knock sensor
MAF: mass airflow sensor
MAP: manifold absolute pressure
MAT: manifold air temperature
MCT: manifold charge temperature (Ford)
OS: oxygen sensor
PA: pressure air (Honda)
PIP : profile ignition pickup (Ford)
SS: speed sensor (Honda)
TA: temperature air (Honda)

ACRONYMS

TP: throttle position sensor (Ford)
TPP: throttle position potentiometer
TPS: throttle position sensor
TPT: throttle position transducer (Chrysler)
TVS: thermal vacuum switch (GM)
VAF: vane airflow sensor
VSS: vehicle speed sensor
WOT: wide open throttle switch (GM)
WSS: wheel speed sensor

ELECTRONIC COMPONENTS

ALCL: assembly line communications link (GM)
ALDL: assembly line data link (GM)
ASDM: airbag system diagnostic module (Chrysler)
BCM: body control module (GM)
CALPAK: calibration pack
CECU: central electronic control unit (Nissan)
CPU: central processing unit
DERM: diagnostic energy reserve module (GM)
DLC: data link connector (GM)
EACV: electronic air control valve (Honda)
EBCM electronic brake control module (GM)
EBM: electronic body module (GM)
ECA: electronic control assembly
ECM electronic control module (GM)
ECU: electronic control unit (Ford, Honda & Toyota)
EEPROM: electronically erasable programmable read only memory chip
EI: electronic ignition (GM)
EPROM: erasable programmable read only memory chip
IC: integrated circuit
ICS: idle control solenoid (GM)
ISC: idle speed control (GM)

LCD: liquid crystal display
LED: light emitting diode
MIL: malfunction indicator lamp (GM)
PCM: powertrain control module (GM & Chrysler)
PROM: program read only memory computer chip
SES: service engine soon indicator (GM)
TCC: torque converter clutch (GM)
VCC: viscous converter clutch (GM)

ELECTRONIC IGNITIONS

BID: Breakerless Inductive Discharge
CDI: Capacitor Discharge Ignition
C3I: Computer Controlled Coil Ignition (GM)
CSSA: Cold Start Spark Advance
CSSH: Cold Start Spark Hold (Ford)
DIS: Distributorless Ignition System
DIS: Direct Ignition System (GM)
EDIS: Electronic Distributorless Ignition System (Ford)
HEI: High Energy Ignition (GM)
ITCS: Ignition Timing Control System (Honda)
SDI: Saab Direct Ignition
SSI: Solid State Ignition (Ford)

SPARK CONTROL SYSTEMS AND DEVICES

CCEVS: Coolant Controlled Engine Vacuum Switch (Chrysler)
CSC: Coolant Spark Control (Ford)
CSSA: Cold Start Spark Advance (Ford)
CTAV: Cold Temperature Actuated Vacuum (Ford)
CTO: Coolant Temperature Override Switch (AMC)
DRCV: distributor retard control valve

DSSA: Dual Signal Spark Advance (Ford)
DVDSV: differential vacuum delay and separator valve
DVDV: distributor vacuum delay valve
ESA: Electronic Spark Advance (Chrysler)
ESC: Electronic Spark Control (GM)
ESS: Electronic Spark Selection (GM)
EST: Electronic Spark Timing (GM)
OSAC: Orifice Spark Advance Control (Chrysler)
PVA: ported vacuum advance
PVS: ported vacuum switch
SAVM: spark advance vacuum modulator
SPOUT: Spark Output signal (Ford)
SRDV: spark retard delay valve
TAV: temperature actuated vacuum
TCS: Transmission Controlled Spark (GM)
TIC: thermal ignition control (Chrysler)
TRS: Transmission Regulated Spark (Ford)
VDV: vacuum delay valve

FUEL SYSTEM

AIS: automatic idle speed motor (Chrysler)
ABSV: air bypass solenoid valve
ASD: automatic shutdown relay (Chrysler)
CANP: canister purge solenoid valve
CCEI: Coolant Controlled Idle Enrichment (Chrysler)
CER: cold enrichment rod (Ford)
CIS: Continuous Injection System (Bosch)
CPI: Central Port Injection (GM)
CVR: control vacuum regulator (Ford)
DEFI: Digital Electronic Fuel

ACRONYMS

Injection (Cadillac)
DFS: deceleration fuel shutoff (Ford)
EFC: electronic fuel control
EFC: electronic feedback carburetor
EFCA: electronic fuel control assembly (Ford)
EFI: electronic fuel injection
FBC: feedback carburetor system (Ford & Mitsubishi)
FBCA: feedback carburetor actuator (Ford)
FCA: fuel control assembly (Chrysler)
FCS: fuel control solenoid (Ford)
FDC: fuel deceleration valve (Ford)
FI: fuel injection
IAC: idle air control (GM)
ISC: idle speed control (GM)
ITS: idle tracking switch (Ford)
JAS: Jet Air System (Mitsubishi)
MCS: mixture control solenoid (GM)
MFI: multi-port fuel injection
MPFI: multi-point fuel injection
MPI: multi-port injection
PECV: power enrichment control valve
PFI: port fuel injection (GM)
SFI: Sequential Fuel Injection (GM)
SMPI: Sequential Multiport Fuel Injection (Chrysler)
TABPV: throttle air bypass valve (Ford)
TBI: throttle body injection
TIV: Thermactor idle vacuum valve (Ford)
TKS: throttle kicker solenoid (Ford))
TPI: Tuned Port Injection (GM
TSP: throttle solenoid positioner (Ford)
TV: throttle valve

GENERAL

A/C: air conditioning
AC: alternating current
A/F: air/fuel ratio
A/T: automatic transmission
ATC: after top center
ATDC: after top dead center
ATF: automatic transmission fluid
AWD: all-wheel drive
BAT: battery
BTC: before top center
BTDC: before top dead center
Btu: British thermal units
C: Celsius
cc: cubic centimeters
cfm: cubic feet per minute
CID: cubic inch displacement
dB: decibels
DC: direct current
DOHC: dual overhead cams
DVOM: digital volt ohm meter
EMI: electromagnetic interference
FWD: front-wheel drive
GND: ground
GPM: grams per mile
Hg: mercury
ID: inside diameter
IGN: ignition
I/P: instrument panel
kHz: kilohertz
Km: kilometers
kPa: kilopascals
KV: kilovolts
ms: millisecond
mV: millivolts
Nm: Newton meters
OD: outside diameter
OE: original equipment
OEM: original equipment manufacturer
OHC: overhead cam
P/B: power brakes
P/N: part number
PPM: parts per million
PS: power steering
PSI: pounds per square inch
RFI: radio frequency interference
RPM: revolutions per minute
RPO: regular production option
RWD: rear-wheel drive
SOHC: single overhead cam
TDC: top dead center
VAC: volts alternating current
VDC: volts direct current
VIN: vehicle identification number
WOT: wide open throttle

MISCELLANEOUS

ABS: antilock brake system
API: American Petroleum Institute
ASE: Automotive Service Excellence
CAFE: corporate average fuel economy
CARB: California Air Resources Board
CCOT: clutch cycling orifice tube
DOT: Department of Transportation
DRBII: Diagnostic Readout Box
IM240: inspection/maintenance 240 program
ISO: International Standards Organization
NHTSA: National Highway Traffic Safety Administration
OBD: onboard diagnostics
RABS: Rear wheel Antilock Brake System (Ford)
RWAL: Rear Wheel Antilock brakes
SAE: Society of Automotive Engineers
SRS: Supplemental Restraint System (air bag)

INDEX

A

Air Injection Reaction-GM (AIR), 58
Air injection systems, 57-63
   aspirators, 63
   components, 59-61
      diverter valve, 59-60
      gulp valve, 60
      pumps, 59
      vacuum differential valve (VDV), 60-61
   GM's Pulsair, 63
   new catalytic converter requirements, 61-63
   smog pump, 58-61
Air intake, heated, 52-56
   components, 54
   operation, 54-55
   the problem, 52-53
   the solution, 53
   without vacuum, 55-56
Air/fuel ratio and emissions, 75-85
   carburetor basics, 76-78
   combustion basics, 75-76
   fuel injection, 81-84
   Rochester Central Port Injection (CPI), 85
   throttle body injection (TBI), 84-85
   wide open throttle, 78-81
      computer-controlled carburetion, 80-81
      leaner and later strategy, 79-80
      super rich, 79
Aldehydes, 44
Aspirators, 63
Automotive emissions control, history, 1-8
Autos, computer, 5-7
Autos, dirty old, 121

C

California
   and emissions controls, 3
   warranty rules, 29-30
Camshafts, 99-101
   aftermarket cams, 99-100
   stock cam problems, 100-101
CARB (California Air Resources Board), 3-5
   emissions inspections, 4-5
   executive orders (EOs), 23
   I/M (inspection and maintenance), 4
   periodic motor vehicle inspection (PMVI), 4
Carbon monoxide (CO), 34-35
Carburetors, 145-147
   cruising, 77-78
   curing dieseling, 145-146
   feedback, 80
   idle and up, 77
   kickers, 146-147
   propane, 145
   reservoir, 77
Catalytic converter requirements, new, 61-63
   Chrysler's version, 62-63
   Ford's early system, 61-62
   operation, 62
Catalytic converters, 2-3, 64-67, 127-128
   causes, 127
   checking air injection, 128
   cold engine operation, 67
   construction, 65
   diagnosis, 127-128
   heat test, 128
   hot engine operation, 67
   operation, 64-67
   purchasing new, 30
   and smog pump, 58-59
   test pipes, 21
   three-way converters, 65-67
   thump test, 128
Corporate average fuel economy (CAFE) standards, 2

CFCs (chlorofluorocarbons), 12, 14-19
   finding and stopping leaks, 18-19
   refrigerant laws, 17-18
Charcoal canisters, 49-51
   computer-controlled units, 51
   open bottom, 50
   sealed bottom, 50-51
Checklist, tune-up, 147
CIS (Continuous Injection System), 81
Clean Air Act (1970), 2, 26
Clean Air Act (1990), 5
Clunker laws, 112
Clunker legislation, 8
Cold weather modulator (Ford), 55
Computer cars, 5-7
Computerized engine controls, 103-111
   airflow sensors, 110-111
   analog and digital, 105-106
   applications, 104-105
   coolant temperature sensor, 110
   evolution, 103-105
   knock sensors, 111
   memory, 106-107
   operation, 107-108
   oxygen sensors, 108-109
   programming, 105-108
   sensors, 108-111
   switches, 111
   system overview, 104
   throttle position sensor (TPS), 109-110
Coolant control engine vacuum switch (CCEVS), 62-63
Crankcase ventilation, positive, 39-43
CVCC (controlled vortex combustion chamber - Honda), 57, 99

D

Defect warranty, 27-29
Dirty old cars, 121

INDEX

Diverter valve, 59-60

E
ECA (electronic control assembly - Ford), 61-62
ECS (evaporation control system - Chrysler), 46
ECU (electronic control unit - Ford), 62, 104
EEC (evaporative emissions control - Ford), 46
EECS (evaporative emissions control system - GM), 46
EFE (early fuel evaporation control system - GM & Nissan), 46
EGR; see Exhaust gas recirculation
Electronic engine control systems, 134-140
    Bosch hot wire air mass sensor, 140
    coolant temp sensor checks, 137-138
    Delco MAF sensor checks, 140
    knock sensor checks, 136-137
    MAF sensor checks, 138-139
    MAP sensor checks, 138
    O2 sensor checks, 136
    system checks, 135-140
    test equipment, 135
    TPS sensor checks, 137
    trouble codes, 139
    vane-type air meter checks, 139-140
Electronic fuel injection (EFI), 6
Electronic leak detector, 19
Electronic spark timing (EST), 93-94
Emissions; see Exhaust emissions
Emissions
    and air/fuel ratio, 75-85
    certification, 5, 23-25
    control history, 1-8
    exhaust, 31-38
    and ignition, 86-94
Emissions, controlling, 91-93
    blocking vacuum advance, 92
    ported vacuum switches, 92
    spark delay valve, 92-93
Emissions controls
    benefits, 3
    evaporative, 44-51
    future, 7-8
Emissions controls, evaporative, 44-51
    anti-percolator valve, 51
    charcoal canister, 49-51
    components, 46-51
    controlling, 45
    fuel tank, 46
    gas cap, 47
    liquid-vapor separator, 47-49
    sealed systems, 45-46
Emissions, history of automotive, 1-8
    California Air Resources Board (CARB), 3-5
    catalytic converters, 2-3
    Clean Air Act (1970), 2
    Clean Air Act (1990), 5
    clunker legislation, 8
    computer cars, 5-7
    corporate average fuel economy (CAFE), 2
    electronic fuel injection (EFI), 6
    emissions control benefits, 3
    Environmental Protection Agency, 2
    future, 7-8
    gross polluters, 8
    loaded mode testing, 8
    multi-point injection (MPI), 6
    on-board diagnosis (OBD), 6
    roadside sniffers, 8
    side stepping the law, 5
    throttle body injection (TBI), 6
    waiver limit, 8
Emissions testing, 112-123
    actual test, 115-119
    carbon monoxide, 117
    emissions violations, 120-123
    exemptions, 113-114
    failing a test, 120
    four-gas analyzers, 116-117
    high readings, 117-119
    hydrocarbon, 117
    I/M 240 test, 115
    OBD II & I/M 240 diagnostic systems, 116
    oxides of nitrogen, 119
    oxygen, 117-119
    retesting, 122
    test facilities, 114-115
    test procedures & equipment, 116-119
    test standards, 115-116
    testing overview, 114-115
    tips for passing, 123
    vehicle emission test report forms, 118-119
    waivers, 122-123
    warranty coverage, 121-122
Emissions-controlled engines, tuning, 141-147
    carburetors, 145-147
    emissions-control devices, 147
    fuel injection, 147
    history, 141-142
    primary ignition, 142-143
    secondary ignition, 143-145
    tune-up checklist, 147
Engine wear, 95-98
    compression & blowby, 96-97
    ring sealing, 96-97
    valve seating, 96
Environmental Protection Agency, 2
Evaporative emissions controls (EEC), 44-51
Exhaust emissions, 31-38
Exhaust emissions, carbon monoxide (CO), 34-35
Exhaust emissions, excessive, 31-34
Exhaust emissions, oxides of nitrogen (NOx), 31-38
Exhaust emissions, unburned hydrocarbons (HC), 31-38

INDEX

Exhaust gas recirculation (EGR), 68-74
    aftermarket exhaust systems, 72
    backpressure, 71-72
    bypassing, 70
    and computer controls, 73-74
    controlling, 72-73
    digital control, 73-74
    knocks and, 70
    linear valve, 74
    negative backpressure, 71-72
    and NOx, 70
    operation, 69-70
    other modulating methods, 73
    positive backpressure, 71
    problems, 74
    pulse-width control, 73
    regulating EGR flow, 73-74
    valve replacement, 74
    valves, 70-72
    wide open throttle (W.O.T.), 72
Exhaust gas recirculation (EGR) problems, 128-132
    checking TVS, 131
    checking vacuum, 131
    diagnosis, 129-130
    reconnecting the TVS hose, 131
    throttle test, 130-131
    vacuum amplifier diagnosis, 130-131

F

Feedback carburetors, 80
FTVC (fuel tank vapor control - AMC), 46
Fuel considerations, 101-102
    octane, 102
    reformulated gasolines, 101-102
Fuel evaporation, early, 52-56
Fuel injection, 81-84
Fuel injection, electronic, 82-84
    closed-loop EFI, 84
    operation, 83
    sensors, 83-84
Fuel injection, mechanical, 81-82
FVCR (fuel vapor control recovery - most imports), 46

G

Greenhouse effect, 13
Gross polluters, 8, 112
Gulp valve, 60

H

Head configuration, 98-99
HEI (high-energy ignition), 143
HFC134a CFC-free refrigerant, 15
High swirl combustion (HSC), 99
Hydrocarbons (HC) 31-38, 44

I

I/M 240 (inspection and maintenance) inspection program, 7-8
    and OBD II diagnostic systems, 116
    test, 115
Ignition and emissions, 86-94
    advance timing, 87-90
        centrifugal advance, 89
        detonation, 89
        the four strokes, 87-88
        vacuum advance, 89-90
    controlling emissions, 91-93
    electronic spark timing (EST), 93-94
    timing and emissions, 90-93
Instant smog, 44
International Mobile Air Conditioning Society (IMACA), 17
Inversion layer, 3

J

Jet Valve (Mitsubishi), 99

K

K-Jetronic fuel injection, 81-82, 83

L

Laws
    CFC refrigerant, 17-18
    clunker, 112
    rules and regulations, 20-30
    side-stepping, 5
Leak detector, electronic, 19
Lean misfire defined, 87
Legislation, clunker, 8
Liquid vapor separator, 47-49
    checking venting, 48-49
    operation, 48
Loaded mode testing, 8
Los Angeles, 3

M

MAF (mass air flow), 85
Maintenance and repairs, 9-10
    changing coolant, 9-10
    changing oil, 9
    performing a tuneup, 10
    recharging the air conditioner, 10
    replacing a battery, 10
    replacing tires, 10
MAP (manifold absolute pressure), 85
Mobile Air Conditioning Society (MACS), 17
Motor vehicle requirements, 25
Multi-point injection (MPI), 6

O

OBD II & I/M 240 diagnostic systems, 116
Octane, 102
On-board diagnosis (OBD), 6
OPEC, 2
Open loop mode, 51
Operating temperatures, 101
Ozone, 11-19
    alerts, 12
    CFC refrigerant laws, 17-18
    and CFCs (chlorofluorocarbons), 12, 14-19

INDEX

defined, 11-14
disappearing, 12-14
and Greenhouse effect, 13
and phytoplankton, 13
R-134a alternative, 15-17
replacing R-12, 16-17
terms, 14
Ozone layer, hole in, 13

P

Performance warranty, 27
Periodic motor vehicle inspection (PMVI), 4
Phytoplankton, 13
Polluters, gross, 8, 112
Pollution control valve, 39
Polyalkylene glycol (PAG), 15
Positive crankcase ventilation (PCV), 39-43
    and blowby gases, 97
    failure, 43
    old system, 40-42
        closed system, 41-42
        open system, 40-41
    valve, 42-43
    without valves, 42-43
Pulsair (GM) aspirator system, 63
Pulse width defined, 83

Q

Quench area, 98

R

R-12 CFC, 14-19
R-134a, 14-19, 15
Reduction catalysis, 65
Reference voltage, 104
Regulations, rules and laws, 20-30
Repairs
    and maintenance, 9-10
    qualifying for free, 29
Roadside sniffers, 8
Robert Bosch K-Jetronic, 81-82, 83
Rules, regulations and laws, 20-30

California warranty rules, 29-30
compliance and enforcement, 25-26
defect warranty, 27-29
emissions certification, 23-25
emissions restrictions, 21
installing certified parts, 23-24
installing non-certified parts, 24-25
loopholes plugged, 22
misc. warranty coverage, 30
parts affected, 22-23
performance warranty, 27
warranties, 26-30

S

SEMA emissions product labeling, 23
SEMA (Specialty Equipment Market Association), 23
Smog, instant, 44
Smog pump, 58-61
    and catalytic converters, 58-59
    thermal reactor, 58
Spark plugs, 143-145
Spark timing, electronic, 93-94
    modifying, 94
    operation, 94
    troubleshooting, 94
Stoichiometric ratio, 76
SUVA CFC-free refrigerant, 15

T

Temperature, operating, 101
Thermactor air pump (Ford), 58, 62
Throttle body injection (TBI), 6
TPS (throttle position sensor), 109-110
Troubleshooting, 124-140
    air injection, 126-127
    catalytic converters, 127-128
        causes, 127
        diagnosis, 127-128
    continuous injection, 132
        adjustments, 132
        diagnosis, 132
    EFE (early fuel evaporation control system), 126

EGR problems, 128-132
    diagnosis, 129-130
    unclogging, 131-132
    vacuum amplifier diagnosis, 130-131
electronic engine control systems, 134-140
electronic fuel injection, 132-134
    diagnosis, 133
    Rochester Central Point Injection (CPI), 133-134
EVAP (evaporative emissions control systems), 125
heated air intake, 125-126
    diagnosis, 125-126
    non-vacuum type, 126
overall system checks, 135-140
PCV (positive crankcase ventilation), 124
test equipment, 135
Tune-up checklist, 147

V

Vacuum motor defined, 54
Valve guides and seals, 97-98
VDV (vacuum differential valve), 60-61
Vehicle emission test report forms, 118-119

W

Waiver limit, 8
Warranties, 26-30
    coverage, 30
    defect, 27-29
    performance, 27
Wide open throttle (W.O.T.), 72, 78-81
    computer-controlled carburetion, 80-81
    leaner and later strategy, 79-80
    super rich, 79

ABOUT THE AUTHORS

LARRY CARLEY

Larry Carley is a well-known automotive technical writer, and has written over 900 technical articles and eleven automotive-related books. He is currently the technical editor of Babcox Communications, which publishes *Automotive Rebuilder*, *Import Car & Truck* and a number of other automotive trade publications. He has been certified by the National Institute for Automotive Service Excellence (ASE) in all areas except automatic transmissions, and is a recognized expert in many areas of automotive service and repair. He lives in Clarendon Hills, Illinois.

BOB FREUDENBERGER

Bob Freudenberger is a veteran freelance automotive writer, editor and photographer with over 1,000 articles published. Since 1973, he has served as the Senior Technical Editor of *Motor Service*, the world's largest magazine for professional auto repair technicians, and he is a regular contributor to *Automobile & Truck International*, which is distributed in 140 countries. In the early 1970s, Bob was the editor of *Speed Shop* magazine, and he has also written for *Popular Mechanics*, *Home Mechanix*, *Back Home* and *Reader's Digest*. He has also written training books and video scripts for NAPA, and he serves as a consultant for several tool and parts manufacturers. He lives in Marlboro, New Jersey.

HP AUTOMOTIVE BOOKS

HANDBOOK SERIES

Auto Electrical Handbook
Auto Math Handbook
Automotive Paint Handbook
Baja Bugs & Buggies
Brake Handbook
Camaro Restoration Handbook
Classic Car Restorer's Handbook
Clutch & Flywheel Handbook
Metal Fabricator's Handbook
Mustang Restoration Handbook
Off-Roader's Handbook
Paint & Body Handbook
Sheet Metal Handbook
Small Trucks
Street Rodder's Handbook
Turbochargers
Turbo Hydra-matic 350
Welder's Handbook

CARBURETORS

Holley 4150
Holley Carburetors, Manifolds & Fuel Injection
Rochester Carburetors
Weber Carburetors

To Order Call: 1-800-223-0510
HPBooks
The Berkley Publishing Group
200 Madison Avenue
New York, NY 10016

PERFORMANCE SERIES

Camaro Performance
Chassis Engineering
Chevrolet Power
How to Hot Rod Big-Block Chevys
How to Hot Rod Small-Block Chevys
How to Hot Rod Small-Block Mopar Engines
How to Hot Rod VW Engines
How to Make Your Car Handle
Mustang Performance Handbook
Mustang Performance Handbook 2
Race Car Engineering & Mechanics
Small-Block Chevy Performance
1001 High Performance Tech Tips

REBUILD SERIES

How to Rebuild Air-Cooled VW Engines
How to Rebuild Big-Block Chevy Engines
How to Rebuild Big-Block Ford Engines
How to Rebuild Big-Block Mopar Engines
How to Rebuild Small-Block Chevy Engines
How to Rebuild Small-Block Ford Engines
How to Rebuild Small-Block Mopar Engines
How to Rebuild Your Ford V-8

GENERAL INTEREST

Auto Dictionary
Auto Repair Shams & Scams
Car Collector's Handbook
Fabulous Funny Cars
Fast Fords
Guide to GM Muscle Cars
Understanding Automotive Emissions Control